Fuzzy Model Identification

Springer
Berlin
Heidelberg
New York
Barcelona
Budapest
Hong Kong
London
Milan
Paris
Santa Clara
Singapore
Tokyo

Hans Hellendoorn
Dimiter Driankov (Eds.)

Fuzzy Model Identification

Selected Approaches

With 129 Figures and 13 Tables

 Springer

Dr. Hans Hellendoorn
Siemens Corporate R&D
D-81730 München, Germany

Prof. Dr. Dimiter Driankov
University of Linköping
S-58183 Linköping, Sweden

ISBN-13: 978-3-540-62721-0 Springer-Verlag Berlin Heidelberg New York

Library of Congress Cataloging–in–Publication Data

Fuzzy model identification : selected approaches / Hans Hellendoorn,
    Dimiter Driankov, eds.
            p.   cm.
        Includes bibliographical references.
        ISBN-13: 978-3-540-62721-0   e-ISBN-13: 978-3-642-60767-7
        DOI:10.1007/978-3-642-60767-7
        1. Automatic control-- Mathematical models. 2. Fuzzy systems.
    3. Neural networks (Computer science) 4. System identification.
    5. Soft computing. I. Hellendoorn, Hans. II. Driankov, Dimiter.
    TJ213.F893   1997
    629.8--dc21

Cover Design: design & production GmbH, Heidelberg
Typesetting: Camera-ready by the authors
SPIN 105773453   45/3142-5 4 3 2 1 – Printed on acid-free paper

Preface

During the past few years two principally different approaches to the design of fuzzy controllers have emerged: heuristics-based design and model-based design.

The main motivation for the heuristics-based design is given by the fact that many industrial processes are still controlled in one of the following two ways:

− The process is controlled manually by an experienced operator.
− The process is controlled by an automatic control system which needs manual, on-line 'trimming' of its parameters by an experienced operator.

In both cases it is enough to translate in terms of a set of fuzzy if-then rules the operator's manual control algorithm or manual on-line 'trimming' strategy in order to obtain an equally good, or even better, wholly automatic fuzzy control system. This implies that the design of a fuzzy controller can only be done *after* a manual control algorithm or trimming strategy exists.

It is admitted in the literature on fuzzy control that the heuristics-based approach to the design of fuzzy controllers is very difficult to apply to multiple-input/multiple-output control problems which represent the largest part of challenging industrial process control applications. Furthermore, the heuristics-based design lacks systematic and formally verifiable tuning techniques. Also, studies of the stability, performance, and robustness of a closed loop system incorporating a heuristics-based fuzzy controller can only be done via extensive simulations. Last but not least, there is a lack of systematic and easily verifiable knowledge acquisition techniques via which the heuristic knowledge constituting the basis of a manual control algorithm or a manual on-line trimming strategy can be extracted.

These difficulties of the heuristics-based approach explain the recent surge of interest in the model-based design of fuzzy controllers. Model-based fuzzy control uses a given conventional or a fuzzy open loop model of the plant under control in order to derive the set of fuzzy if-then rules constituting the corresponding fuzzy controller. Furthermore, of central interest are the consequent stability, performance, and robustness analysis of the resulting closed loop system involving a conventional model and a fuzzy controller, or a fuzzy model and a fuzzy controller. The major objective of the model-based fuzzy control is to use the full available range of existing linear and nonlinear design and analysis methods for the design of such fuzzy controllers which have better stability, performance, and robustness properties than the corresponding non-fuzzy controllers designed by the use of these same techniques.

This interest in the model-based design of fuzzy controllers is paralleled by a similar surge of interest in fuzzy identification. That is, the derivation of *black box* fuzzy models of the plant under control, in terms of the identification of a set of fuzzy if-then rules (encoding structural knowledge) together with their corresponding parameters, by the use of clustering techniques, neural networks, genetic algorithms, or a mixture of these techniques. Fuzzy identification is a young discipline, and many publications on the subject can only be found in specialized conference proceedings and journal articles of insufficient size, presentation style, and depth.

We believe that research in fuzzy identification should be suitably reported to a much larger group of practitioners and researchers, and out of this belief the present edited volume was born. The collection of papers represented in the volume does not cover all methods for fuzzy identification that can be found in the literature. Instead, only methods that constitute the major trends in fuzzy identification, that have shown their relevance in solving practical problems, and whose implementation can be easily automated have been selected. Furthermore, the presentation of the methods presented in this volume is not attempted as a comprehensive discussion of all theoretical and implementation issues characteristic for a particular method. Instead, only important issues in connection with the principal technique (e.g., clustering, neural networks, genetic algorithms, etc.) underlying a particular method and the application of this technique for the purpose of fuzzy identification are discussed. In particular, each presentation highlights the algorithmic aspects of the application of a particular technique for the identification of a particular type of fuzzy model. Finally, each presentation summarizes the advantages and drawbacks of the particular method it considers, focusing on its practical aspects.

It is our hope that by collecting such contributions in a single volume we will not only present the major themes in fuzzy identification, but also encourage their use in practice and possibly call attention to still open research problems.

Munich, Germany, 1997   *Hans Hellendoorn*
                                                                Dimiter Driankov

Table of Contents

Preface V

List of Contributors XI

Introduction XIII

General Overview

Fuzzy Identification from a Grey Box Modeling Point of View
P. Lindskog 3

1. Introduction 3
2. System Identification 5
3. Fuzzy Modeling Framework 15
4. Fuzzy Identification Based on Prior Knowledge 25
5. Example – Tank Level Modeling 38
6. Practical Aspects 44
7. Conclusions and Future Work 46
References 47

Clustering Methods

Constructing Fuzzy Models by Product Space Clustering
R. Babuška and H.B. Verbruggen 53

1. Introduction 53
2. Overview of Fuzzy Models 55
3. Structure Selection for Modeling of Dynamic Systems 57
4. Fuzzy Clustering 61
5. Deriving Takagi–Sugeno Fuzzy Models 68
6. Example: pH Neutralization 79
7. Practical Considerations and Concluding Remarks 83
A. The Gustafson–Kessel Algorithm – MATLAB Implementation 85
References 87

**Identification of Takagi-Sugeno Fuzzy Models
via Clustering and Hough Transform**
M.-K. Park, S.-H. Ji, E.-T. Kim, and M. Park 91

1. Introduction 91
2. The Identification Method 93
3. Example 1 108
4. Example 2 110
5. Summary of the Identification Procedure 115
6. Practical Considerations and Concluding Remarks 116
References 119

**Rapid Prototyping of Fuzzy Models
Based on Hierarchical Clustering**
M. Delgado, M.A. Vila, and A.F. Gomez-Skarmeta 121

1. Introduction 121
2. The Fuzzy C-Means Algorithm 123
3. Using Hierarchical Clustering to Preprocess Data 125
4. Rapid Prototyping of Approximative Fuzzy Models 138
5. Rapid Prototyping of Descriptive Fuzzy Models 144
6. Examples 150
7. Practical Considerations and Concluding Remarks 156
A. Proofs of Propositions 157
References 158

Neural Networks

Fuzzy Identification Using Methods of Intelligent Data Analysis
J. Hollatz 165

1. Introduction 165
2. Neuro-Fuzzy Methods 167
3. Density Estimation 176
4. Fuzzy Clustering 181
5. Conclusion 186
A. From Rules to Networks 187
B. Learning Rule for RBF Networks 188
C. Update Equations for Gaussian Mixtures 189
D. Adaptation Algorithm for Fuzzy Clustering 189
References 191

**Identification of Singleton Fuzzy Models
via Fuzzy Hyperrectangular Composite NN**
M.-C. Su 193

1. Introduction 193
2. Classification of Fuzzy Models 194
3. Fuzzy Neural Networks 201
4. Identification of Singleton Fuzzy Models 204
5. Simulation Results 206
6. Practical Considerations and Concluding Remarks 209
References 211

Genetic Algorithms

**Identification of Linguistic Fuzzy Models
by Means of Genetic Algorithms**
O. Cordón and F. Herrera 215

1. Introduction 215
2. Evolutionary Algorithms and Genetic Fuzzy Systems 216
3. The Fuzzy Model Identification Problem 223
4. The Genetic Fuzzy Identification Method 228
5. Example 242
6. Practical Considerations and Concluding Remarks 247
References 248

Optimization of Fuzzy Models by Global Numeric Optimization
V. Vergara and C. Moraga 251

1. Introduction 251
2. Theoretical Aspects of Fuzzy Models 251
3. The Fuzzy Identification Method 260
4. Simulation Results 267
5. Practical Aspects 275
References 277

Artificial Intelligence

Identification of Linguistic Fuzzy Models Based on Learning
Y. Nakoula, S. Galichet, and L. Foulloy 281

1. Introduction 281
2. Basic Concepts and Notation 282
3. The Identification Problem 284
4. The Fuzzy Identification Method 295
5. Numeric Examples 307
6. Practical Aspects and Concluding Remarks 314
References 317

List of Contributors

R. Babuška
Delft University of Technology
Dept. of Electrical Engineering
P.O. Box 5031
NL-2600 GA Delft, The Netherlands

O. Cordón
University of Granada
Dept. of Computer Science and
Artificial Intelligence
E.T.S.I. Informatica
E-18071 Granada, Spain

M. Delgado
University of Granada
Dept. of Computer Science and
Artificial Intelligence
E.T.S.I. Informatica
E-18071 Granada, Spain

L. Foulloy
Lab. d'Automatique et de Micro
Informatique Industrielle
Université de Savoie, BP 806
F-74016 Annecy Cedex, France

S. Galichet
Lab. d'Automatique et de Micro
Informatique Industrielle
Université de Savoie, BP 806
F-74016 Annecy Cedex, France

A.F. Gomez-Skarmeta
University of Murcia
Dept. Informatics and Systems
E-Murcia, Spain

F. Herrera
University of Granada
Dept. of Computer Science and
Artificial Intelligence
E.T.S.I. Informatica
E-18071 Granada, Spain

J. Hollatz
Siemens AG
Corporate Technology
D-81730 Munich, Germany

Seung-Hwan Ji
Yonsei University
134 Shinchon-dong, Seodaemun-gu
Seoul, 120-749, Korea

Eun-Tai Kim
Yonsei University
134 Shinchon-dong, Seodaemun-gu
Seoul, 120-749, Korea

P. Lindskog
Linköping University
Department of Electrical Engineering
S-581 83 Linköping, Sweden

C. Moraga
University of Dortmund
Dept. of Computer Science I
D-44221 Dortmund, Germany

Y. Nakoula
Lab. d'Automatique et de Micro
Informatique Industrielle
Université de Savoie, BP 806
F-74016 Annecy Cedex, France

Mignon Park
Yonsei University
134 Shinchon-dong, Seodaemun-gu
Seoul, 120-749, Korea

Min-Kee Park
National Polytechnic University
172 Kongneung-dong, Nowon-gu
Seoul, 139-743, Korea

Mu-Chun Su
Tamkang University
Tamsui, Taipei Hsien
Taiwan, 25137, R.O.C.

H.B. Verbruggen
Delft University of Technology
Dept. of Electrical Engineering
P.O. Box 5031
NL-2600 GA Delft, The Netherlands

V. Vergara
University of Dortmund
Dept. of Computer Science I
D-44221 Dortmund, Germany

M.A. Vila
University of Granada
Dept. of Computer Science and
Artificial Intelligence
E.T.S.I. Informatica
E-18071 Granada, Spain

Introduction

This edited volume assembles a collection of recent works in the field of fuzzy model identification. Our goal is twofold:

1. To expose the field of fuzzy identification to conventional control theorists as a complement to the existing approaches to identification of nonlinear systems and provide practicing control engineers with the algorithmic and practical aspects of a set of new identification techniques.
2. To emphasize the need for a more systematic and coherent theory of fuzzy identification by bringing together methods based on different techniques but aiming at the identification of the same types of fuzzy models.

In what follows we will describe the general control engineering context in which fuzzy identification has a useful role to play.

Fuzzy Models

In the theory and practice of control engineering mathematical models are often constructed, for examples based on differential or difference equations, derived from physical laws without any use of measurements of the system (i.e., white box models). Mathematical models can also be derived entirely from data using no physical insight whatsoever (i.e., black box models). The structure of such a model is chosen from families of structures that are known to be flexible and successful in applications.

Assume that one has already specified the type of model to use. This means that one has made a selection between linear and nonlinear structures, between black box and physically parameterized approaches, etc. Furthermore, let us assume that one has decided on the size of the model set. That is, one has chosen the possible variables and combinations of these to be used in the model. Also, one has fixed orders and degrees of the chosen model types. With these assumptions fulfilled one has in principle determined a model set over which the search for the model can be carried out. The next item one has to consider is how to parameterize the model set so that the estimation algorithms can find reasonable parameter values. Thus we assume that the members of the already determined model set can be parameterized

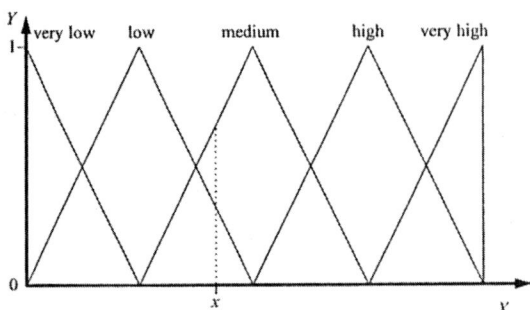

Fig. 1. The partitioning of a continuous domain by fuzzy sets

by a parameter vector $\boldsymbol{\theta}$ of finite dimension. Then we look for one particular model corresponding to $\boldsymbol{\theta}$. The family of model structures to which such a model belongs is

$$\hat{y}(t, \boldsymbol{\theta}) = g(t, \boldsymbol{\theta}, \varphi(t, \boldsymbol{\theta})),$$

where $\hat{y}(\cdot)$ emphasizes that the function $g(\cdot)$ is a predictor, i.e., it is based on signals that are known at time t. The predictor structure is captured by the regressor $\varphi(\cdot)$ which maps output signals up to an index $t-1$ ($\mathbf{y}^{t-1}$) and input signals up to an index t ($\mathbf{u}^t$) to an r-dimensional regression vector

$$\varphi(t, \boldsymbol{\theta}) = \varphi(\mathbf{y}^{t-1}, \mathbf{u}^t, \boldsymbol{\theta}).$$

With this, the choice of model structure splits into two subproblems (both possibly nonlinear):

1. The choice of the dynamic regression vector $\varphi(t, \boldsymbol{\theta})$.
2. The choice of static mapping $g(t, \boldsymbol{\theta}, \varphi(t, \boldsymbol{\theta}))$.

The papers collected in this edited volume consider one of the following three types of static mapping (or fuzzy model) represented as a set of if-then fuzzy rules:

1. Mamdani-type fuzzy rules defining a linguistic fuzzy model.
2. Takagi–Sugeno-type fuzzy rules defining a Takagi–Sugeno (TS) fuzzy model.
3. Singleton-type fuzzy rules defining a singleton fuzzy model.

In general, any one of the above types of fuzzy models employs fuzzy sets for the partitioning of the continuous domains of the input and output variables into a smaller number of overlapping regions which constitute the so-called linguistic (or symbolic) values (e.g., Low, Medium, High) of the input and output variables (see Fig. 1).

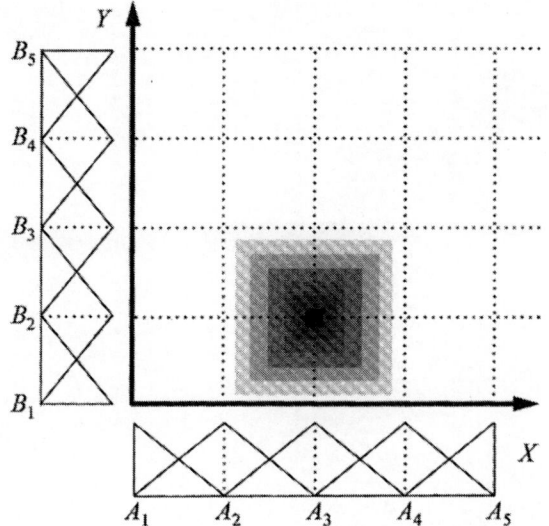

Fig. 2. The input-output space divided in fuzzy regions, the rule **If** x is A_3 **then** y is B_2

Then a fuzzy model describes the system under consideration by establishing relationships between the linguistic values of the input and output variables. These relationships are expressed in the form of if-then fuzzy rules. Each fuzzy rule maps a fuzzy region from the antecedent space (if-part) of a fuzzy rule to the fuzzy region from the consequent space (then-part) of this same fuzzy rule.

Mamdani-type models

A Mamdani-type (SISO) fuzzy model consists, for example, of the following fuzzy rules:

R_1: **If** x is A_1 **then** y is B_1
R_2: **If** x is A_2 **then** y is B_2
R_3: **If** x is A_3 **then** y is B_1
R_4: **If** x is A_4 **then** y is B_2

In the above, x is an input variable which takes its crisp (pointwise) values on the domain X while A_1, A_2, A_3, and A_4 are the linguistic (fuzzy) values of x defined as fuzzy sets on X via membership functions $\mu_{A_i}(x)$.

Furthermore, y is an input variable which takes its crisp (pointwise) values on the domain Y while B_1 and B_2 are the linguistic (fuzzy) values of y defined as fuzzy sets on Y via membership functions $\mu_{B_i}(y)$.

Each rule of the above model defines a fuzzy region in the input-output space. The set of all rules then partitions the input-output space into a number of overlapping fuzzy regions. See Fig. 2.

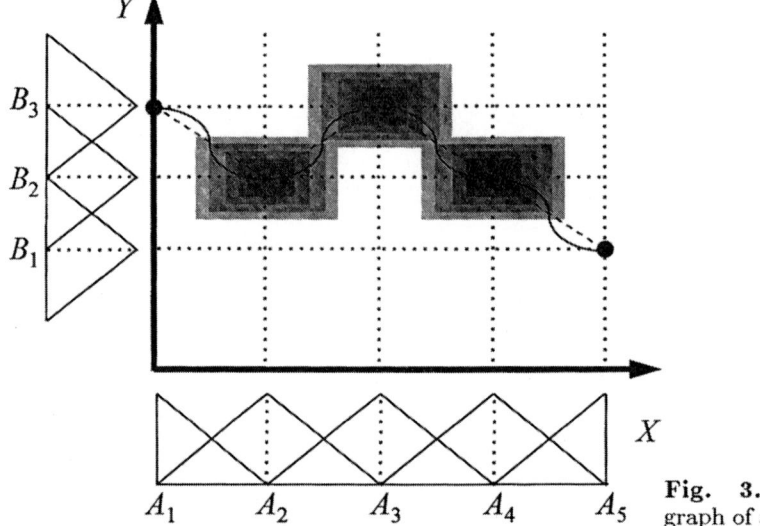

Fig. 3. The fuzzy graph of a fuzzy model

As a result of computation with the above fuzzy model one obtains the fuzzy graph of the fuzzy model as illustrated in Fig. 3.

Dynamics is reflected in a Mamdani-type of fuzzy model in exactly the same way as in conventional mathematical models. For the continuous case and for a first-order model one may have, for example,

R_1: **If** x is A_1 **and** u is B_2 **then** $\dot{x}$ is C_1
R_2: **If** x is A_2 **and** u is B_1 **then** $\dot{x}$ is C_2

while for the time-discrete case one may have (for a first-order system), for example,

R_1: **If** $y(k)$ is A_1 **and** $u(k)$ is B_2 **then** $\dot{y}(k+1)$ is A_3
R_2: **If** $y(k)$ is A_4 **and** $u(k)$ is A_1 **then** $\dot{y}(k)$ is A_1.

When the latter two rules are looked upon in the context of model identification, they map the input and output regressors to the new (next in time) output. The black-box identification of Mamdani-type fuzzy models (or linguistic models) is the subject of the contributions of Cordón and Herrera (pp. 215–250), of Delgado, Vila, and Gomez-Skarmeta (pp. 121–161), and of Vergara and Moraga (pp. 251–278). Lindskog (pp. 3–50) describes a grey-box approach to the identification of this type of fuzzy models.

Takagi–Sugeno fuzzy models

The affine TS-fuzzy model combines a global rule-based description with a local functional description which in the context of black-box identification is chosen as a linear regression model. Thus, an affine TS-fuzzy model is less

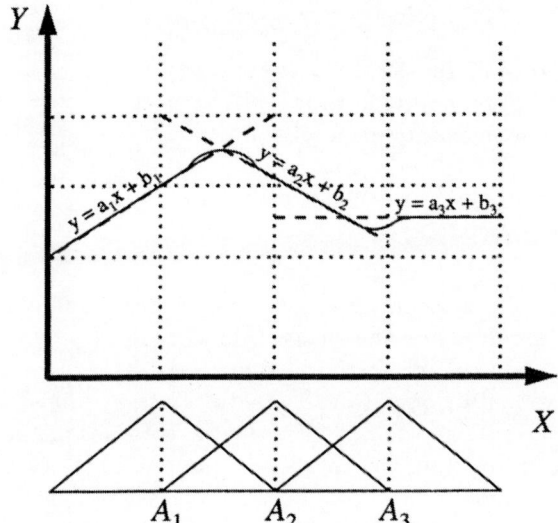

Fig. 4. The graph of an affine TS-fuzzy model

general than a Mamdani-type fuzzy model with respect to modeling arbitrary nonlinear static mappings. On the other hand, it is easier to identify, because each rule describes a different fuzzy region where the outputs depend on the inputs in a linear manner. Also, the consequent parameters can be easily estimated, e.g., with least squares, and less input-output data is needed.

For example, the following three fuzzy rules describe an affine TS-fuzzy model:

R_1: **If** x is A_1 **then** $y = a_1 x + b_1$
R_2: **If** x is A_2 **then** $y = a_2 x + b_2$
R_3: **If** x is A_3 **then** $y = a_3 x + b_3$.

After computation with these fuzzy rules one obtains the graph from Fig. 4.

In the context of black-box identification, the antecedent of an affine TS-fuzzy model defines a fuzzy region on the regressors and the consequent is (in most cases) an autoregressive model with an exogenous input. For example,

R_1: **If** $y(k)$ is A_1 **and** $u(k)$ is B_1 **then** $y(k+1) = a_1 y(k) + b_1 u(k) + c_1$
R_2: **If** $y(k)$ is A_2 **and** $u(k)$ is B_2 **then** $y(k+2) = a_2 y(k) + b_2 u(k) + c_2$.

The black-box identification of affine TS-fuzzy models is the subject of the contributions of Babuška and Verbruggen (pp. 53–90), of Park, Ji, Kim, and Park (pp. 91–119), and of Hollatz (pp. 165–191).

Singleton fuzzy models

When the consequent parameters a_i in the TS-fuzzy model are constant, the resulting model is called a fuzzy singleton model, since a constant is a singleton fuzzy set. For the previous example one obtains

R_1: **If** x is A_1 **then** $y = b_1$
R_2: **If** x is A_2 **then** $y = b_2$
R_3: **If** x is A_3 **then** $y = b_3$

However, this type of fuzzy model can also be seen as a special case of the Mamdani-type fuzzy model where the fuzzy sets corresponding to the linguistic values of the consequent are singleton fuzzy sets. It has been shown that this type of fuzzy model belongs to the general class of function approximators, called basis function expansion, where the basis functions are given (normalized) degrees of fulfillment of the antecedents of the fuzzy rules constituting the singleton fuzzy model.

The black-box identification of singleton fuzzy models is the subject of the contribution of Su (pp. 193–212).

The learning method of Nakoula, Galichet, and Foulloy is applied to the identification of Mamdani, TS-, and singleton fuzzy models.

The Identification Methods

We have considered the goal of the identification method, viz. the achievement of either a Mamdani fuzzy model, a TS-fuzzy model, or a singleton fuzzy model. We will now deal with the methods for achieving these models, viz. clustering, neural networks, genetic algorithms, and classical learning methods. Note that we have sorted the contributions in this volume with regard to these methods.

The first paper in this volume is a contribution by **Lindskog** (pp. 3–50). We have put it at the beginning because it gives a general introduction and overview to modeling and identification. Anyone who is interested in a thorough theoretical foundation for fuzzy identification should read this contribution.

Clustering

The first method is *clustering*. Clustering means the grouping of data points into a number of classes. The number or even the location of these classes can be predetermined but usually they are not. It is also possible to mix the input and output data into one vector and then do the clustering of these vectors, but usually input and output data are clustered separately. **Babuška and Verbruggen** (pp. 53–90) deal extensively with clustering methods, and present methods to construct clusters, to determine the number of clusters, to derive consequent parameters in the form of Takagi–Sugeno expressions for each cluster, to derive the antecedent membership functions, and to simplify the resulting rules. Finally, they deal with the validation of the derived fuzzy model.

Park et al. (pp. 91–119) also derive Takagi–Sugeno fuzzy models with the help of clustering methods. Their focus is a two-step approach, first course tuning, then fine tuning. It is interesting to compare this approach with the one mentioned above. Park et al. carefully describe similarities and differences throughout the whole paper! The reader is recommended to compare hyperplane–based c-means with the Gustafson–Kessel algorithm as well as the Hough transform with iteration-based methods. The Hough-transform method turns out to be a helpful tool in the determination of the number of clusters.

Another more explicit two-step clustering method is presented by **Delgado** et al. (pp. 121–161). They follow an approach between descriptive and approximative approaches by first establishing a rough description to achieve an initial fuzzy model that can afterwards be tuned by neural networks, genetic algorithms, etc. This *rapid prototyping approach* shows some similarities with the approach of Park et al., but differs with respect to the chosen learning methods, e.g., hierarchical clustering and special measures, and the chosen fuzzy model, viz. Mamdani instead of Takagi–Sugeno.

Neural Networks

Actually, the contribution of **Hollatz** (pp. 165–191) is on the border between neural networks and clustering techniques. Hollatz is an experienced practitioner in the area of data analysis with noisy and incomprehensible data, with thin data sets in high dimensional spaces, and with large, redundant, and inconsistent data sets. He especially stresses the use of prior knowledge to start with. Fuzzy logic is an ideal methodology to express prior, vague, and incomplete expert knowledge. In this contribution the use of neural networks and fuzzy clustering methods are shown in parallel and hence can be compared with respect to their methodological aspects and to the expressive power and transparency of the resulting rule base.

In a certain sense the contribution of **Su** (pp. 193–212) also belongs on the boundary of clustering and neural networks, because he uses neural networks to cluster the input data into more or less hyperrectangular entities. These entities can be found explicitly in his identified fuzzy rules. The outputs are singleton fuzzy sets. Su gives some very interesting examples from system identification and time series prediction to illustrate his theory.

Genetic Algorithms

Cordón and Herrera (pp. 215–250) use genetic algorithms or, more generally, evolutionary algorithms to identify an initial set of fuzzy rules *and* to simplify this fuzzy rulebase. A general problem of automatic generation of fuzzy rules – with genetic algorithms as well as with neural networks or conventional learning techniques – is the often chaotic structure of the input fuzzy sets, the rule weights, rule structure, etc. Genetic algorithms are particularly appropriate to simplify and 'straighten' such rulebases. In a final step they can tune the resulting rulebase by shifting and reshaping the membership functions.

In the contribution of **Vergara and Moraga** (pp. 251–278) the authors use a combination of a self-organizing map, a fuzzy associative memory, a neural network, and genetic algorithms to identify and tune the fuzzy model. The choice of the parameters to be tuned and the fitness function play an important role in this process. The reader should be aware of complexity problems using this approach.

Artificial Intelligence

Artificial intelligence is a broad field and many scientists judge that also fuzzy systems, neural networks, and genetic algorithms belong to it, although they are better described by the notion computational intelligence. **Nakoula, Galichet, and Foulloy** (pp. 281–319) propose an identification approach based on learning techniques that are different from clustering, neural networks, and genetic algorithms. The learning method could be described as ad hoc; however, this should not be understood in its (common) pejorative sense, but positively, as an appropriate method to solve the identification problem, i.e., the partitioning of the input fuzzy sets, the generation of the initial and additional output fuzzy sets and the rules, and the evaluation of the derived fuzzy model.

General Remarks

We have asked all authors to use the same notation and the same interpretation of technical terms. Therefore we have checked all papers carefully and sent them back to the authors to change sometimes significant parts of them. We honestly thank the authors for their patience and their willingness to make all the changes, which in some cases meant a partial rewriting of the papers. We believe that the presentation style, the terminology, and the notation as a whole has significantly profited from this iteration loop.

With regard to the structure of the paper we have asked each author to adhere generally to the following leitmotiv: The introduction should give a general overview of the paper. Then there should be a section dealing with the model to be identified. Then comes the identification technique, which should be accompanied by algorithms. These *algorithms* play an important role in this volume. We believe that in particular they will help the reader to use the presented methods in his or her own environment to solve real application problems. If this goal were achieved we would be very satisfied. We also asked each author to give a practical example to show how the presented method works. Furthermore, in the last section, each author has been asked to say something about

– the use of prior knowledge,
– the model complexity,
– the robustness of the identification method, and
– real-world applications.

General Overview

Fuzzy Identification from a Grey Box Modeling Point of View

P. Lindskog

Linköping University, S-581 83 Linköping, Sweden

1. Introduction

The design of mathematical models of complex real-world (and typically non-linear) systems is essential in many fields of science and engineering. The developed models can be used, e.g., to explain the behavior of the underlying system as well as for prediction and control purposes.

A common approach for building mathematical models is so-called *black box modeling* (Ljung, 1987; Söderström and Stoica, 1989), as opposed to more traditional physical modeling (or white box modeling), where everything is considered known a priori from physics. Strictly speaking, a black box model is designed entirely from data using no physical or verbal insight whatsoever. The structure of the model is chosen from families that are known to be very flexible and successful in past applications. This also means that the model parameters lack physical or verbal significance; they are tuned just to fit the observed data as well as possible.

The term "black box modeling" is sometimes used almost as a synonym for *system identification*, although a much more convenient definition, and the one often used, is that system identification is the theory of designing mathematical models of dynamical systems from observed data. Hence, by combining the black box approach with physical or verbal modeling in such a way that certain prior knowledge from the system is taken into account, we end up with special identification procedures that commonly are referred to as *grey box modeling* approaches, see, e.g., (Bohlin, 1991, Hangos, 1995). Two important facts make such methods intuitively appealing.

1. In a real-world modeling situation, we never have complete process knowledge. There are always uncertain factors affecting the system, thus indicating that a complete physical model can hardly ever be constructed. However, uncertain factors can be revealed through experiments and, at least partly, taken care of by employing sufficiently flexible model families.

2. The modeling procedure on the other hand allows us to restrict the flexibility to comply with prior knowledge. This makes it possible to follow, at least partly, another basic identification principle, namely to only estimate what is still unknown.

Traditional grey box approaches assume that the structure of the model is given directly as a parameterized mathematical function, which is based (at least partly) on physical principles. However, for many real-world systems a great deal of information is provided by human experts, who do not reason in terms of mathematics but instead describe the system verbally through vague or imprecise statements. For example, in case it is hard to design a suitable mathematical model of a heating system, an important part of its behavior can still be characterized, e.g., through

If *more energy is supplied to the heater element*
then *the temperature will increase.*   (1.1)

Because so much human knowledge and expertise comes in terms of verbal rules, a sound engineering approach is to try to integrate such linguistic information into the identification process. A convenient and common way of doing this is to use fuzzy logic concepts in order to cast the verbal knowledge into a conventional mathematical representation (a model structure), which subsequently can be fine-tuned using input-output data. It turns out that the structure so obtained can be viewed very well as a layered network having much in common with an ordinary neural network, see, e.g., (Brown and Harris, 1994; Haykin, 1994; Roger Jang and Sun, 1995; Lin and Lee, 1996; Chen, 1996). As a matter of fact, the kinship is so evident that many researchers refer to this approach as *neuro-fuzzy modeling*.

With this in mind, the palpable question is what is conceptually gained by this approach compared to standard black box neural network modeling?

– Firstly and contrary to neural networks, neuro-fuzzy modeling, or just *fuzzy modeling*, offers a high-level, structured, and convenient way of incorporating linguistic prior knowledge into the models.
– Secondly, the basic linguistic knowledge entered is of the form "SPEED$(t-1)$ is HIGH." In fuzzy modeling, such a proposition is given a precise mathematical meaning through a basis function (membership function) having parameters associated with the property "HIGH", thus meaning that the parameters can be assigned reasonable initial values. This is important in that the parameter estimation algorithm (which often is iterative) can be started from a point where the risk of getting stuck in an undesired local minimum is reduced compared to that if the initial parameters are chosen at random (which often is the case for neural networks).
– Thirdly, physically unsound regions can be avoided. By randomly choosing initial parameter values in a neural network this cannot be guaranteed, and although regularization (see below) is applied in the estimation phase, basis functions corresponding to unsound regions are seldom removed from the final model, which then becomes more complex than necessary.
– A fourth potential advantage comes in terms of extrapolation capabilities. While data can be used to explain certain system features, the linguis-

tic expert knowledge (here the rules) can be employed to pick up other phenomena that are not revealed in the available data.
− Finally, the human expert who supplied the verbal knowledge can always be consulted for model validation.

This contribution concentrates on how to maintain these advantages when fuzzy modeling is complemented with system identification techniques. More precisely, the aim is to provide answers to a number of central grey-box-type questions:

1. What kind of mathematical rule base interpretation is suited when system identification aspects are also taken into account?
2. What parameter estimation algorithms should be used?
3. How can the knowledge provided by the domain expert, i.e., the meaning of the rule base, be preserved throughout the parameter estimation step?
4. How can different non-structural system features be built into the models? By non-structural knowledge we mean, e.g., that the step response is known to be monotone, or that the steady-state gain curve is monotonic in certain input variables, or some other qualitative property.

To be able to address these issues we first give a brief introduction to the field of parametric system identification, focusing mainly on basic concepts, ideas, and algorithms from which the following sections can depart. Section 3 addresses various fuzzy modeling matters. It is argued that a Mamdani type of rule base interpretation[1] (Mamdani and Assilian, 1975; Roger Jang and Sun, 1995) is suited when the rules are of the form (1.1) and when identification aspects are also accounted for. The remaining main three questions from above are then considered and answered in Sect. 4, whereupon Sect. 5 illustrates the usefulness of the suggested framework on a real-world laboratory-scale application example. Some practical aspects of the proposed modeling approach are thereafter discussed in Sect. 6, and in Sect. 7 we finally put forward some concluding remarks and give a few directions for further research within the fuzzy identification area.

2. System Identification

2.1 Basic Ingredients and Notation

System identification deals with the problem of how to infer relationships between past input-output measurements and future outputs (Ljung, 1987; Söderström and Stoica, 1989). In practice, this is a procedure that is highly

---

[1] In fact the considered model representation turns out to be structurally equivalent to a zero-order Takagi–Sugeno fuzzy structure (Takagi and Sugeno, 1985; Sugeno and Kang, 1988), which is just a special case of the general Takagi–Sugeno fuzzy model family.

iterative in nature and is made up from three main ingredients: the data, the model structure, and the selection criterion, all of which include choices that are subject to personal judgments.

The data $\mathbf{Z}_N$. By the row vector

$$\mathbf{z}(t) = [\, y(t) \quad u_1(t) \quad \ldots \quad u_m(t) \,] \in \mathbb{R}^{1+m}, \tag{2.1}$$

we denote one particular data sample at time t collected from a system having one output and m input signals, i.e., we consider a multi-input single-output (MISO) system. This restriction is mainly for ease of notation and the extension to multi-output (MIMO) systems is fairly straightforward; see (Ljung, 1987; Lee, 1990; Wang, 1994). Stacking N consecutive samples on top of each other gives the data matrix

$$\mathbf{Z}_N = [\, \mathbf{z}(1)^T \quad \mathbf{z}(2)^T \quad \ldots \quad \mathbf{z}(N)^T \,]^T \in \mathbb{R}^{N \times (1+m)}. \tag{2.2}$$

It is of course crucial that the data reflect the important features of the underlying system. This will typically be the case if the input signals are "sufficiently" exciting and if large "enough" data sets are collected. However, such a situation is unrealistic in many real-world applications, since, firstly, the experimental time is limited and, secondly, many of the inputs are restricted to certain signal classes. Having to live with this reality, it is worth stressing that the problem of having incomplete data can be alleviated considerably by building various prior system properties into the models (or rather into the applied model structure).

The model structure $g(\boldsymbol{\varphi}(t), \boldsymbol{\theta})$. It is generally agreed that the single most difficult step in identification is that of model structure selection. Roughly speaking, the problem can be divided into three subproblems. The first one is to specify the *type* of model set to use. This involves the selection between linear and nonlinear representations, between black box, grey box and physically parameterized approaches, and so forth. The next issue is to decide the *size* of the model set. This includes the choice of possible variables (inputs and outputs) and combinations of variables to use in the models. It also involves fixing orders and degrees of the chosen model types, often to some intervals. The last item to consider is how to *parameterize* the model set: what basis functions should be used and how should these be parameterized, etc. With the type already determined here, in the sequel we will focus on the latter two issues.

Mathematically speaking, a quite general MISO predictor family or *model structure* is

$$\hat{y}(t|\boldsymbol{\theta}) = g(\boldsymbol{\varphi}(t), \boldsymbol{\theta}) \in \mathbb{R}, \tag{2.3}$$

where $\hat{y}(t|\boldsymbol{\theta})$ accentuates that the function $g(\cdot, \cdot)$ is a predictor, i.e., it is based on signals that are known at time t. The predictor structure is ensured by the regressor $\boldsymbol{\varphi}(t)$, which maps output signals up to index $t-1$ and input signals up to index t to an r-dimensional regression vector. This vector is often of the form

$$\boldsymbol{\varphi}(t) = [y(t-1) \quad \ldots \quad y(t-k) \quad u_1(t) \quad \ldots \quad u_1(t-k_1) \quad \ldots$$
$$u_m(t) \quad \ldots \quad u_m(t-k_m)]^T, \qquad (2.4)$$

although in general its entries can be any combinations of input-output signals known at time t. The mapping $g(\cdot,\cdot)$ (from $\mathbb{R}^r$ to $\mathbb{R}$) is parameterized by $\boldsymbol{\theta} \in \mathcal{D} \subset \mathbb{R}^d$, with the set $\mathcal{D}$ denoting the set of values over which $\boldsymbol{\theta}$ is allowed to range due to parameter restrictions. With this formulation, the work of finding a suitable model structure splits naturally into two subtasks, both possibly nonlinear in nature:

1. the choice of dynamics, i.e., the choice of regression vector $\boldsymbol{\varphi}(t)$, followed by
2. the choice of static mapping $g(\cdot,\cdot)$.

The selection criterion $V_N(\mathbf{Z}_N, \boldsymbol{\theta})$. Measured and model outputs never match perfectly in practice, but differ as

$$\varepsilon(t|\boldsymbol{\theta}) = y(t) - \hat{y}(t|\boldsymbol{\theta}), \qquad (2.5)$$

where $\varepsilon(t|\boldsymbol{\theta})$ is an error term reflecting unmodeled dynamics on one hand and noise on the other hand. An obvious modeling goal must be that this discrepancy is "small" in some sense. This is achieved by the selection criterion, which ranks different models according to some pre-determined cost function. The selection criterion can come in several shapes, although we will here start off with the usual quadratic measure of the fit between measured and predicted values, i.e., with

$$V_N(\mathbf{Z}_N, \boldsymbol{\theta}) = \frac{1}{N}\sum_{t=1}^{N}\frac{1}{2}\left(y(t) - \hat{y}(t|\boldsymbol{\theta})\right)^2 = \frac{1}{N}\sum_{t=1}^{N}\frac{1}{2}\varepsilon(t|\boldsymbol{\theta})^2. \qquad (2.6)$$

Once these three issues are settled we have in principle defined the searched for model. It then "only" remains to estimate the parameters $\boldsymbol{\theta}$ and to decide whether the model is good enough or not. If the model cannot be accepted some or even all of the entities above have to be reconsidered; in the worst conceivable case one must start from the very beginning and collect new data.

2.2 Nonlinear Model Structures – The Series Expansion Approach

In the introduction we stated that fuzzy models have much in common with neural networks. As a matter of fact, these and many other nonlinear modeling approaches can be viewed as series expansions. Adopting the ideas of the comprehensive and unifying work of (Sjöberg et al., 1995), such a function expansion can be written:

$$\hat{y}(t|\boldsymbol{\theta}) = g(\boldsymbol{\varphi}(t), \boldsymbol{\theta}) = \sum_{j=1}^{n} \alpha_j g_j(\boldsymbol{\varphi}(t), \boldsymbol{\beta}_j, \boldsymbol{\gamma}_j), \tag{2.7}$$

$$\boldsymbol{\theta}^T = [\,\boldsymbol{\alpha}^T \quad \boldsymbol{\beta}^T \quad \boldsymbol{\gamma}^T\,], \tag{2.8}$$

where, sometimes with abuse of notation, we call a $g_j(\cdot, \cdot, \cdot)$ a *basis function*. These are usually rather simple and typically they are all of one single type. The basis functions are also local in the sense that each $g_j(\cdot, \cdot, \cdot)$ essentially covers a certain part of the total regression space. Which part is specified by the parameters $\boldsymbol{\beta}_j$ and $\boldsymbol{\gamma}_j$, where $\boldsymbol{\beta}_j$ is related to the scale or direction of the basis function and $\boldsymbol{\gamma}_j$ specify the position or translation of it. The remaining α_j parameter is a "coordinate" parameter – a weight – giving the basis function its final amplitude shape.

The basic difference from one series expansion approach to another is the choice of basis functions. In principle, there are three fundamentally different ways of generalizing simple univariate basis functions to multi-variate ones:

Ridge construction. A ridge basis function has the form

$$g_j(\boldsymbol{\varphi}(t), \boldsymbol{\beta}_j, \boldsymbol{\gamma}_j) = \kappa\left(\boldsymbol{\beta}_j^T \boldsymbol{\varphi}(t) - \gamma_j\right), \tag{2.9}$$

where $\kappa(\cdot)$ is a function in one variable having parameters $\boldsymbol{\beta}_j \in \mathbb{R}^r$ and $\gamma_j \in \mathbb{R}$. Notice that the ridge nature has nothing to do with the choice of $\kappa(\cdot)$, but it is due to the fact that $\kappa(\cdot)$ is constant for all regression vectors in the sub-space where $\boldsymbol{\beta}_j^T \boldsymbol{\varphi}(t)$ is constant, thus forming a ridge along that direction; see Fig. 2.1. With n weighted ridge basis functions the dimension of $\boldsymbol{\theta}$ becomes $n(r+2)$. Typical examples of this family are, e.g., *feed-forward neural networks* with one hidden layer (Kung, 1993; Haykin, 1994; Ljung et al., 1996) and *hinging hyperplane models* (Breiman, 1993; Pucar and Sjöberg, 1995a; Pucar and Sjöberg, 1995b).

Radial construction. Radial basis functions do not show the ridge directional property but have true local support as illustrated in Fig. 2.1. Such a radial support can be obtained by using basis functions of the form

$$g_j(\boldsymbol{\varphi}(t), \boldsymbol{\beta}_j, \boldsymbol{\gamma}_j) = \kappa\left(\|\boldsymbol{\varphi}(t) - \boldsymbol{\gamma}_j\|_{\boldsymbol{\beta}_j}^2\right), \tag{2.10}$$

where the weighted norm $\boldsymbol{\beta}_j$ specify the scaling of the basis function. In general, $\boldsymbol{\beta}_j$ is a positive semi-definite and symmetric matrix of dimension $r \times r$, although quite often it is chosen to be a scaled identity matrix. This means that the dimension of $\boldsymbol{\theta}$ is at least $n(r+2)$ and at most $n(r^2+r+1)$. Popular choices within this category are *kernel estimators* (Watson, 1969), *radial basis function networks* (RBFN) (Poggio and Girosi, 1990; Chen and Billings, 1992), and *wavelet networks* (Zhang and Benveniste, 1992).

Composition. A composition (tensor product in (Sjöberg et al., 1995)) is obtained whenever ridge and radial constructions are combined when forming the basis functions. A typical example is shown in the rightmost plot of Fig. 2.1. The most extreme composition is

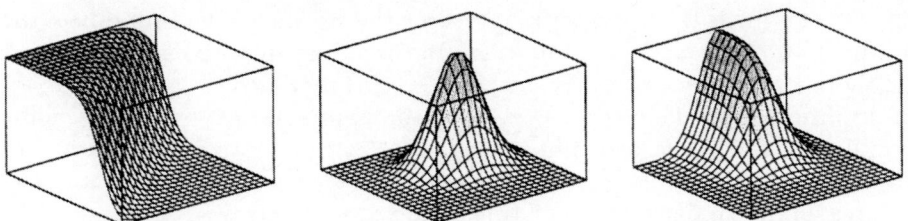

Fig. 2.1. From left to right: ridge construction, radial construction, and composition

$$g_j(\boldsymbol{\varphi}(t), \boldsymbol{\beta}_j, \boldsymbol{\gamma}_j) = \prod_{k=1}^{r} g_{j,k}(\varphi_k(t), \beta_{j,k}, \gamma_{j,k}), \tag{2.11}$$

where each $g_{j,k}(\cdot, \cdot, \cdot) \in \mathbb{R}$ is either a ridge or a radial function. In a more general setting, such an element need not live in $\mathbb{R}$ but can be defined in a any sub-space of $\mathbb{R}^r$. If all n basis functions are of the commonly encountered form (2.11), then it is easy to verify that the dimension of $\boldsymbol{\theta}$ becomes $n(2r+1)$. Within this model class we find certain *regression tree* approaches (Breiman et al., 1984; Strömberg et al., 1990) and, as will be discussed in the following section, the kind of *fuzzy identifiers* considered in this contribution.

2.3 General Parameter Estimation Techniques

After having determined the type of basis functions to apply, the next step is to use input-output data to estimate what is still unknown. It is here useful to distinguish the estimation needs by the kind of parameters involved in the models. The following three categories can be identified.

Structure estimation. This is the case when the type of basis functions to use have been decided, but where the size, i.e., the number of basis functions n to employ, is estimated. Selecting the r "best" regressors out of a set of possible regressors is a typical example in this category. It should be noted that structure estimation often can viewed as a combinatorial optimization problem that in complexity grows exponentially, e.g., with the number of regressors. This means that exhaustive algorithms soon become impractical, which has motivated schemes that provide, if not optimal, then at least good enough solutions. However, another way to reduce the complexity is to use prior structural system knowledge as is the case when a grey box approach is adopted.

Nonlinear-in-the-parameters estimation. Having decided the size of the model structure it remains to find reasonable parameter values $\boldsymbol{\theta}$. With the scalar loss function (2.6) as the performance criterion the parameter estimate $\hat{\boldsymbol{\theta}}_N$ is given by

$$\hat{\boldsymbol{\theta}}_N = \arg \min_{\boldsymbol{\theta} \in \mathcal{D}} V_N(\mathbf{Z}_N, \boldsymbol{\theta}), \tag{2.12}$$

where "arg min" is the operator that returns the argument that minimizes the loss function. This is a very important and well-known problem formulation leading to *prediction error minimization* (PEM) methods. The type of PEM algorithm to apply depends on whether the parameters θ enter the model structure in a linear or a nonlinear way. The latter situation leads to a *nonlinear least-squares* problem, and appears whenever the model structure (2.7) contains unknown direction β or translation γ parameters.

Linear-in-the-parameters estimation. In case all parameters enter the structure in a linear fashion one usually talks about a *linear least-squares* problem. For the series expansion (2.7) such an approach is applicable if only coordinate parameters α_j are to be estimated.

It should be emphasized that the complexity of the estimation problem decreases in the listed order, yet at the price that the amount of prior knowledge needed to arrive at a useful model typically increases. With these preliminary observations, we next present some different minimization algorithms, unconstrained as well as constrained ones.

2.3.1 Unconstrained linear least-squares algorithms. The parameters of an unconstrained linear least-squares structure (a linear regression) can be estimated efficiently and analytically by solving the normal equations

$$\varphi(t)\varphi(t)^T \hat{\theta}_N = \varphi(t)y(t) \qquad (2.13)$$

for $t = 1, \ldots, N$. The optimal parameter estimate is

$$\hat{\theta}_N = \left[\sum_{t=1}^{N} \varphi(t)\varphi(t)^T \right]^{-1} \sum_{t=1}^{N} \varphi(t)y(t) = \boldsymbol{R}_N^{-1} \boldsymbol{f}_N, \qquad (2.14)$$

provided that the inverse of the $d \times d$ regression matrix $\boldsymbol{R}_N$ exists. For numerical reasons this inverse is rarely formed, but instead the estimate is computed via so-called QR- or singular value decomposition (SVD) (Golub and Van Loan, 1989; Björk, 1996), both of which are able to handle rank deficient regression matrices.

2.3.2 Unconstrained nonlinear least-squares algorithms. When the parameters appear in a nonlinear fashion the typical situation is that the minimum of the loss function cannot be computed analytically. Instead we have to resort to certain iterative search routines, most of which can be seen as special cases of *Newton's algorithm* (see among many others (Dennis and Schnabel, 1983; Scales, 1985; Fletcher, 1987))

$$\hat{\theta}_N^{(i+1)} = \hat{\theta}_N^{(i)} - \left[V_N''(\mathbf{Z}_N, \hat{\theta}_N^{(i)}) \right]^{-1} V_N'(\mathbf{Z}_N, \hat{\theta}_N^{(i)})$$

$$= \hat{\theta}_N^{(i)} - \hat{\boldsymbol{\Delta}}_N^{(i)}(\mathbf{Z}_N, \hat{\theta}_N^{(i)}), \qquad (2.15)$$

where $\hat{\theta}_N^{(i)} \in \mathbb{R}^d$ is the parameter estimate at the i-th iteration, $V_N'(\cdot, \cdot) \in \mathbb{R}^d$ is the gradient of the loss function, and $V_N''(\cdot, \cdot) \in \mathbb{R}^{d \times d}$ the Hessian of it, both computed with respect to the current parameter vector. More specifically, the gradient is given by

$$V_N'(\mathbf{Z}_N, \hat{\boldsymbol{\theta}}_N^{(i)}) = -\frac{1}{N}\sum_{t=1}^{N} \boldsymbol{J}(t|\hat{\boldsymbol{\theta}}_N^{(i)})\varepsilon(t|\hat{\boldsymbol{\theta}}_N^{(i)}), \tag{2.16}$$

with $\boldsymbol{J}(t|\hat{\boldsymbol{\theta}}_N^{(i)}) \in \mathbb{R}^d$ being the *Jacobian* vector

$$\boldsymbol{J}(t|\hat{\boldsymbol{\theta}}_N^{(i)}) = \left[\frac{\partial \hat{y}(t|\hat{\boldsymbol{\theta}}_N^{(i)})}{\partial \hat{\theta}_1^{(i)}} \quad \cdots \quad \frac{\partial \hat{y}(t|\hat{\boldsymbol{\theta}}_N^{(i)})}{\partial \hat{\theta}_d^{(i)}} \right]^T. \tag{2.17}$$

Differentiating the gradient with respect to the parameters yields the *Hessian*

$$V_N''(\mathbf{Z}_N, \hat{\boldsymbol{\theta}}_N^{(i)}) = \frac{1}{N}\sum_{t=1}^{N} \left(\boldsymbol{J}(t|\hat{\boldsymbol{\theta}}_N^{(i)})\boldsymbol{J}(t|\hat{\boldsymbol{\theta}}_N^{(i)})^T - \frac{\partial \boldsymbol{J}(t|\hat{\boldsymbol{\theta}}_N^{(i)})}{\partial \hat{\boldsymbol{\theta}}_N^{(i)}}\varepsilon(t|\hat{\boldsymbol{\theta}}_N^{(i)}) \right), \tag{2.18}$$

which thus means that the second derivative of the loss function is needed in (2.15). Simply put, Newton's algorithm searches for the new parameter vector along a Hessian modified gradient of the current loss function.

The availability of derivatives of the loss function with respect to the parameters is of course of paramount importance in all Newton-based estimation schemes. In case arbitrary (though differentiable) predictor structures are considered these may very well be too hard to obtain analytically or too expensive to compute. One way around this difficulty is to numerically approximate the derivatives by finite differences. The simplest such method is just to replace each of the d elements of the Jacobian by the forward difference

$$J_j(t|\hat{\boldsymbol{\theta}}_N^{(i)}) = \frac{\partial \hat{y}(t|\hat{\boldsymbol{\theta}}_N^{(i)})}{\partial \hat{\theta}_j^{(i)}} \approx \frac{\hat{y}(t|\hat{\boldsymbol{\theta}}_N^{(i)} + h_j e_j) - \hat{y}(t|\hat{\boldsymbol{\theta}}_N^{(i)})}{h_j}, \tag{2.19}$$

where e_j is a column vector with a one in the j-th position and zeros elsewhere, and where h_j is a small positive scalar perturbation. Because the parameters may differ substantially in magnitude it is expedient here to choose these perturbations individually. A typical choice is $h_j = \sqrt{\epsilon}\max(h_{\min}, |\hat{\theta}_j^{(i)}|)$, where ϵ is the relative machine precision and $h_{\min} > 0$ is the smallest perturbation allowed; consult (Dennis and Schnabel, 1983; Scales, 1985) for further details on this. If a more accurate approximation is deemed necessary one can employ the central difference

$$J_j(t, \hat{\boldsymbol{\theta}}_N^{(i)}) = \frac{\partial \hat{y}(t|\hat{\boldsymbol{\theta}}_N^{(i)})}{\partial \hat{\theta}_j^{(i)}} \approx \frac{\hat{y}(t|\hat{\boldsymbol{\theta}}_N^{(i)} + h_j e_j) - \hat{y}(t|\hat{\boldsymbol{\theta}}_N^{(i)} - h_j e_j)}{2h_j}, \tag{2.20}$$

at the cost of d additional function evaluations.

It now turns out that the Newton update (2.15) has some severe drawbacks, most of which are associated with the computation of the Hessian (2.18). Firstly, it is in general expensive to compute the derivative of the Jacobian. It may also happen that the inverse of the Hessian does not exist, so if further progress towards a minimum is to be made the update vector must be constructed in a different way. Furthermore, even if the inverse exists it is not guaranteed to be positive definite and it may therefore happen that the parameter update vector is such that the loss function actually becomes larger. Finally, although the parameter update vector is a descent one it might be much too large, locating the new parameters at a point with higher loss than what is currently the case. To avoid these problems other search directions than (2.15) are much more common in practice:

Gradient method. Simply replace the Hessian by an identity matrix of appropriate size. This, however, does not prevent the update vector from being so large that also $V_N(\cdot,\cdot)$ becomes larger. To avoid such behavior the updating is often complemented with a line search technique

$$\hat{\Delta}_N^{(i)}(\mathbf{Z}_N,\hat{\theta}_N^{(i)}) = \mu^{(i)} V_N'(\mathbf{Z}_N,\hat{\theta}_N^{(i)}), \qquad (2.21)$$

where $0 < \mu^{(i)} \leq 1$, thereby giving a *damped gradient algorithm*. The choice of step length $\mu^{(i)}$ is not critical, and the procedure often used is to start with $\mu^{(i)} = 1$ and then repeatedly halve it until a lower value of the loss function is obtained.

Gauss–Newton method. By neglecting the second derivative term of the Hessian (2.18) and including line search as above we arrive at a *damped Gauss–Newton algorithm* with update vector

$$\hat{\Delta}_N^{(i)}(\mathbf{Z}_N,\hat{\theta}_N^{(i)}) =$$
$$\mu^{(i)} \left[\sum_{t=1}^N \boldsymbol{J}(t|\hat{\theta}_N^{(i)})\boldsymbol{J}(t|\hat{\theta}_N^{(i)})^T \right]^{-1} \sum_{t=1}^N \boldsymbol{J}(t|\hat{\theta}_N^{(i)})\varepsilon(t|\hat{\theta}_N^{(i)}), \qquad (2.22)$$

which is of the same form as the linear least-squares formula (2.14). To cope with a singular or near to singular Hessian approximation the inverse is normally replaced by the so-called pseudo-inverse, which can be obtained easily by computing the SVD of the Jacobian (Golub and Van Loan, 1989).

Levenberg–Marquardt method. The *Levenberg–Marquardt algorithm* handles simultaneously the update step size and the singularity problems through the update

$$\hat{\Delta}_N^{(i)}(\mathbf{Z}_N,\hat{\theta}_N^{(i)}) =$$
$$\left[\left(\sum_{t=1}^N \boldsymbol{J}(t|\hat{\theta}_N^{(i)})\boldsymbol{J}(t|\hat{\theta}_N^{(i)})^T \right) + \mu^{(i)}\boldsymbol{I} \right]^{-1} \sum_{t=1}^N \boldsymbol{J}(t|\hat{\theta}_N^{(i)})\varepsilon(t|\hat{\theta}_N^{(i)}), \quad (2.23)$$

where the Hessian is guaranteed to be positive definite since $\mu^{(i)} > 0$. As is the case for the above procedures, it can be shown that this update is in a descent direction. However, $\mu^{(i)}$ must be carefully chosen so that the loss function also decreases. The method by Marquardt (Scales, 1985) achieves this by starting with a $\mu^{(i)} > 0$, whereupon it is reduced (typically a factor 10) at the beginning of each iteration, thereby resembling a Gauss–Newton update step. If this results in an increased loss, then the step $\mu^{(i)}$ is repeatedly increased (typically a factor 10) until $V_N(\mathbf{Z}_N, \hat{\boldsymbol{\theta}}_N^{(i+1)}) < V_N(\mathbf{Z}_N, \hat{\boldsymbol{\theta}}_N^{(i)})$, which means that the update is forced towards a scaled gradient direction. Other and more elaborate choices of $\mu^{(i)}$ are discussed in, e.g., (Fletcher, 1987).

Although simple, a major drawback with the gradient method is that the convergence rate can be fairly poor close to the minimum. This fact favors the latter two methods, which, especially near the minimum, show convergence properties similar to those of the full Newton algorithm (Dennis and Schnabel, 1983). For ill-conditioned problems (Dennis and Schnabel, 1983) recommend the Levenberg–Marquardt modification. However, this choice is far less obvious when the pseudo-inverse is used in the Gauss–Newton update. In such a case both methods try to update the parameters that really influence the criterion fit most, whereas the remaining parameters are kept unchanged. This means that so-called *regularization* is built into the algorithms (see below).

A last algorithmic issue to consider here is when to terminate the search. In theory, $V_N'(\cdot, \cdot)$ is zero at a minimum, so an obvious practical test is to terminate once $|V_N'(\cdot, \cdot)|$ is sufficiently small. Another useful test is to investigate the relative change in parameters from one iteration to another and terminate if this quantity falls below some tolerance level. The algorithms will also terminate when a certain number of maximum iterations has been carried out, or if the line search algorithm fails to decrease the loss function in a predetermined number of iterations.

It is worth stressing that the three schemes above all return estimates that are at least as good as the starting point. Nonetheless, should the algorithm converge to a minimum, then it is important to remember that convergence needs not be to a global minimum but can be to a local one.

2.3.3 Constrained minimization algorithms. In a grey box modeling situation the parameters usually have physical or linguistic significance. To really maintain such a property it is necessary to take the corresponding parameter restrictions into account in the estimation procedure, i.e., *constrained optimization* methods are needed.

Therefore assume that there are ℓ parameter constraints collected in a vector

$$c(\boldsymbol{\theta}) = [\, c_1(\boldsymbol{\theta}) \quad c_2(\boldsymbol{\theta}) \quad \dots \quad c_\ell(\boldsymbol{\theta}) \,]^T \in \mathbb{R}^\ell, \tag{2.24}$$

where each $c_j(\boldsymbol{\theta})$ is a well-defined function such that $c_j(\boldsymbol{\theta}) > 0$ for $j = 1, \dots, \ell$, hence specifying a feasible parameter region $\mathcal{D}$. There exist quite a few schemes that handle such constraints, see, e.g., (Scales, 1985). An old but simple and versatile idea is to rephrase the original problem into a sequence of unconstrained minimization problems for which a Newton-type method (like the gradient one) can be applied without too much extra coding effort.

This is the basic idea behind the *barrier function* estimation procedure. Algorithmically, the method starts with a feasible parameter vector $\hat{\boldsymbol{\theta}}_N^{(0)}$, whereupon the parameter estimate is iteratively obtained by solving (each iteration is started with the estimate of the previous iteration)

$$
\begin{aligned}
\hat{\boldsymbol{\theta}}_N^{(k+1)} &= \arg\min_{\boldsymbol{\theta}\in\mathcal{D}} W_N(\mathbf{Z}_N,\boldsymbol{\theta}) \\
&= \arg\min_{\boldsymbol{\theta}\in\mathcal{D}} \left(V_N(\mathbf{Z}_N,\boldsymbol{\theta}) + \rho^{(k)}\sum_{j=1}^{\ell}\vartheta(c_j(\boldsymbol{\theta})) \right),
\end{aligned}
\tag{2.25}
$$

where, typically, $\rho^{(k)} = 10^{-k}$ with k starting from 0 and then increasing by 1 for each iteration until convergence is obtained. In order to maintain a feasible estimate the barrier function $\vartheta(\cdot)$ is chosen so that an increasingly larger value is added to the objective function $W_N(\cdot,\cdot)$ as the boundary of the feasibility region $\mathcal{D}$ is approached from the interior; at the boundary itself this quantity should be infinite. A good choice of barrier function for many kinds of problems seems to be the *log barrier function* $\vartheta(c_j(\boldsymbol{\theta})) = -\ln(c_j(\boldsymbol{\theta}))$; see (Scales, 1985) for further details on this.

At this stage one may wonder why it is not sufficient to set $\rho^{(k)}$ to a much smaller value initially. One reason is that if the true minimum is near the boundary, then it could be difficult to minimize the overall cost function because of its rapidly changing curvature near this minimum, thus giving rise to an ill-conditioned problem. One could also argue that the method is too complex as an outer iteration is added. This is only partially true as the inner estimate (especially at the first few outer iterations) need not be that accurate. A rule of thumb is to perform only some five iterations in the inner loop. Finally, the outer loop is terminated once the parameter update is sufficiently small or when a number of maximum outer iterations has been carried out.

2.4 The Bias-Variance Trade-Off

The series expansion approach has been widely used in nonlinear black box identification, where the idea is to employ a parameterization that covers a system class as broad as possible. In practice, however, the typical situation is that merely a fraction of the available flexibility is really needed, i.e., the applied model structures are often over-parameterized. This fact, possibly in combination with an insufficiently informative data set $\mathbf{Z}_N$, leads to ill-conditioning of the Jacobian and the Hessian. This observation also suggests that the parameters should be divided into two sets: the set of *spurious* parameters, which do not influence the criterion fit that much, and the set of *efficient* parameters, which do affect the fit. Having such a decomposition it is intuitively reasonable to treat the spurious or redundant parameters as constants that are not estimated. The problem with this is now that it is in general hard to make this decomposition beforehand.

However, using data one can overcome the ill-conditioning problem and automatically unveil an efficient parameterization by incorporating regularization techniques (or trust region techniques) (Dennis and Schnabel, 1983). When such an effect is built into the estimation procedure, as in the Levenberg–Marquardt algorithm, we get so-called implicit regularization, as opposed to explicit regularization, which is obtained by adding a penalty term to the criterion function, e.g.,

$$
\begin{aligned}
W_N(\mathbf{Z}_N, \boldsymbol{\theta}) &= \left(\frac{1}{N} \sum_{t=1}^{N} \frac{1}{2} \varepsilon(t|\boldsymbol{\theta})^2 \right) + \frac{\mu}{2} |\boldsymbol{\theta} - \boldsymbol{\theta}^{\sharp}|^2 \\
&= V_N(\mathbf{Z}_N, \boldsymbol{\theta}) + \frac{\mu}{2} |\boldsymbol{\theta} - \boldsymbol{\theta}^{\sharp}|^2,
\end{aligned} \tag{2.26}
$$

where $\mu > 0$ is a small user-tunable parameter ensuring a positive definite Hessian and $\boldsymbol{\theta}^{\sharp} \in \mathbb{R}^d$ is some a priori determined parameter vector (possibly representing prior parameter knowledge). Here the important point is that a parameter not affecting the first term that much will be kept close to $\boldsymbol{\theta}^{\sharp}$ by the second term. This means that the regularization parameter μ can be viewed as a threshold that labels the parameters to be either efficient or spurious (Sjöberg et al., 1995). A large μ simply means that the number of efficient parameters $d^{\sharp}$ becomes small.

From a system identification viewpoint regularization is a very important means for addressing the ever present bias-variance trade-off, as is emphasized in (Ljung, 1987; Ljung et al., 1996). There it is shown, under fairly general assumptions, that the asymptotic criterion misfit essentially depends on two factors that can be affected by the choice of model structure. First we have the *bias error*, which reflects the misfit between the true system and the best possible approximation of it, given a certain model structure. Typically, this error decreases when the number of parameters d increases. The other term is the parameter *variance error*, which usually grows with d but decreases with N. There is thus a clear trade-off between the bias and the variance contributions.

At this point, suppose that a flexible enough model structure has been decided upon. Decreasing the number of parameters that are actually updated ($d^{\sharp}$) by increasing μ is beneficial for the total misfit as long as the decrease in variance error is larger than the increase in bias error. In other words, the purpose of regularization is to decrease the variance error contribution to a level where it balances the bias misfit.

3. Fuzzy Modeling Framework

The history of methods based on fuzzy concepts is rather short. It all started in the mid 1960s with Zadeh's pioneering article (Zadeh, 1965), in which a new way of characterizing non-probabilistic uncertainties via so-called fuzzy sets was suggested. Since then, and especially in the last ten years or so, there has been a dramatic growth of sub-disciplines in science and engineering that have adopted fuzzy ideas. To a great extent this development is due to a large number of successful industrial applications, spanning such diverse fields as robotics, consumer electronics, signal processing, bioengineering, image processing, pattern recognition, management and control. See the comprehensive compilations (Marks II, 1994, Chen, 1996).

The fields of fuzzy control and fuzzy identification have been developed largely in parallel. A good first book on fuzzy control is (Driankov et al., 1993), and a shorter but informative overview is given by (Lee, 1990). Various fuzzy identification methods have been proposed by several authors. The work by Sugeno and coworkers (Takagi and Sugeno, 1985, Sugeno and Kang, 1988, Sugeno and Yasukawa, 1993) and by (Wang, 1995) constitute some of the most influential contributions. The merging of fuzzy control and fuzzy identification is discussed, e.g., in (Wang, 1994) and in (Roger Jang and Sun, 1995). Many of the ideas detailed in this section can be found in the latter reference, which is exceptionally well written and highly recommended. With these sources as a basis, the aim of the section is to derive and motivate the use of one particular fuzzy rule base interpretation that is suited for identification purposes.

3.1 Components of a Fuzzy Model

The basic configuration of a *fuzzy model* is shown in Fig. 3.1. The model involves six components, of which the four lowermost ones are fuzzy model specific.

Scaling. The physical values of the actual inputs and outputs may differ significantly in magnitude. By mapping these to proper normalized domains via scaling, one can instead work with signals that roughly are of the same magnitude, which is desirable from an estimation point of view. However, the need for scaling is highly problem dependent and therefore not considered any further here, i.e., from now on we assume that $\varphi(t)$ is formed directly from $\mathbf{z}(t)$ and that $\hat{y}_s(t|\boldsymbol{\theta}) = \hat{y}(t|\boldsymbol{\theta})$.

Regressor generator. The kind of dynamics to include in a fuzzy model is embodied in the regressor generator. The regression vector $\varphi(t)$ can contain any at time t known combinations of input-output measurements $\mathbf{z}(t)$, although for such a combination to make sense it ought to have a linguistic interpretation. This is due to the fact that the entries of $\varphi(t)$ are specified by the linguistic database or, actually, by so-called linguistic variables (see below). Such a typical variable is SPEED$(t-1)$, which in terms of input-output data may be interpreted as $z_1(t-1)$. This also means that the mathematical purpose of the remaining components of a fuzzy model is to provide a static map from $\varphi(t) \in \mathbb{R}^r$ to $\hat{y}(t|\boldsymbol{\theta}) \in \mathbb{R}$.

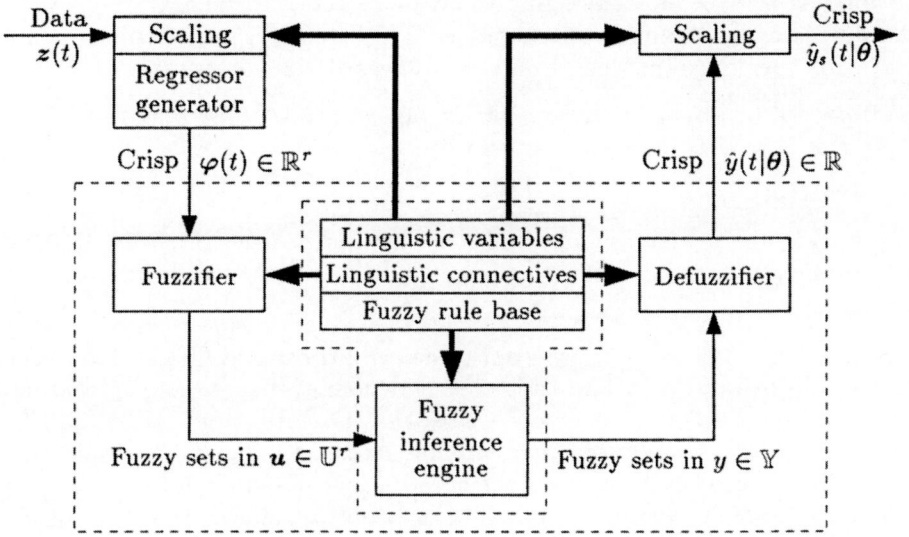

Fig. 3.1. Structure of a MISO fuzzy model. Thin arrows indicate the computational flow and thick arrows the information flow. The grey box is a linguistic database, reflecting prior knowledge.

Linguistic database. The linguistic database is the heart of a fuzzy model. The expert knowledge, which is assumed to be given as a number of if-then rules, is stored in a fuzzy rule base. These rules are subsequently given a precise mathematical meaning through user-supplied definitions of the linguistic variables and connectives employed (and, or, etc.).

Fuzzifier. The fuzzifier maps the crisp values of $\varphi(t)$ into suitable fuzzy sets (discussed below).

Fuzzy inference engine. The fuzzy sets provided by the fuzzifier are then interpreted by the fuzzy inference engine, which uses the fuzzy rule base knowledge in order to produce some fuzzy sets in the output y.

Defuzzifier. As a last step the defuzzifier converts the output fuzzy sets to a standard crisp signal $\hat{y}(t|\theta) \in \mathbb{R}$.

From this short description it should be clear that fuzzy sets are vital objects to comprehend in order to understand how a fuzzy model operates. Let us therefore discuss such sets in more detail.

3.2 Fuzzy Sets and Membership Functions

An ordinary set is a set with a crisp boundary, i.e., an element can either be or not be a member of that set. A *fuzzy set* on the other hand does not show this absolute "either-or" membership property. The transition from "belonging to" to "not belonging to" a fuzzy set is instead gradual, where the degree of belonging is characterized by a *membership function*. Mathematically speaking, the definition is as follows (Driankov et al., 1993).

Definition 3.1. *If u is an element in the universe of discourse $\mathbb{U}$, then a fuzzy set A in $\mathbb{U}$ is the set of ordered pairs*

$$A = \{(u, \mu_A(u)) : u \in \mathbb{U}\}, \tag{3.1}$$

where $\mu_A(u)$ is a membership function carrying an element from $\mathbb{U}$ into a membership value between 0 (no degree of membership) and 1 (full degree of membership).

Example 3.1. Suppose that we want to describe a car traveling at high speed on the motorway. As a first step, let u denote the speed of any car and introduce $\mathbb{U} = [0, 300] \subset \mathbb{R}$, which states that no car can go faster than 300 km/h. By Nordic standards, a car running at, say, 140 km/h is considered to have a high speed, but not when the speed is, say, 80 km/h. Moreover, when a car is running at around 110 km/h most people would say that the speed is neither low nor high. Based on this information, a fuzzy set describing that the speed of a car is HIGH is, e.g.,

$$\text{HIGH} = \left\{ \left(u, \mu_{\text{HIGH}}(u) = \frac{1}{1 + e^{-0.1(u-110)}} \right) : u \in [0, 300] \right\}. \tag{3.2}$$

This subjective choice of membership function gives that cars running at 80, 110 and 140 km/h are considered to go at a high speed to a degree of 0.05, 0.50 and 0.95, respectively.

An important point illustrated in this example is that the fuzziness does not emanate from the fuzzy set itself but rather from the vagueness of what it describes. This is manifested by the subjective and non-random nature of the choice of membership function, which may vary considerably depending on who determined it. This is also the main philosophical difference between fuzzy memberships and probabilities (which convey objective information about random phenomena).

As noted above, the membership function (MF) can be any function producing a value between 0 and 1. Here we will focus on three common classes of MFs, all convex in nature, i.e., the membership functions are of the form "increasing", "decreasing", or "bell-shaped"; see (Driankov et al., 1993) for the mathematical definition. First we have what may be called the *network-classic* MFs, which because of their smoothness are becoming increasingly popular in fuzzy modeling.

Definition 3.2 (Network-classic MFs). *This class consists of the sigmoidal and the Gaussian membership functions defined as*

$$\texttt{mfsig}(u, \beta, \gamma) : \qquad \mu_A(u, \beta, \gamma) = \frac{1}{1 + e^{-\beta(u-\gamma)}},$$

$$\texttt{mfgauss}(u, \beta, \gamma) : \qquad \mu_A(u, \beta, \gamma) = e^{-\frac{1}{2}\left(\frac{u-\gamma}{\beta}\right)^2},$$

where β and γ are related to the scale and the position of the membership function, respectively.

The second class, widely used in fuzzy logic theory, was originally suggested by Zadeh, thus meriting the label *Zadeh-formed MFs*.

Definition 3.3 (Zadeh-formed MFs). *The Zadeh-formed MFs are the Z-, the S-, and the π-functions (named after their shape) in order defined as* $(\gamma_1 \leq \gamma_2 \leq \gamma_3 \leq \gamma_4)$

$$\mathtt{mfz}(u, \gamma_1, \gamma_2) : \mu_A(u, \gamma) = \begin{cases} 1 & u \leq \gamma_1 \\ 1 - 2\left(\frac{u-\gamma_1}{\gamma_2-\gamma_1}\right)^2 & \gamma_1 < u \leq \frac{\gamma_1+\gamma_2}{2} \\ 2\left(\frac{u-\gamma_2}{\gamma_2-\gamma_1}\right)^2 & \frac{\gamma_1+\gamma_2}{2} < u \leq \gamma_2 \\ 0 & u > \gamma_2, \end{cases}$$

$$\mathtt{mfs}(u, \gamma_1, \gamma_2) : \mu_A(u, \gamma) = 1 - \mathtt{mfz}(u, \gamma_1, \gamma_2),$$

$$\mathtt{mfpi}(u, \gamma_1, \gamma_2, \gamma_3, \gamma_4) : \mu_A(u, \gamma) = \begin{cases} \mathtt{mfs}(u, \gamma_1, \gamma_2) & u \leq \gamma_2 \\ 1 & \gamma_2 < u \leq \gamma_3 \\ \mathtt{mfz}(u, \gamma_3, \gamma_4) & u > \gamma_3. \end{cases}$$

The last category is that of *piecewise linear MFs*, which, primarily because of real-time aspects, have been extensively used in various fuzzy control applications (Driankov et al., 1993).

Definition 3.4 (Piecewise linear MFs). *The piecewise linear MFs are the open left, the open right, the triangular, and the trapezoidal functions* $(\gamma_1 \leq \gamma_2 \leq \gamma_3 \leq \gamma_4)$:

$$\mathtt{mfl}(u, \gamma_1, \gamma_2) : \qquad \mu_A(u, \gamma) = \max\left(\min\left(\tfrac{\gamma_2-u}{\gamma_2-\gamma_1}, 1\right), 0\right),$$

$$\mathtt{mfr}(u, \gamma_1, \gamma_2) : \qquad \mu_A(u, \gamma) = \max\left(\min\left(\tfrac{u-\gamma_1}{\gamma_2-\gamma_1}, 1\right), 0\right),$$

$$\mathtt{mftri}(u, \gamma_1, \gamma_2, \gamma_3) : \qquad \mu_A(u, \gamma) = \max\left(\min\left(\tfrac{u-\gamma_1}{\gamma_2-\gamma_1}, \tfrac{\gamma_3-u}{\gamma_3-\gamma_2}\right), 0\right),$$

$$\mathtt{mftrap}(u, \gamma_1, \gamma_2, \gamma_3, \gamma_4) : \quad \mu_A(u, \gamma) = \max\left(\min\left(\tfrac{u-\gamma_1}{\gamma_2-\gamma_1}, 1, \tfrac{\gamma_4-u}{\gamma_4-\gamma_3}\right), 0\right).$$

Notice that with the terminology from the previous section an MF is really nothing but a basis function, and since it involves one variable (u) only it is of composition type. As will be evident in the following section, fuzzy sets constitute the main building block of a linguistic variable.

3.3 Linguistic Variables and Fuzzy Propositions

Linguistic variables are fundamental in approximate or fuzzy reasoning. In a generalized form, cf. (Driankov et al., 1993), such a variable is conveniently described by a triple:

$$\langle U, \boldsymbol{A}(\mathbb{U}, \boldsymbol{\theta}), D \rangle, \tag{3.3}$$

where U is the name of the variable, $\boldsymbol{A}(\cdot, \cdot)$ is a set of linguistic values, each of which is characterized by a fuzzy set, that can be assigned to U, and D

provides information on how to connect the linguistic domain to the physical measurement domain.

Example 3.2. The linguistic variable SPEED$(t-1)$ of a car on a motorway, is, e.g.,

$$\langle \text{SPEED}(t-1), \boldsymbol{A}(\mathbb{U}, \boldsymbol{\theta}) = \{\text{LOW}, \text{MEDIUM}, \text{HIGH}\}, D : \varphi_1(t) = z_1(t-1)\rangle$$
$$\text{LOW} = \{(u, \mu_{\text{LOW}}(u, \boldsymbol{\gamma}) = \texttt{mfl}(u, 60, 90)) : u \in \mathbb{U}\}$$
$$\text{MEDIUM} = \{(u, \mu_{\text{MEDIUM}}(u, \boldsymbol{\gamma}) = \texttt{mftrap}(u, 60, 90, 110, 140)) : u \in \mathbb{U}\}$$
$$\text{HIGH} = \{(u, \mu_{\text{HIGH}}(u, \boldsymbol{\gamma}) = \texttt{mfr}(u, 110, 140)) : u \in \mathbb{U}\},$$

where $\mathbb{U} = [0, 300]$.

The assignment of values to a linguistic variable is simply achieved by an atomic fuzzy proposition using the syntax "U is PROPERTY", e.g., "SPEED$(t-1)$ is LOW."

Several atomic fuzzy propositions can now be combined using linguistic connectives such as 'not', 'and', and 'or', thus forming more complex propositions such as

$$(U_1 \text{ is } \mathbf{not} \ A_1) \ \mathbf{and} \ (U_2 \text{ is } A_2), \tag{3.4}$$

where A_1 and A_2 refer to two different fuzzy sets which normally are defined in different universes $\mathbb{U}_1$ and $\mathbb{U}_2$, respectively. While it is mathematically natural to interpret $(U_1 \text{ is } \mathbf{not} \ A_1)$ as the fuzzy set $1 - \mu_{A_1}(u_1, \cdot)$ with $\mu_{A_1}(u_1, \cdot)$ being the MF associated with A_1, there are many different ways of interpreting 'and' and 'or'. Often, however, a fuzzy conjunction (and) is defined in terms of a *triangular norm* $\star$, which combines MFs as $\mu_{A_1}(u_1, \cdot) \star \mu_{A_2}(u_2, \cdot)$; see (Driankov et al., 1993) for the details. The most widely used triangular norms are intersection (the min operator) and algebraic product (multiplication). Similarly, a fuzzy disjunction (or) is usually defined as a *triangular co-norm* $\dot{+}$, syntactically written $\mu_{A_1}(u_1, \cdot) \dot{+} \mu_{A_2}(u_2, \cdot)$. The most commonly encountered co-norms are union (the max operator) and algebraic sum $(\mu_{A_1}(u_1, \cdot) + \mu_{A_2}(u_2, \cdot) - \mu_{A_1}(u_1, \cdot) \cdot \mu_{A_2}(u_2, \cdot))$.

If, in the above operations, u_1 and u_2 are defined in different universes, then a triangular norm or co-norm performs a mapping from $[0, 1] \times [0, 1]$ to $[0, 1]$. Otherwise, the mapping is from $[0, 1]$ to $[0, 1]$. By combining several atomic fuzzy expressions using suitable connectives (others, different than those above, can of course also be defined) it is possible to construct arbitrarily complex fuzzy sets. In doing so the important point is that the result always is a new fuzzy set, although the space in which it is defined is not restricted to one or two dimensions.

3.4 Fuzzy Model Structure

We are now in a position to discuss the computational units of Fig. 3.1. As with the interpretation of the basic connectives, there are also here a number of choices to be made. However, various grey box identification aspects, e.g., the complexity of computing predictors, Jacobians, and possibly Hessians, the approximation capability, and the interpretability of the estimated models, naturally lead to one particular type of fuzzy model structure, as will be derived next.

Fuzzification. The fuzzification unit is conceptually quite simple. For each value of $\varphi_k(t)$, $k = 1, 2, \ldots, r$, it returns a fuzzy set (a fuzzy fact) denoted $A_k^!$ with MF $\mu_{A_k^!}(u_k) \in \mathbb{U}_k$. Thus, if we are given N measurements and use r regressors, then this implies that $N \cdot r$ fuzzy sets will be generated by the fuzzifier. These sets can in principle be constructed either by a singleton fuzzifier or by a non-singleton one. The latter approach may be used to capture noise properties of the regressors (Wang, 1994), though at a rather high computational cost. This fact motivates the use of a *singleton fuzzifier*, which simply returns a 1 in a single point in $\mathbb{U}_k$:

$$\mu_{A_k^!}(u_k) = \begin{cases} 1 & \text{if } u_k = \varphi_k(t) \\ 0 & \text{otherwise.} \end{cases} \tag{3.5}$$

Fuzzy rule base. A *fuzzy rule base* $\boldsymbol{R}$ consists of a set of, say, n fuzzy rules. Using a somewhat unorthodox grid-oriented multi-indexing labeling system, such a rule base often takes the form:

$$
\begin{aligned}
&R_{1,\ldots,1}: && \textbf{If } (U_1 \text{ is } A_{1,1}) \textbf{ and } \ldots \textbf{ and } (U_r \text{ is } A_{1,r}) \\
& && \textbf{then } (Y \text{ is } B_{1,\ldots,1}) \\
&R_{1,\ldots,2}: && \textbf{If } (U_1 \text{ is } A_{1,1}) \textbf{ and } \ldots \textbf{ and } (U_r \text{ is } A_{2,r}) \\
& && \textbf{then } (Y \text{ is } B_{1,\ldots,2}) \\
& && \qquad\qquad \vdots \\
&R_{1,\ldots,n_r}: && \textbf{If } (U_1 \text{ is } A_{1,1}) \textbf{ and } \ldots \textbf{ and } (U_r \text{ is } A_{n_r,r}) \\
& && \textbf{then } (Y \text{ is } B_{1,\ldots,n_r}) \\
& && \qquad\qquad \vdots \\
&R_{n_1,\ldots,1}: && \textbf{If } (U_1 \text{ is } A_{n_1,1}) \textbf{ and } \ldots \textbf{ and } (U_r \text{ is } A_{1,r}) \\
& && \textbf{then } (Y \text{ is } B_{n_1,\ldots,1}) \\
&R_{n_1,\ldots,2}: && \textbf{If } (U_1 \text{ is } A_{n_1,1}) \textbf{ and } \ldots \textbf{ and } (U_r \text{ is } A_{2,r}) \\
& && \textbf{then } (Y \text{ is } B_{n_1,\ldots,2}) \\
& && \qquad\qquad \vdots \\
&R_{n_1,\ldots,n_r}: && \textbf{If } (U_1 \text{ is } A_{n_1,1}) \textbf{ and } \ldots \textbf{ and } (U_r \text{ is } A_{n_r,r}) \\
& && \textbf{then } (Y \text{ is } B_{n_1,\ldots,n_r}),
\end{aligned}
\tag{3.6}
$$

where $A_{1,1}, \ldots, A_{n_r,r}$ are the linguistic values that can be assigned to the linguistic variables $U_1, \ldots, U_r$, while $B_{1,\ldots,1}, \ldots, B_{n_1,\ldots,n_r}$ denote the linguistic values that can be allotted to the linguistic output Y. As usual, the mathematical meaning of any $A_{j_k,k}$ and $B_{j_1,\ldots,j_r}$ are given by suitable membership functions, denoted $\mu_{A_{j_k,k}}(u_k, \cdot)$ and $\mu_{B_{j_1,\ldots,j_r}}(y, \cdot)$, respectively.

With the rule base (3.6) it is worthwhile stressing that each $B_{j_1,\ldots,j_r}$ needs not be linguistically unique. On the contrary, the typical situation is that many rules share the same linguistic consequence, yet they have different antecedents. Furthermore, each U_j can be assigned to n_j different linguistic values, which means that the number of rules in a *complete* fuzzy rule base becomes $n = \prod_{k=1}^{r} n_k$. In case the rule base is incomplete there will be regions (e.g., physically impossible ones) in the overall regression space for which no output can be inferred. Notice also that a rule base like (3.6) with 'and' connectives only is not as restricted as one might first suspect. The reason is that many other constructs (e.g., 'not' and 'or') can be logically converted (e.g., by De Morgan's Law) to the desired form, as is shown in (Wang, 1994; Lindskog, 1996).

Interpreting fuzzy if-then rules. Each of the rules of (3.6) describes essentially a relation between $U_1, \ldots, U_r$ on one side and Y on the other. This suggests that a rule $R_{j_1, \ldots, j_r}$ should be defined as a fuzzy relation with MF $\mu_{A_j \to B_j}(\boldsymbol{u}, y, \cdot) = \mu_{A_{j_1,1} \times \ldots \times A_{j_r,r} \to B_{j_1, \ldots, j_r}}(u_1, \ldots, u_r, y, \cdot)$ defined on $\mathbb{U}_1 \times \ldots \times \mathbb{U}_r \times \mathbb{Y}$. There now exist some 40 different ways of interpreting implication (Lee, 1990), but striving for computational simplicity we also here choose the simplest and most widely used translation, namely

$$
\begin{aligned}
\mu_{A_j \to B_j}(\boldsymbol{u}, y, \cdot) &= \mu_{A_j}(\boldsymbol{u}, \cdot) \star \mu_{B_j}(y, \cdot) \\
&= \mu_{A_{j_1},1}(u_1, \cdot) \star \ldots \star \mu_{A_{j_r},r}(u_r, \cdot) \star \mu_{B_j}(y, \cdot),
\end{aligned}
\tag{3.7}
$$

which is sometimes referred to as *Mamdani implication*.

Inference engine. Faced with some facts from the fuzzifier and a fuzzy rule base (3.6), the fuzzy inference engine is responsible for inferring conclusions in terms of output fuzzy sets. The most widely used inference mechanism is *generalized modus ponens* (GMP), which, as the name indicates, is a generalization of the classical modus ponens rule to the fuzzy domain; see, e.g., (Driankov et al., 1993).

More specifically, the GMP inference scheme takes some fuzzy facts as input, maps it via a fuzzy relation – the rule representation – and returns a fuzzy conclusion:

$$
\frac{(U_1 \text{ is } A_1^!) \text{ and } \ldots \text{ and } (U_r \text{ is } A_r^!)}{\mathbf{If} \ (U_1 \text{ is } A_{j_1,1}) \text{ and } \ldots \text{ and } (U_r \text{ is } A_{j_r,r}) \ \mathbf{then} \ (Y \text{ is } B_{j_1, \ldots, j_r})}
$$
$$
(Y \text{ is } B_{j_1, \ldots, j_r}^?)
$$

Mathematically speaking (Dubois and Prade, 1992), this scheme first combines the overall MFs of the given fact and the rule. Projecting the MF so generated onto the y-axis gives a new MF $\mu_{B_{j_1, \ldots, j_r}^?}(y, \cdot) \in \mathbb{Y}$, which can be thought of as the possibility distribution of the fuzzy output given the fuzzy fact. Representing the fuzzy fact by a MF $\mu_{\boldsymbol{A}^!}(\boldsymbol{u}) = \mu_{A_1^!}(u_1) \star \ldots \star \mu_{A_r^!}(u_r)$ defined on $\mathbb{U}_1 \times \ldots \times \mathbb{U}_r$, we thus get

$$
\begin{aligned}
\mu_{B_{j_1, \ldots, j_r}^?}(y, \cdot) &= \text{proj}_{\boldsymbol{u} \in \mathbb{U}^r} \left(\mu_{\boldsymbol{A}^!}(\boldsymbol{u}) \text{ and } \mu_{A_j \to B_j}(\boldsymbol{u}, y, \cdot) \right) \\
&= \sup_{\boldsymbol{u} \in \mathbb{U}^r} \left(\mu_{\boldsymbol{A}^!}(\boldsymbol{u}) \star \mu_{A_j \to B_j}(\boldsymbol{u}, y, \cdot) \right).
\end{aligned}
\tag{3.8}
$$

Assuming now that the facts are represented by fuzzy singletons, $\mu_{A_j'}(u)$ will be 1 in a single point in $\mathbb{U}^r$, namely when $u = \varphi(t)$, and zero elsewhere. Because a triangular norm $\star$ returns a 0 membership degree when one of its operands is 0, (3.8) simplifies to

$$\mu_{B_j^?}(y,\cdot) = \mu_{B_{j_1,\ldots,j_r}^?}(y,\cdot) = \mu_{A_j \to B_j}(\varphi(t), y, \cdot), \qquad (3.9)$$

which means that $\mu_{B_j^?}(y,\cdot)$ is obtained by slicing $\mu_{A_j \to B_j}(u, y, \cdot)$ along the u-coordinate specified by $\varphi(t)$. Compare this with a standard function evaluation. Notice also that the sup-star computation becomes much more involved in case a non-singleton fuzzifier is used.

Finally, by using the Mamdani rule interpretation (3.7), (3.9) can be written

$$
\begin{aligned}
\mu_{B_j^?}(y,\cdot) &= \mu_{A_j}(\varphi(t),\cdot) \star \mu_{B_j}(y,\cdot) \\
&= w_{A_{j_1,1}} \star \ldots w_{A_{j_r,r}} \star \mu_{B_j}(y,\cdot) = w_{A_j}\mu_{B_j}(y,\cdot), \qquad (3.10)
\end{aligned}
$$

where $w_{A_j} \in [0,1]$ is a weight known as the *degree of fulfillment*, or *firing strength*, of the rule. Clearly, the higher the value of this weight the higher the value of $\mu_{B_j^?}(y,\cdot)$. In particular, with $w_{A_j} = 1$ we get the intuitively reasonable result that $\mu_{B_j^?}(y,\cdot)$ equals $\mu_{B_j}(y,\cdot)$.

Defuzzification. The last issue to consider is how to aggregate the generated fuzzy sets into a form that can be converted into a crisp output $\hat{y}(t|\theta) \in \mathbb{R}$. For this purpose we use a *defuzzifier*, which returns the crisp value that in some sense best corresponds to the possibility distribution of the combined output fuzzy sets. Because there is no universally correct way of doing this, quite a few different defuzzification schemes have been suggested in the literature (Driankov et al., 1993).

Dominating in the fuzzy control genre is center of area (COA) or center of sums (COS) defuzzification, which are chosen largely for performance reasons (Lee, 1990). The *center of sums* method operates on the rules on an individual basis:

$$\hat{y}(t|\theta) = \frac{\displaystyle\int_{\mathbb{Y}} y \cdot \sum_{j_1=1}^{n_1} \cdots \sum_{j_r=1}^{n_r} \mu_{B_{j_1,\ldots,j_r}^?}(y,\theta)\,dy}{\displaystyle\int_{\mathbb{Y}} \sum_{j_1=1}^{n_1} \cdots \sum_{j_r=1}^{n_r} \mu_{B_{j_1,\ldots,j_r}^?}(y,\theta)\,dy} \qquad (3.11)$$

and is here preferred to the COA method since it does not involve a complex rule aggregation part.

Network representation. To predict $y(t)$ using (3.11) it now only remains to specify which $\star$-operator to use. Although max is often employed for $\star$ in control applications (Lee, 1990), it is not so suitable from an estimation point of view. The main reason is that it introduces discontinuities, which may lead to problems when computing gradients. Another problem that may

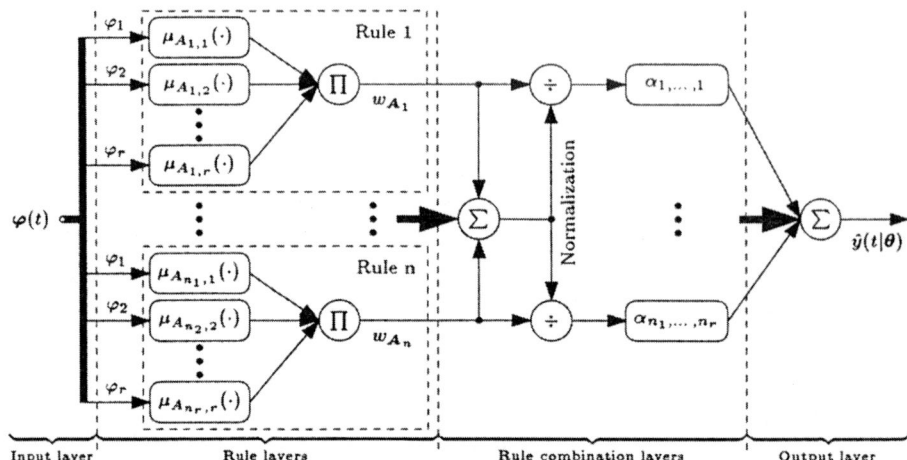

Fig. 3.2. Fuzzy model structure suitable for identification purposes

occur when updating MF parameters is that certain MFs can affect the overall mapping at one iteration, while at the next iteration they have no impact on the mapping whatsoever. It is not hard to see that such behavior can confuse the estimation algorithm quite a bit. Algebraic product (multiplication) on the other hand shows none of these deficiencies, thus making it better suited for identification purposes.

The integrals of (3.11) are in general also rather costly to compute. This is serious as the number of integrals that must be computed is at least the number of parameters d times the number of iterations needed in the estimation procedure. However, by restricting the MFs of the rule consequences to be fuzzy singletons, i.e., each $\mu_{B_j}(y, \cdot)$ is 1 in a single, as yet unknown, point $\alpha_j \in \mathbb{Y}$, the predictor structure (3.11) simplifies to the *Mamdani fuzzy model structure*

$$
\begin{aligned}
\hat{y}(t|\boldsymbol{\theta}) \;&=\; \frac{\displaystyle\sum_{j_1=1}^{n_1} \cdots \sum_{j_r=1}^{n_r} \alpha_{j_1,\dots,j_r} \prod_{k=1}^{r} \mu_{A_{j_k,k}}(\varphi_k(t), \boldsymbol{\beta}_{j_k,k}, \boldsymbol{\gamma}_{j_k,k})}{\displaystyle\sum_{j_1=1}^{n_1} \cdots \sum_{j_r=1}^{n_r} \prod_{k=1}^{r} \mu_{A_{j_k,k}}(\varphi_k(t), \boldsymbol{\beta}_{j_k,k}, \boldsymbol{\gamma}_{j_k,k})} \\[2ex]
&=\; \frac{\displaystyle\sum_{j_1=1}^{n_1} \cdots \sum_{j_r=1}^{n_r} \alpha_j w_{A_j}}{\displaystyle\sum_{j_1=1}^{n_1} \cdots \sum_{j_r=1}^{n_r} w_{A_j}}.
\end{aligned} \qquad (3.12)
$$

The corresponding network of (3.12) is reproduced in Fig. 3.2, from which it is evident that it belongs to the composition type of series expansions that was discussed in Sect. 2.2, although here all parameters have linguistic meanings. The simplicity of structure (3.12) is the main reason we focus on it in the following sections. Another and more theoretical reason is the fact that it is actually capable of approximating any real continuous function on a compact (closed and bounded) domain to any degree of accuracy (Kosko, 1992; Wang, 1992; Wang and Mendel, 1992).

To this end, observe that each α_j can be viewed as a local model (a leaf in the model regression tree terminology), which is active to a degree w_{A_j}. An obvious generalization is to replace the constants α_j with more complex local models, such as linear regressions. This gives what is known as the Takagi–Sugeno model structure (Takagi and Sugeno, 1985; Sugeno and Kang, 1988; Sugeno and Yasukawa, 1993), which also has been suggested by (Johansen, 1994; Johansen and Foss, 1994) (who suppressed the fuzziness and instead referred to the approach as "operating regime based identification"). The concluding part of each fuzzy rule is thus replaced by a linear-in-the-parameters predictor structure for which it is hard to find a linguistic interpretation.

4. Fuzzy Identification Based on Prior Knowledge

This section considers various system identification issues based on the model structure (3.12). In Sect. 4.1 we further discuss and motivate the use of a rule base provided by an expert, as opposed to a pure black box approach where the rule base itself is estimated. The following three subsections constitute the core of the section. Their purpose is to provide answers to three fundamental grey-box-type questions: how to estimate the parameters of structure (3.12) – Sect. 4.2; how to preserve the expert knowledge when updating these parameters – Sect. 4.3; and lastly, how to ensure certain non-structural system features, in this case how to guarantee a monotone steady-state gain curve – Sect. 4.4. Finally, Sect. 4.5 presents three fuzzy hybrid modeling approaches, of which two aim at reducing the number of parameters to estimate, and one aims at modeling dynamics that could not be captured by the fuzzy model.

4.1 Black Versus Grey Box Approaches

There are two main ways of determining the regressors, the number of MFs, and the number of rules to use in the model structure (3.12). The black box way implies that the structure is estimated from data, the modeling (grey box) way that it is provided by an expert.

Structure estimation. *Structure estimation* can be split into two separate tasks. The first one is to determine what regressors $\varphi(t)$ to construct from the data $z(t)$. This is clearly not a fuzzy specific problem but is present in all black box approaches. The matter is usually resolved by restricting the regressors to delayed versions of the measurements. What regressors to include can be determined in many different ways, e.g., by statistical screening tests (Draper and Smith, 1981), by certain clustering techniques (Aguirre and Billings, 1995), by trial and error, and so forth. Here the key thing to recognize is that regressor selection is a combinatorial problem, growing rapidly in complexity with the number of possible regressors.

Having chosen r regressors the second task is to find the number of different membership functions associated with each regressor, the number of different centers α_j, and the number of rules n. As mentioned earlier, the number of rules in a complete rule base is

$$n = \prod_{k=1}^{r} n_k,$$ (4.1)

which clearly suffers from the *curse of dimensionality* problem, i.e., the number of possible rules increases exponentially with r. Observe that even moderate values of r and n_k give a large n, e.g., $r = n_1 = \ldots = n_r = 4$ result in $n = 256$. Moreover, a large n often leads to many MF parameters to estimate, which in turn easily leads to overfit problems (recall the bias-variance trade-off). For many complex modeling problems the typical situation is that only some of the n rules are important for describing the underlying system. The modeling of glycemic variations in the human body detailed in (Sjöberg et al., 1995), e.g., includes only 64 out of 1620 possible rules. Hence we should keep the m out of n rules that are most influential, i.e., we should again solve a combinatorial optimization problem.

Now, in the light of the fact that iterative search schemes typically must be employed for MF parameter estimation (see the next section) it is highly impractical to apply exhaustive search algorithms for rule base structure estimation. This has motivated methods that try to explore the most promising alternatives in one way or another. It seems that three main classes of structure estimation algorithms have been suggested in the literature: those that successively try to partition the regression space in a tree-like manner (Takagi and Sugeno, 1985; Sugeno and Kang, 1988; Higgins and Goodman, 1994; Sun, 1994), those that provide non-optimal but often good enough solutions (simulated annealing, genetic algorithms, etc.) (Kirkpatrick et al., 1983; Ishigami et al., 1995; Lin and Lee, 1996), and those that apply clustering techniques (Sugeno and Yasukawa, 1993; Yoshinari et al., 1993; Babuška and Verbruggen, 1994; Kaymak and Babuška, 1995).

Even if these methods usually are far less computationally demanding than exhaustive search, they are still quite complex, chiefly because structure estimation in general cannot be separated from the task of adjusting MF parameters. Another problem is that it is not so easy to say how many different output centers α_j to estimate. The fuzzy modeling advantages listed in the first section are also weakened or even lost if a black box approach is followed.

Modeling. The modeling approach overcomes these difficulties, at least partly. The computational cost is reduced at the price of the time it takes for the expert to provide the knowledge. At the linguistic level the knowledge required is the name of the linguistic variables, the corresponding linguistic values and the rule base itself. This is later complemented with descriptions of how to create the regressors, the initial shape of the MFs and other numeric information (e.g., parameter restrictions). Because natural language is rather coarse it can here be argued that the rule base often becomes relatively small, and that superfluous parameters therefore can be more easily avoided than if a black box approach is adopted.

Perhaps the most important advantage with the modeling approach is due to the curse of dimensionality and comes in terms of extrapolation capabilities. To see this, recall that the regression vector lives in $\mathbb{R}^r$. Even for a moderate r the observations $\varphi(t)$ are by necessity sparse in any bounded region of $\mathbb{R}^r$. For example, filling up the unit cube in $\mathbb{R}^6$ using a grid of granularity 0.1 requires a million measurements. Since such excessive amount of data cannot be collected in practice there will always be regression regions having no real data support. By using a black box method, rules corresponding to such regions are likely to be removed because they do not improve the optimization criterion. This is in contrast to the modeling approach, which allows the combination of data–expert and purely expert explained regression regions. However, in case it is hard to determine the latter regions, it must here be required that only "data supported MF parameters" are updated by the estimation algorithm.

A potential drawback with the modeling approach is now that it can be quite arduous to capture all the important system features, especially in a complex modeling situation. It is then reasonable to condense what is actually linguistically known into a fuzzy rule base, and then try to describe the remaining dynamics within some other model structure, e.g., by letting a black box structure (a standard neural network, etc.) operate in parallel with the fuzzy rule base. Such hybrid approaches will be further investigated in Sect. 4.5. Altogether these facts and possibilities suggest that the strength of fuzzy identification really shows up when the modeling path is followed.

4.2 Parameter Estimation

The model structure (3.12) is a regular mathematical function with tunable parameters θ of the form (2.8). By adopting the quadratic performance criterion (2.6) and for a moment neglecting that the parameters have linguistic meanings, this merely reduces to a standard unconstrained nonlinear least-squares problem, where the nonlinear nature stems from the fact that the scale β and the position γ parameters enter the predictor in a nonlinear fashion.

The nonlinear least-squares estimation algorithms presented in Sect. 2.3 rely on the assumption that the Jacobian (2.17) can be constructed for any $\boldsymbol{\theta} \in \mathcal{D} \subset \mathbb{R}^d$. A possible complication with the fuzzy predictor (3.12) occurs if $\boldsymbol{\varphi}(t)$ and $\boldsymbol{\theta}$ are such that the denominator of the predictor evaluates to zero. This takes place in the rare situations when also the numerator is zero and is due to the fact that no rule is able to explain the current $\boldsymbol{\varphi}(t)$. The natural way around the difficulty is simply to exclude samples causing an undefined $\hat{y}(t|\boldsymbol{\theta})$.

The Jacobian can now be constructed if the derivatives of the individual MFs with respect to their parameters exist. This is always the case for the α_j parameters because they enter the predictor linearly, and thus it suffices to investigate the derivatives of the MFs at the input side.

Both network-classic (Definition 3.2) and Zadeh-formed (Definition 3.3) MFs have well-defined and continuous derivatives with respect to any of their parameters, at least if pathological cases are excluded (e.g., $\beta = 0$ of a Gaussian, $\gamma_1 = \gamma_2$ of a Z-formed MF, etc.). However, this is not the case for piecewise linear MFs (Definition 3.4), and a rather common misunderstanding is therefore that Jacobian based estimation algorithms cannot be used in such a case. To put an end to this misconception we first observe that the only points without well-defined derivatives correspond to data located at the breakpoints (corners) of these MFs. Since the breakpoints are finitely many in any universe of practical interest, it follows that there will be no data points at these positions in the generic case. Nevertheless, in order to really guarantee a well-defined estimation problem we simply adopt the convention that derivatives at the breakpoints are zero, which means that any such data is excluded from the criterion fit. We thus conclude that the Jacobian (2.17) can be formed regardless of which of the mentioned MFs are used, and that algorithms built on it can be employed to estimate the parameters of the structure (3.12).

Based on these observations a suitable MF parameter estimation procedure is as follows.

1. Fix all β and γ parameters and estimate $\boldsymbol{\alpha}$ using an unconstrained linear least-squares method (see Sect. 2.3). This gives a rough idea of the quality of the rule base and provides further clues on how to choose the initial values of the $\boldsymbol{\alpha}$ components.

2. Next, let all parameters loose and estimate these using either the Levenberg–Marquardt or the damped pseudo-inverse variant of the Gauss–Newton algorithm, which are both equipped with regularization, thus meaning that only data supported MF parameters are updated (see Sect. 2.3). Besides this extrapolation feature it is worth emphasizing that another decisive reason for using one of these schemes instead of a pure gradient procedure (which is quite common in this area) is due to their superior convergence properties near the minimum (Dennis and Schnabel,

1983). Notice though that a gradient descent method can be warranted in on-line situations where real-time aspects are especially important.

The main problem with this straightforward approach is now that nothing hampers the parameters from being updated in such a way that the original meaning of the parameters is lost. In what follows we show how to avoid such behavior.

4.3 Preserving the Meaning of a Fuzzy Rule Base

The MFs and their parameters θ are directly coupled to certain linguistic values. What is important to recognize is that these values are ordered. Consider, e.g., the linguistic variable SPEED(t),

$$\langle \text{SPEED}(t), \boldsymbol{A}(\cdot, \cdot) = \\ \{\text{LOW}, \text{RATHER LOW}, \text{MEDIUM}, \text{RATHER HIGH}, \text{HIGH}\}, D \rangle, \quad (4.2)$$

which can be assigned to five different values, here listed in an order representing higher and higher speed. It is of course crucial that this order is reflected by the corresponding MFs. If this is the case for all the linguistic variables involved, then we say that the corresponding fuzzy model is *linguistically sound*. The important question is now how to relate this soundness concept to the parameters of the membership functions.

In case SPEED(t) is an output linguistic variable, then the MFs characterizing its linguistic values are fuzzy singletons located at positions $\alpha_{j_1}, \ldots, \alpha_{j_5} \in \mathbb{Y}$. Assuming that the order of these centers corresponds to the order of the linguistic values, the interpretation of SPEED(t) is preserved if and only if

$$\alpha_{j_1} < \alpha_{j_2} < \alpha_{j_3} < \alpha_{j_4} < \alpha_{j_5}. \quad (4.3)$$

See plot (a) of Fig. 4.1. For n different linguistic values of the output variable this merely generalizes to the parameter restrictions

$$\alpha_{j_1} < \alpha_{j_2} < \ldots < \alpha_{j_n}. \quad (4.4)$$

The ordering of the linguistic values is also essential for the linguistic variables involved in the rule antecedents, but then the situation becomes more complicated owing to the fact that the corresponding membership functions are more complex.

Suppose that SPEED(t) is a linguistic input variable (coupled to the k-th regressor) with linguistic values described by network-classic MFs. For these MFs to have a linguistic interpretation a first requirement is that the five position parameters $\gamma_{j_k,k}$ are ordered as

$$\gamma_{1,k} < \gamma_{2,k} < \gamma_{3,k} < \gamma_{4,k} < \gamma_{5,k}. \quad (4.5)$$

Furthermore, the scale parameters $\beta_{1,k}, \ldots, \beta_{5,k}$ can be divided into two classes: those that must be positive, β^+, and those that must be negative, β^-. The latter category is applicable for MFs of sigmoidal type and are used

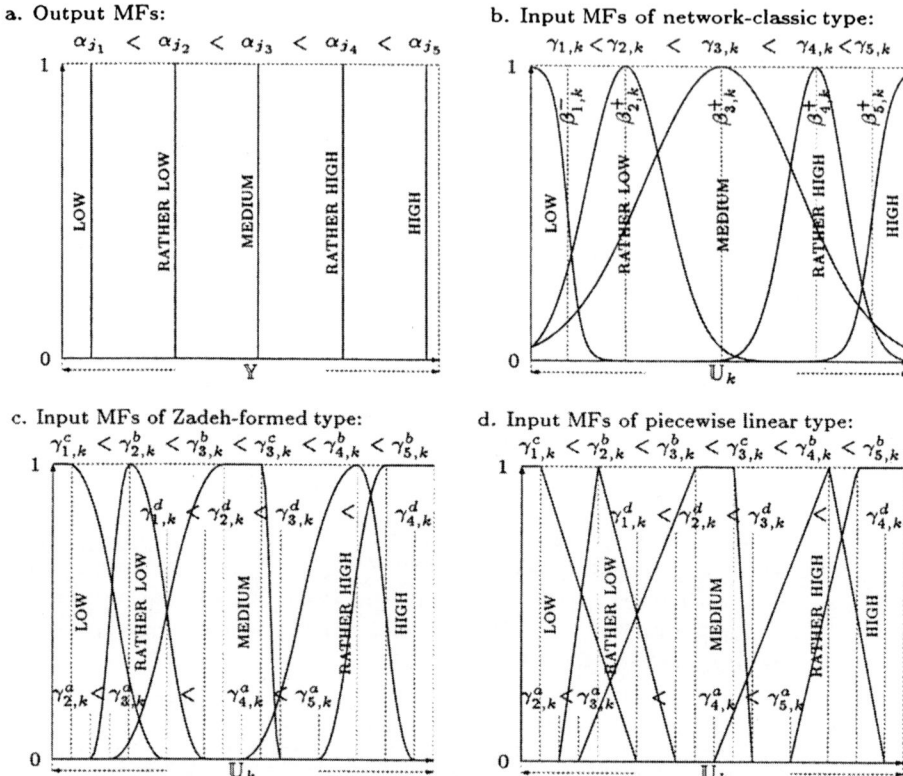

Fig. 4.1. Restrictions on the parameters α, β and γ that are necessary for a linguistic variable (here SPEED(t)) to have a reasonable linguistic interpretation

to reflect that the degree of membership decreases when the input increases. If, on the other hand, the membership degree should increase with the input or if the MF is a Gaussian, then the scale parameter must be positive. See Fig. 4.1b. The generalization to r regressors each with n_k different linguistic values listed in increasing sense is obvious:

$$\gamma_{1,k} < \gamma_{2,k} < \ldots < \gamma_{n_k,k}, \quad k = 1, \ldots, r, \tag{4.6}$$

$$0 < -\beta^-, \quad 0 < \beta^+. \tag{4.7}$$

These restrictions still allow some inconsistencies as is also illustrated in Fig. 4.1b. The problem is that the membership degree of MEDIUM exceeds the membership degree of RATHER HIGH for large input values, which is quite illogical. To cope with this difficulty one idea is to impose further restrictions on β. However, this is not that easy (unless unnecessarily hard restrictions are inflicted) because what can be accepted depends partly on the current universe of discourse and partly on the values of the position parameters involved.

Zadeh-formed or piecewise linear MFs are both able to resolve this dilemma as they do not involve scale parameters. The j_k-th linguistic value of the k-th linguistic variable is instead characterized by an MF with two, three, or four position parameters $\gamma_{j_k,k}$. Let the components of $\gamma_{j_k,k}$ be ordered on $\mathbb{U}_k$ and denoted $\gamma_{j_k,k}^i$ for $i \in \{a, b, c, d\}$, where a denotes the point in $\mathbb{U}_k$ from which the membership degree starts to increase, b is the point where it reaches a full degree of membership, c is the point where the membership degree starts to decrease (a parameter if and only if $b \neq c$), and, finally, d is the point from which the membership degree is zero. For each individual Zadeh-formed or piecewise linear MF it must thus hold that

$$\gamma_{j_k,k}^a < \gamma_{j_k,k}^b < \gamma_{j_k,k}^c < \gamma_{j_k,k}^d, \qquad (4.8)$$

where parameters not present in the current MF are removed from (4.8). In order to maintain a reasonable meaning for the linguistic variables we must also guarantee the ordering of their respective MFs. For the linguistic variable SPEED(t) we obtain a language-consistent meaning if, besides (4.8), it additionally holds that (see Fig. 4.1c or 4.1d):

$$\gamma_{2,k}^a < \gamma_{3,k}^a < \gamma_{4,k}^a < \gamma_{5,k}^a \qquad (4.9)$$

$$\gamma_{1,k}^c < \gamma_{2,k}^b < \gamma_{3,k}^b < \gamma_{3,k}^c < \gamma_{4,k}^b < \gamma_{5,k}^b \qquad (4.10)$$

$$\gamma_{1,k}^d < \gamma_{2,k}^d < \gamma_{3,k}^d < \gamma_{4,k}^d. \qquad (4.11)$$

In case there are r linguistic variables each with n_k, $k = 1, \ldots, r$, different and ordered linguistic values this generalizes to

$$\gamma_{1,k}^a < \gamma_{2,k}^a < \cdots < \gamma_{n_k,k}^a \qquad (4.12)$$

$$\gamma_{1,k}^b < \gamma_{1,k}^c < \gamma_{2,k}^b < \gamma_{2,k}^c < \cdots < \gamma_{n_k,k}^b < \gamma_{n_k,k}^c \qquad (4.13)$$

$$\gamma_{1,k}^d < \gamma_{2,k}^d < \cdots < \gamma_{n_k,k}^d, \qquad (4.14)$$

where, as before, parameters not present in the corresponding MFs are removed from (4.12)–(4.14).

By a proper modeling procedure the initial values of α, β and γ will agree with the applicable restrictions from above, i.e., the initial value of θ can be assumed to correspond to the feasibility region $\mathcal{D}$. In order to also ensure a feasible parameter estimate when using an unconstrained algorithm it must here be required that the parameter update at any stage is such that the constraints are not violated. Otherwise convergence can be to a (local and undesired) minimum where the parameters cannot be linked to the linguistic domain.

Although this problem does not always occur, it still appears surprisingly often in practice, even for simple static systems, as demonstrated in (Lindskog, 1996). There are several plausible reasons for this. One is that certain regression regions may be reflected by few and noisy data points that actually suggest an infeasible update of the parameter vector. Notice that this is likely to happen when many parameters are estimated, i.e., when the model structure is too flexible. Another reason is that the initial parameter values and thereby the initial shape of the corresponding membership functions are just too inaccurate.

The only way to really guarantee that the expert knowledge is preserved is to take the parameter restrictions into account in the estimation phase. Since all restrictions considered thus far are inequalities this is a situation that can be handled straightforwardly through the barrier function estimation approach discussed in Sect. 2.3. More specifically, we introduce a number of simple inequality constraints $c_j(\boldsymbol{\theta})$:

$$\text{if } 0 < \theta_i, \quad \text{then let} \quad c_j(\boldsymbol{\theta}) = \theta_i, \tag{4.15}$$

$$\text{if } \theta_i < \theta_{i+1}, \quad \text{then let} \quad c_j(\boldsymbol{\theta}) = \theta_{i+1} - \theta_i, \tag{4.16}$$

which, when put together, bring about any of the parameter restriction "chains" (4.4), (4.6)–(4.7) or (4.8), (4.12)–(4.14), thereby defining $\mathcal{D}$.

Further parameter restrictions of inequality type can of course also be imposed via barrier functions. For example, we can restrict a position parameter to any interval, thus guaranteeing that an MF stays in a position where the corresponding linguistic value is considered valid.

Another possibility is to allow soft parameter knowledge which is more or less consultative in nature, and whose purpose it to try to balance the expert and the data knowledge by assessing the quality of the initial parameter values. The easiest method to achieve such behavior is to use explicit regularization as was briefly discussed in Sect. 2.4. As for the barrier function method, the basic idea is to add a penalty term to the objective function of (2.25), i.e., to find parameters by iteratively minimizing something like

$$\hat{\boldsymbol{\theta}}_N^{(k+1)} = \arg\min_{\boldsymbol{\theta} \in \mathcal{D}} \left(\frac{1}{N} \sum_{t=1}^{N} \frac{1}{2} \left(y(t) - \hat{y}(t|\boldsymbol{\theta}) \right)^2 + \rho^{(k)} \sum_{j=1}^{l} \vartheta(c_j(\boldsymbol{\theta})) \right.$$

$$\left. + \sum_{j=1}^{i} \mu_i \left(\theta_i - \theta_i^\sharp \right)^2 \right), \tag{4.17}$$

where $\theta_i^\sharp$ denotes the initial value of the i-th parameter, and $\mu_i > 0$ is a user-tunable parameter expressing the relative belief in the value of θ_i. In effect, a large μ_i (compared to the other μ_j) implies that the cost for moving θ_i away from $\theta_i^\sharp$ becomes high, thus expressing that its value is believed to be close to $\theta_i^\sharp$.

4.4 Guaranteeing Certain Non-structural System Properties

Many dynamic processes are known (from simple physics) to have a steady-state gain curve that is monotonically increasing (decreasing) in the inputs. Consider, e.g., a simple tank system where the inflow is the input and the liquid level the output. Here it is known that a certain constant inflow eventually leads to a "constant" liquid level. Starting from such a steady-state condition it is also known that an increase in the inflow causes the liquid level to increase (in a non-oscillatory manner) and settle at a higher level. This is a non-structural system property that is extremely important to retain in certain applications, e.g., when the model is going to be used in a predictive control arrangement (Koivisto, 1995).

The main problem is now that when applying a flexible nonlinear black box model structure (a neural network, etc.) it can be quite hard to ensure this monotonicity feature, especially if there are regression regions with few and noisy data. To remedy this, we will here consider a restricted variant of the fuzzy model structure (3.12) that guarantees an increasing (decreasing) function mapping from the regression space $\mathbb{R}^r$ to the output space $\mathbb{R}$. This structure together with a proper choice of regressors $\varphi(t)$ (delayed inputs and outputs only) result in dynamical models showing the desired monotone behavior.

A conceptually simple way to ensure monotonicity is to first restrict the MFs at the input side to correspond to *fuzzy partitions* (Brown and Harris, 1994; Sjöberg et al., 1995).

Definition 4.1 (fuzzy partition). *Suppose that the k-th linguistic variable can be assigned to n_k different values each described by a membership function $\mu_{A_{j_k,k}}(\varphi_k(t), \beta_{j_k,k}, \gamma_{j_k,k}) \in \mathbb{U}_k$. These MFs form a fuzzy partition if it holds on the entire domain $\mathbb{U}_k$ that*

$$\sum_{j_k=1}^{n_k} \mu_{A_{j_k,k}}(\varphi_k(t), \beta_{j_k,k}, \gamma_{j_k,k}) = 1. \tag{4.18}$$

By imposing this restriction on all the r linguistic variables (regressors) and additionally assuming that the rule base is complete in the sense that it covers the whole input domain $\mathbb{U}^r$, it immediately follows that the model structure (3.12) simplifies to

$$
\begin{aligned}
\hat{y}(t|\theta) &= \sum_{j_1=1}^{n_1} \cdots \sum_{j_r=1}^{n_r} \alpha_{j_1,\ldots,j_r} \prod_{k=1}^{r} \mu_{A_{j_k,k}}(\varphi_k(t), \beta_{j_k,k}, \gamma_{j_k,k}) \\
&= \sum_{j_1=1}^{n_1} \cdots \sum_{j_r=1}^{n_r} \alpha_j w_{A_j}.
\end{aligned}
\tag{4.19}
$$

Before proceeding, notice that a fuzzy partition puts certain demands on the MFs and their parameters. For example, we cannot in general use sigmoidal or Gaussian MFs because of their spreading and curvature. Zadeh-formed or piecewise linear MFs on the other hand can readily be parameterized so that a fuzzy partition is obtained. See Fig. 4.2.

Besides simplifying the predictor structure (no normalization is needed) a fuzzy partition always leads to fewer parameters at the input side. The

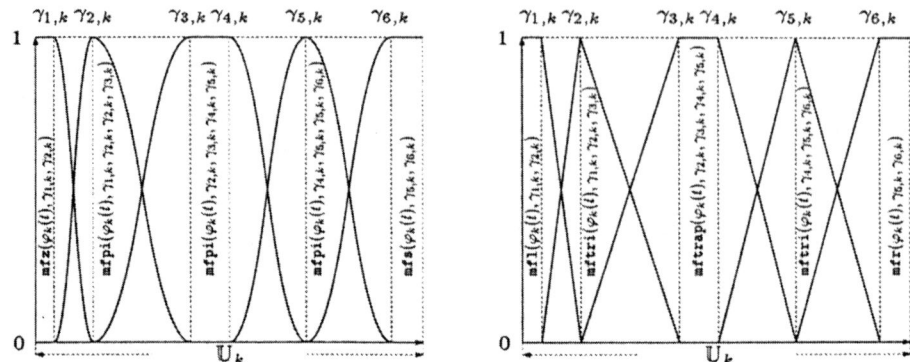

Fig. 4.2. Fuzzy partitions formed by Zadeh-formed MFs (left) and piecewise linear MFs (right)

five MFs shown in Fig. 4.2 are, e.g., described by 6 parameters only, whereas a full degree of freedom parameterization of the MFs implies 14 parameters as shown in Fig. 4.1. Still, of course, the remaining problem is that the complexity of the predictor (4.19) typically increases rapidly with r.

Consider now the case with a single input linguistic variable ($r = 1$). Guaranteeing that the predictor is monotonically increasing in $\varphi(t)$ can be done in many different ways (especially if the origin of (4.19) is neglected), but then it can be quite hard to express restrictions on the parameters that ensure that monotonicity is preserved in the estimation step. However, this is a simple task when the input MFs form a fuzzy partition.

To see that this is the case, assume that all input MFs are ordered on the universe $\mathbb{U}$ in such a way that $\mu_{A_j}(\cdot)$ reaches a full degree of membership for a value of $\varphi(t)$ that is lower than what is the case for $\mu_{A_{j+1}}(\cdot)$. See Fig. 4.2. If the ordered MFs at the input side form a fuzzy partition and the corresponding centers α_j reflecting the output MFs are such that

$$\alpha_1 < \alpha_2 < \ldots < \alpha_n, \qquad (4.20)$$

then $\hat{y}(t|\theta)$ will show a monotonically increasing behavior. In verifying this, we first notice that at intervals where the j-th input MF is fully active then the corresponding output becomes α_j. With fuzzy partitions constructed by Zadeh-formed or piecewise linear MFs we also have that

$$\hat{y}(t|\theta) = \alpha_j \mu_{A_j}(\cdot) + \alpha_{j+1} \mu_{A_{j+1}}(\cdot) = (\alpha_{j+1} - \alpha_j)\mu_{A_{j+1}}(\cdot) + \alpha_j \qquad (4.21)$$

for all intervals $[\gamma_j, \gamma_{j+1}] \subset \mathbb{U}$ such that $\mu_{A_j}(\cdot)$ and $\mu_{A_{j+1}}(\cdot)$ are not always zero. Since $\alpha_{j+1} > \alpha_j$ (equality gives a constant output on the current interval) and $\mu_{A_{j+1}}(\cdot)$ is an increasing function on $[\gamma_j, \gamma_{j+1}]$ it follows that also $\hat{y}(t|\theta)$ is an increasing function on that interval, with values ranging from α_j to α_{j+1}. These facts together give that the overall predictor is a non-decreasing function. To get a strictly increasing mapping it must additionally be required that all the input MFs lack intervals (flat parts) with a full degree of membership.

Two things are worth emphasizing before considering systems with several input linguistic variables. The first is that Zadeh-formed MFs always result in models with local plateaus at each γ_j position. This is a behavior that is quite unrealistic in certain applications, thus favoring piecewise linear MFs as these do not introduce such plateaus (unless trapezoidal MFs are used). The second observation is that the restrictions imposed by a fuzzy partition typically reduce the risk for position changes among the parameters $\boldsymbol{\alpha}$ and $\boldsymbol{\gamma}$, which in terms of estimation algorithms means that it is often sufficient to use an unconstrained procedure instead of a constrained one.

Now, in order to generalize the above result to predictors having r regressors it is instructive to first formally define what is meant by a monotonically increasing predictor in $\boldsymbol{\varphi}(t)$.

Definition 4.2 (regressor ordering). *Let* $\boldsymbol{\varphi}(t)$, $\bar{\boldsymbol{\varphi}}(t) \in \mathbb{R}^r$. *We say that* $\boldsymbol{\varphi}(t) \geq \bar{\boldsymbol{\varphi}}(t)$ *if* $\varphi_j(t) \geq \bar{\varphi}_j(t)$ *for* $j = 1, \ldots, r$.

Definition 4.3 (monotonically increasing predictor). *Let* $\boldsymbol{\varphi}(t), \bar{\boldsymbol{\varphi}}(t) \in \mathbb{R}^r$. *We say that a predictor* $g(\boldsymbol{\varphi}(t), \boldsymbol{\theta})$ *is monotonically increasing in the regressors if whenever* $\boldsymbol{\varphi}(t) \geq \bar{\boldsymbol{\varphi}}(t)$ *it holds that* $g(\boldsymbol{\varphi}(t), \boldsymbol{\theta}) \geq g(\bar{\boldsymbol{\varphi}}(t), \boldsymbol{\theta})$.

Using the latter definition we now have the following central theorem.

Theorem 4.1. *Let the model structure be complete and given by (4.19). If, for all* $k = 1, \ldots, r$, *it holds that*

$$\sum_{j_k=1}^{n_k} \alpha_{j_1,\ldots,j_r} \mu_{A_{j_k,k}}(\varphi_k(t), \beta_{j_k,k}, \gamma_{j_k,k}) \tag{4.22}$$

are monotonically increasing functions in $\varphi_k(t)$ *on* $\mathbb{U}_k$ *for all possible combinations of fixed values of* $j_1, \ldots, j_{k-1}, j_{k+1}, \ldots, j_r$, *then the predictor (4.19) is monotonically increasing in* $\boldsymbol{\varphi}(t)$.

Proof. See (Lindskog, 1996), page 195.   □

The main point with Theorem 4.1 is that it is sufficient to work with one-dimensional functions. A simple way to ensure increasing functions in all $\varphi_k(t)$ is therefore to restrict the input MFs to fuzzy partitions and order the corresponding centers as was done in the one-dimensional case.

Lemma 4.1. *Let the model structure be (4.19) and let* $\varphi_k(t)$ *denote one of its regressors. Assume that the ordered (on* $\mathbb{U}_k$*) MFs associated with* $\varphi_k(t)$ *are either Zadeh-formed or piecewise linear and such that they form a fuzzy partition. If, for all possible combinations of* $j_1, \ldots, j_{k-1}, j_{k+1}, \ldots, j_r$, *it holds that*

$$\alpha_{j_1,\ldots,j_k,\ldots,j_r} \leq \alpha_{j_1,\ldots,j_k+1,\ldots,j_r}, \tag{4.23}$$

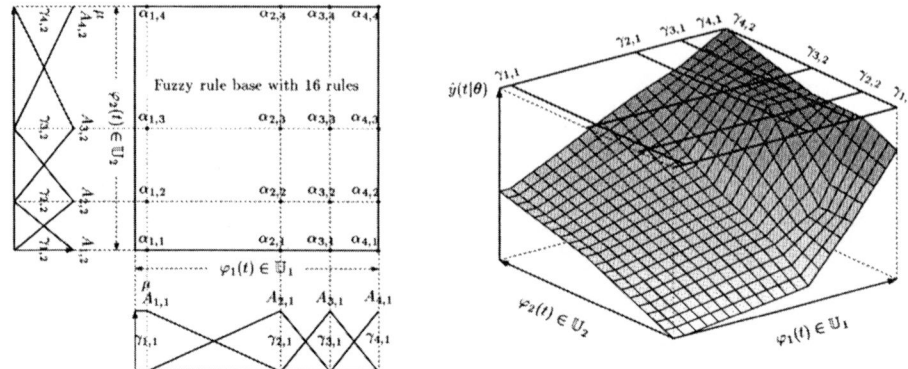

Fig. 4.3. Graphical representation of a complete fuzzy rule base containing 16 rules (left). Both linguistic variables at the input side have MFs forming fuzzy partitions. Ordering the centers as $\alpha_{j_1,j_2} \leq \alpha_{j_1,j_2+1}$ for $j_1 = 1,\ldots,4$ and $j_2 = 1,\ldots,3$ and as $\alpha_{j_1,j_2} \leq \alpha_{j_1+1,j_2}$ for $j_1 = 1,\ldots,3$ and $j_2 = 1,\ldots,4$ gives an increasing function mapping as is shown to the right.

for all $j_k = 1,\ldots,n_k - 1$, then every

$$\sum_{j_k=1}^{n_k} \alpha_{j_1,\ldots,j_r} \mu_{A_{j_k},k}(\varphi_k(t), \gamma_{j_k,k}) \tag{4.24}$$

is a monotonically increasing function in $\varphi_k(t)$ on $\mathbb{U}_k$.

Proof. Follows directly from the one-dimensional case discussed above.   □

The requirements for Theorem 4.1 to hold are fulfilled if all the MFs are chosen according to Lemma 4.1. This is the case for the fuzzy rule base shown in Fig. 4.3, from which it is clear that the resulting predictor returns a larger (or unchanged) output if one or more of the regressors become larger. Notice that this property continues to hold if only the original orders among the parameters $\boldsymbol{\alpha}$ and $\boldsymbol{\gamma}$ are maintained. Since this will be the case when pursuing constrained estimation subject to these order constraints, we conclude that the monotonicity property can be preserved throughout the estimation phase.

At this point, assume that the regressors include dynamics

$$\varphi(t) = [\, y(t-1) \quad y(t-2) \quad \ldots \quad u(t) \quad u(t-1) \quad \ldots]^T, \tag{4.25}$$

where, without loss of generality, only one input signal is present. A globally asymptotically stable predictor $g(\cdot,\cdot)$ in $\varphi(t)$ implies that a constant input $u^* = u(t) = u(t-1) = \ldots$ leads to a constant output y^* as $t \to \infty$. Plotting y^* for each value of u^* gives the so-called steady-state gain curve.

Lemma 4.2. *Let u^*, y^* and $\bar{u}^*, \bar{y}^*$ be two steady-state solutions to a globally asymptotically stable predictor $g(\varphi(t), \boldsymbol{\theta})$, i.e.,*

$$y^* = g([\, y^* \quad y^* \quad \dots \quad u^* \quad u^* \quad \dots]^T, \boldsymbol{\theta}) = g(\boldsymbol{\varphi}^*, \boldsymbol{\theta}), \tag{4.26}$$

$$\bar{y}^* = g([\, \bar{y}^* \quad \bar{y}^* \quad \dots \quad \bar{u}^* \quad \bar{u}^* \quad \dots]^T, \boldsymbol{\theta}) = g(\bar{\boldsymbol{\varphi}}^*, \boldsymbol{\theta}). \tag{4.27}$$

If $g(\boldsymbol{\varphi}(t), \boldsymbol{\theta})$ is monotonically increasing in $\boldsymbol{\varphi}(t)$ and $u^* \geq \bar{u}^*$, then $y^* \geq \bar{y}^*$.

Proof. See (Lindskog, 1996), page 196. □

If the requirements of Lemma 4.2 are fulfilled, then we get a predictor with a monotonically increasing steady-state gain curve in the input. Moreover, starting from a steady-state solution and increasing the input in a stepwise fashion, it follows by simple induction that $\hat{y}(t|\boldsymbol{\theta})$ increases monotonically with t. This in particular means that the predictor shows a non-oscillatory step response behavior, which is a restriction but also a property that is valid for many industrial processes (e.g., thermal systems). We will apply fuzzy identification based on (4.19) to one such process in Sect. 5.

4.5 Fuzzy Hybrid Modeling – Some Possibilities

Although a fuzzy model of the form (3.12) is able to theoretically approximate any "well-behaved" system to a desired degree of accuracy, this may require a "too" complex rule base, particularly if r is large. With the aim of reducing the complexity of the models, while also maintaining (or even enhancing) their performance, it is interesting to marry together fuzzy and other identification approaches. This can be done in many different ways.

Fuzzy modeling based on physically induced regressors. A novel first idea is to keep structure (3.12), but apply more involved and physically motivated regressors (linguistic variables) than just delayed in- and outputs. Parts of the important system nonlinearities can then be captured directly in the regressors, thus typically implying that fewer MFs (parameters) and/or regressors r are needed in the resulting models. For example, in order to model the power delivered by a heater element (a resistor of some kind), an obvious physically motivated regressor to use would be the squared voltage applied to the heater. In other and more sophisticated modeling situations, suitable regressors can be implicitly given in terms of some dynamic and/or static equations. To then arrive at a set of physically induced regressors requires both symbolic and numeric computations, as stressed in (Lindskog and Ljung, 1995, Lindskog, 1996).

Combining fuzzy and traditional grey box modeling. Many real-world systems are composed of several subcomponents, some of which are well described directly in terms of physical principles (conservation laws, etc.) and some of which are better described in linguistic terms. This fact strongly motivates the use of several "small" and interacting fuzzy and grey box model structures, which when combined give the overall predictor. It is our opinion that such a model decomposition always should be considered in a complex modeling situation.

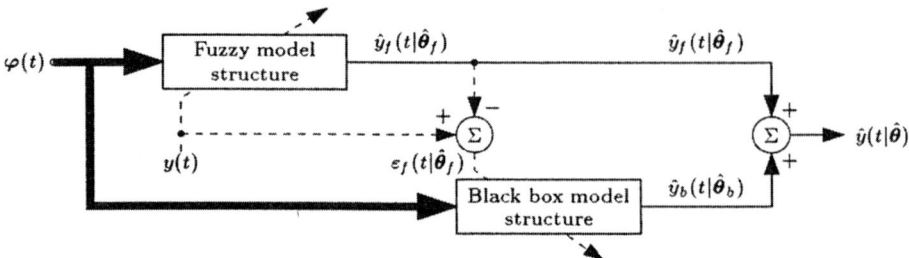

Fig. 4.4. Combined two-stage fuzzy and black box estimation procedure. The fuzzy model is first obtained, whereupon the residuals $\epsilon_f(t|\hat{\theta}_f)$ are used for the tuning of the black box parameters θ_b.

Combining fuzzy and black box modeling. Even if the above possibilities are contemplated there can still be important system phenomena that are hard to reflect within the fuzzy framework. As mentioned earlier, it is then appealing to complement the expert determined fuzzy structure by a sufficiently flexible black box structure (typically a neural network), which is solely responsible for picking up the remaining dynamics. This naturally leads to the setup shown in Fig. 4.4, which in structure is similar to what is discussed and proved useful in (Forssell and Lindskog, 1997).

Notice that the parameters of the fuzzy structure θ_f are first estimated based on the measurements $y(t)$. The residuals $\varepsilon_f(t|\hat{\theta}_f) = y(t) - \hat{y}_f(t|\hat{\theta}_f)$ are then formed and used for the tuning of the black box parameters θ_b. The main reason for using this particular scheme is that the fuzzy model obtained gives useful insight into the choice of black box structure, its size, and so forth.

5. Example – Tank Level Modeling

The objective in this application example is to model how the liquid level $h(t)$ of a simple laboratory-scale tank system, shown in Fig. 5.1, changes with the inflow that is generated by the voltage $u(t)$ applied to the pump. We see that the measured estimation data (1000 samples) cover the modeling domain of interest rather well.

To get a feeling for the nonlinearities it is useful to first take a closer look at the simulation behavior of a simple linear regression model, and compare this with experimental (validation) data. One of the best model structures having regressors of the form (2.4) involves three parameters only:

$$\hat{h}(t|\theta) = \theta_1 h(t-1) + \theta_2 u(t-1) + \theta_3. \tag{5.1}$$

Simulated outputs from the corresponding linear least-squares fitted model are compared to real tank measurements (1000 new samples) in the left plot

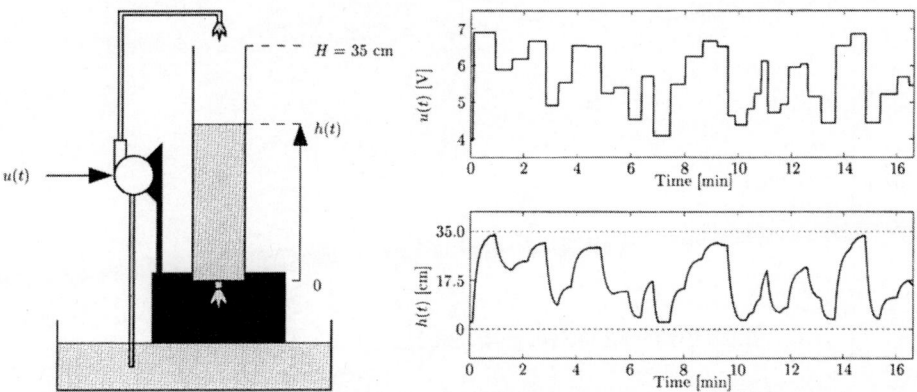

Fig. 5.1. Schematic picture of the laboratory-scale tank system (left), experimental data used for estimation (right)

of Fig. 5.2. The fit is clearly not that bad, yet the model output is physically impossible since it is sometimes negative. This is of course a nontrivial complication if we are going to use the model to study the behavior of the real system. In fact, all linear regression models with delayed in- and outputs as regressors show this defect.

A simple idea to overcome this difficulty is now to try some semi-physical modeling. The tank level change depends on the difference between in- and outflow (conservation of mass). While the inflow is roughly proportional to $u(t)$, the outflow can be approximated using *Bernoulli's law*, which for a small outlet hole states that the outflow is proportional to $\sqrt{h(t)}$. By combining these facts, it is pretty straightforward to arrive at the nonlinear model structure (a linear regression):

$$\hat{h}(t|\boldsymbol{\theta}) = \theta_1 h(t-1) + \theta_2 u(t-1) + \theta_3 + \theta_4 \sqrt{h(t-1)}. \qquad (5.2)$$

By tuning the four parameters of structure (5.2) we obtain a model whose simulation behavior is detailed in the right plot of Fig. 5.2. Compared to the model of the form (5.1), the semi-physical model gives a physically sound response and is seemingly better except for large tank levels. Still, however, there is no guarantee that the model outputs are physically sound for other input values.

Before trying to counteract this it is expedient to list what is actually known about the process.

1. First of all, we know that the more inflow the higher will the liquid level be. The steady-state gain curve of the model should thus be monotonically increasing in $u(t) = u^*$.
2. The input $u(t)$ may vary from 3.5 to 7.5 V, which means that there is always a flow across the tank. Even if the estimation data set is of rather high quality it still shows some gaps. The tank is, e.g., never emptied nor

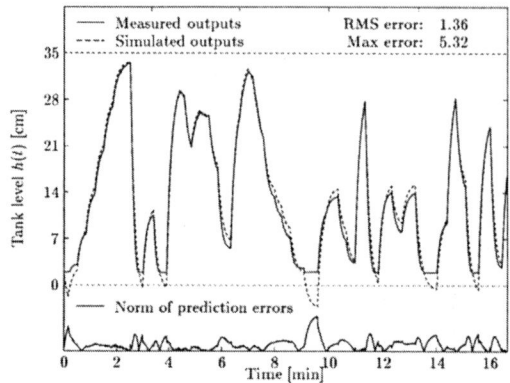

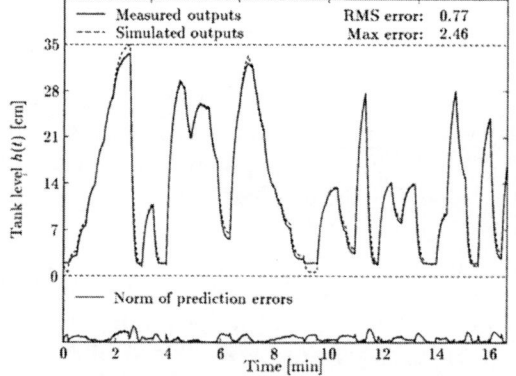

Fig. 5.2. Simulation behavior (based on validation data) of typical ARX (top) and semi-physical (bottom) models describing the level of the tank depicted in Fig. 5.1

is it completely filled up. However, we know for sure that these situations can occur for $u(t) \in [3.5, 7.5]$ V. A good model should be equipped with these extrapolation capabilities.

3. The true function mapping shows no intermediate local plateaus.

Since these features can be captured using the fuzzy framework discussed earlier we next turn to some fuzzy identification. The ARX and the semi-physical models (structures (5.1) and (5.2), respectively) indicate that $h(t-1)$ and $u(t-1)$ are useful signals (regressors). Taking the ARX model as the starting point and noticing its good performance at high levels (above 7 cm) it is reasonable to put further modeling effort into regions where $h(t)$ is low. Desiring also a low complexity model it is sensible to describe each linguistic variable with few linguistic values:

$\langle \text{LEVEL}(t), \boldsymbol{A}(\cdot, \cdot) =$

$\quad \{\text{ZERO, VERY LOW, LOW, RATHER LOW, HIGH, MAX}\}, D : \hat{h}(t|\boldsymbol{\theta})\rangle,$

$\langle \text{LEVEL}(t-1), \boldsymbol{A}(\cdot, \cdot) =$

$\quad \{\text{ZERO, LOW, HIGH}\}, D : \varphi_1(t) = h(t-1) = z_1(t-1)\rangle,$

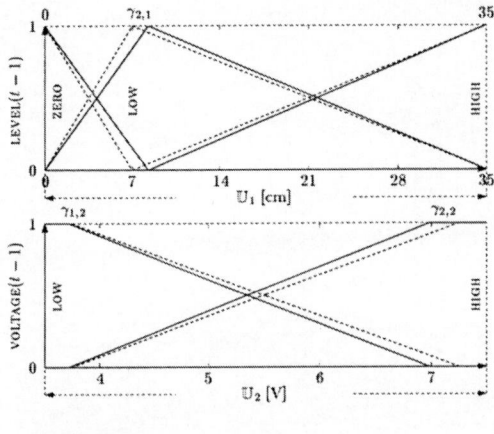

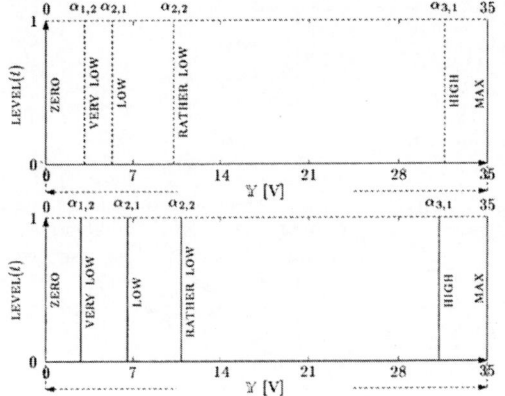

Fig. 5.3. Premise (top) and consequence (bottom) MFs for describing the liquid level of the tank system. Dotted curves show the situation when only the centers $\boldsymbol{\alpha}$ are estimated. Solid curves show the situation after constrained estimation subject to the constraints (5.8).

$$\langle \text{VOLTAGE}(t-1), \boldsymbol{A}(\cdot, \cdot) =$$
$$\{\text{LOW}, \text{HIGH}\}, D : \varphi_2(t) = u(t-1) = z_2(t-1)\rangle,$$

$$\hat{h}(t|\boldsymbol{\theta}) \in \mathbb{Y} = [0, 35], \quad \varphi_1(t) \in \mathbb{U}_1 = [0, 35], \quad \varphi_2(t) \in \mathbb{U}_2 = [3.5, 7.5], \quad (5.3)$$

where $\mathbf{z}(t) = [\,h(t) \quad u(t)\,]$. The listed system properties can now be guaranteed if the MFs

$$
\begin{array}{ll}
\mu_{\text{ZERO}}(\hat{h}(t|\boldsymbol{\theta})) = \alpha_{1,1} = 0, & \mu_{\text{VERY LOW}}(\hat{h}(t|\boldsymbol{\theta})) = \alpha_{1,2}, \\
\mu_{\text{LOW}}(\hat{h}(t|\boldsymbol{\theta})) = \alpha_{2,1}, & \mu_{\text{RATHER LOW}}(\hat{h}(t|\boldsymbol{\theta})) = \alpha_{2,2}, \\
\mu_{\text{HIGH}}(\hat{h}(t|\boldsymbol{\theta})) = \alpha_{3,1}, & \mu_{\text{MAX}}(\hat{h}(t|\boldsymbol{\theta})) = \alpha_{3,2} = 35,
\end{array}
\qquad (5.4)
$$

and

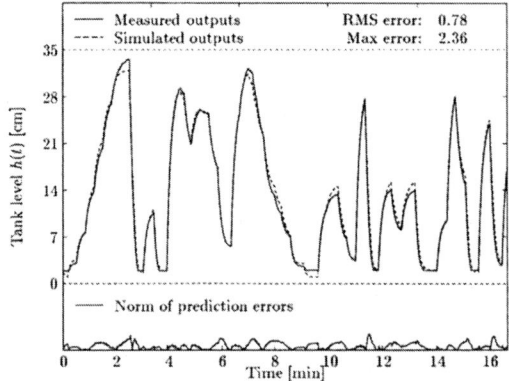

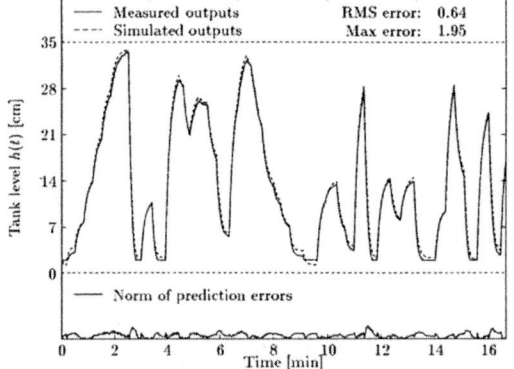

Fig. 5.4. Simulation (based on validation data) of unconstrained linear least-squares (top) and constrained (bottom) estimated fuzzy models reflecting the liquid level of the tank from Fig. 5.1

$$\mu_{\text{ZERO}}(\varphi_1(t), \gamma_{1,1}, \gamma_{2,1}) = \mu_{A_{1,1}}(\varphi_1(t), 0, \gamma_{2,1}) = \texttt{mfl}(\varphi_1(t), 0, \gamma_{2,1}),$$

$$\mu_{\text{LOW}}(\varphi_1(t), \gamma_{1,1}, \gamma_{2,1}, \gamma_{3,1}) = \mu_{A_{2,1}}(\varphi_1(t), 0, \gamma_{2,1}, 35)$$
$$= \texttt{mftri}(\varphi_1(t), 0, \gamma_{2,1}, 35),$$

$$\mu_{\text{HIGH}}(\varphi_1(t), \gamma_{2,1}, \gamma_{3,1}) = \mu_{A_{3,1}}(\varphi_1(t), \gamma_{2,1}) = \texttt{mfr}(\varphi_1(t), \gamma_{2,1}, 35),$$

$$\mu_{\text{LOW}}(\varphi_2(t), \gamma_{1,2}, \gamma_{2,2}) = \mu_{A_{1,2}}(\varphi_2(t), \gamma_{1,2}, \gamma_{2,2})$$
$$= \texttt{mfl}(\varphi_1(t), \gamma_{1,2}, \gamma_{2,2}),$$

$$\mu_{\text{HIGH}}(\varphi_2(t), \gamma_{1,2}, \gamma_{2,2}) = \mu_{A_{2,2}}(\varphi_2(t), \gamma_{1,2}, \gamma_{2,2})$$
$$= \texttt{mfr}(\varphi_2(t), \gamma_{1,2}, \gamma_{2,2})$$

$$(5.5)$$

are used in the fuzzy predictor

$$\hat{h}(t|\boldsymbol{\theta}) = \sum_{j_1=1}^{3} \sum_{j_2=1}^{2} \alpha_{j_1,j_2} \prod_{k=1}^{2} \mu_{A_{j_1,j_2}}(\varphi_k(t), \boldsymbol{\gamma}), \qquad (5.6)$$

which contains seven free parameters

$$\boldsymbol{\theta} = [\, \alpha_{1,2} \quad \alpha_{2,1} \quad \alpha_{2,2} \quad \alpha_{3,1} \quad \gamma_{2,1} \quad \gamma_{1,2} \quad \gamma_{2,2} \,]^T \qquad (5.7)$$

chosen so that

$$0 = \alpha_{1,1} < \alpha_{1,2} < \alpha_{2,1} < \alpha_{2,2} < \alpha_{3,1} < \alpha_{3,2} = 35,$$
$$0 = \gamma_{1,1} < \gamma_{2,1} < \gamma_{3,1} = 35, \tag{5.8}$$
$$3.5 < \gamma_{1,2} < \gamma_{2,2} < 7.5.$$

A graphical representation of the corresponding complete fuzzy rule base is shown in Fig. 5.3. Notice that the MFs associated with each regressor form a fuzzy partition and that this fact together with the restrictions (5.8) guarantee a monotonically increasing predictor in $\varphi(t)$. By Lemma 4.2 we also get a steady-state gain curve that is monotonically increasing in $u(t) = u^*$ (the first property). Furthermore, the extrapolation property is ensured by fixing some of the parameters: $\alpha_{1,1} = \gamma_{1,1} = 0$ and $\alpha_{3,2} = \gamma_{3,1} = 35$, thereby assuring that the predictor is able to return values in the whole output universe $\mathbb{Y}$. The third property is finally guaranteed by the use of piecewise linear MFs in accordance with (5.5).

With γ fixed according to the left plot of Fig. 5.3 (dotted curves), unconstrained linear least-squares estimation of the four free centers α yields a feasible parameter estimate; see the upper right plot of Fig. 5.3. The simulation detailed to the left of Fig. 5.4 indicates also that this first model is rather good. Starting from this point, it is now true that unconstrained estimation of all seven parameters renders a model with a lower root mean square (RMS) error (0.70 compared to 0.78 for the first model), but then it becomes difficult to linguistically interpret the model obtained. Resolving this dilemma by performing constrained estimation subject to the constraints (5.8) gives linguistically sound MFs as shown in Fig. 5.3 (solid curves). On top of that, this final model shows the best simulation performance of all models derived. Compare Figs. 5.2 and 5.4. The built-in increasing nature of the final predictor is now evident from the left plot of Fig. 5.5, and although this mapping is at first sight quite similar to a linear one, it is clear from the steady-state gain curve of Fig. 5.5 that the model shows important nonlinear behavior in the operating region of interest. Notice also that this steady-state gain curve is monotonically increasing in $u(t) = u^*$.

From this discussion we conclude that the tank system can be accurately described by a fuzzy model having few (7) estimated parameters. Sound physical behavior is guaranteed by applying the model structure (4.19), which allows inclusion of certain extrapolation and steady-state gain monotonicity features. The latter property is especially important to reflect in certain predictive control applications, as stressed in (Koivisto, 1995). Even if models that are good from a loss function point of view are used, it is shown there that without such a property (when known from physics) severe stability problems often arise. This behavior is indeed related to the difficulties occurring when performing standard adaptive control based on linear models for which the sign of the first $B(q)$ parameter is incorrect (Åström and Wittenmark, 1995).

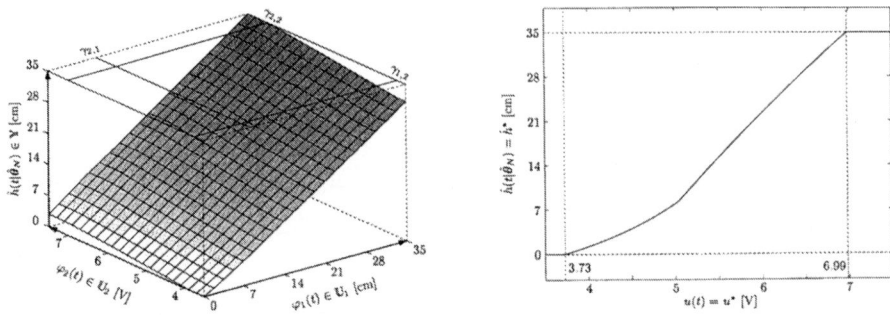

Fig. 5.5. Function mapping (left) and steady-state gain curve (right) of the fuzzy tank model obtained using constrained estimation

6. Practical Aspects

This section addresses a number of practical issues that ought to be considered in connection with fuzzy grey box identification.

6.1 Model Complexity

We have earlier stressed that the model complexity typically increases rapidly with the number of possible linguistic variables r, particularly if these can be assigned to many different linguistic values. While the number of variables can be reduced by a fuzzy hybrid approach (see Sect. 4.5), the number of linguistic values can be kept down by using a coarse description language.

In a way this is in conflict with the experts' attempt to pursue accurate linguistic modeling, yet the use of "few" MFs is often desirable from an estimation point of view, especially in complex modeling situations where the data are sparse in the regression space. The main reason for this is that a "too" dense MF configuration implies that the corresponding parameters are fitted to few data, which typically leads to models that perform rather poorly for other data records.

To overcome these difficulties a good practice is often to start with a rather coarse rule base, and, if necessary, successively refine it. This can be accomplished by lumping together similar linguistic values into one single notion: treat, e.g., VERY LOW, LOW, and SOMEWHAT LOW as one linguistic value described by one MF. If the estimated coarse model is not good enough, then introduce new MFs based partly on the expert knowledge and partly on the performance of the coarse model. The use of the coarse model is foremost motivated by the fact that it provides local performance information, i.e., it gives useful data guided refinement information. It may also provide information about phenomena that were overlooked in the modeling phase. This fine-grain procedure is now iterated until (hopefully) a good enough model is found. Although user interaction is good for identifying and avoiding pitfalls, the major drawback with the approach is of course that it is rather time-consuming compared to pure black box modeling.

6.2 Robustness of the Identification Method

The success of any identification method relies on the descriptive power of the model structure as well as on the quality of the estimation data. Compared to a pure data driven identification approach, it is easier to avoid data caused pitfalls by using an expert determined model structure.

Concerning estimation algorithms, it is worth stressing that the preferred schemes of Sect. 2.3 are all robust in the sense that the fit of the tuned model is at least as good as what is obtained with the initial parameters. With a constrained estimation procedure, we can in addition guarantee that the estimated models are linguistically sound. Notice, though, that this does not imply that a "better" model is obtained, as this only can be assessed after a careful validation procedure.

A distinct advantage with expert modeling is that redundancy in terms of similar MFs as well as physically unsound regions can be avoided. Apart from reducing the model complexity, this also leads to fewer ill-conditioning problems. To fully handle ill-conditioning we used regularization, having the nice add-on property that it enables extrapolation into more or less expert explained regression regions. Such a regularizing effect is mediated in the algorithms through SVD computations, which are known to be numerically robust to carry out (Golub and Van Loan, 1989).

6.3 Software

An always present and relevant issue in system identification is the availability of software tools. The prototype package used for the above experiments consists of a number of MATLAB (The MathWorks, Inc., 1992) m-files, which can be downloaded from the library

`ftp://ftp.control.isy.liu.se/pub/Software/Fuzzy/`

Because the model structure as well as the constraints are represented as strings, this package can be used for rather general predictors of the form (2.3), and not just fuzzy ones. For example, it can be applied directly to the fuzzy hybrid approaches suggested in Sect. 4.5.

However, working with strings on a textual basis is a bit awkward and prone to error. This is a problem that can be relaxed significantly through a graphical user interface (GUI) of the kind provided by MathWorks' fuzzy logic toolbox (Roger Jang and Gulley, 1995). The design of such GUI means is an obvious project for the future. Other important software projects for the future include the implementation of more efficient constrained estimation algorithms as well as the development of general and versatile validation procedures.

7. Conclusions and Future Work

After experiment design and data collection, a typical system identification session involves two main issues: model structure determination followed by parameter estimation. In this contribution we have considered fuzzy grey box identification, which assumes that the former problem is addressed, at least partly, by a human domain expert who indirectly describes the model structure in terms of a number of if-then rules. Taking various fuzzy and identification aspects into account we arrived at the Mamdani fuzzy model structure (3.12), which, in a more traditional identification setting, is nothing but a series expansion of composition type having much in common with feed-forward neural networks, RBFN networks, model regression trees, etc.

This kinship in particular means that efficient Newton-type algorithms (the pseudo-inverse Gauss–Newton or the Levenberg–Marquardt procedures) can be applied for MF parameter estimation. Since these schemes are equipped with regularization it is to some extent possible to preserve expert knowledge having minor data support. However, the series expansions are usually rich in terms of the number of parameters. This fact, especially in combination with few and noisy data, sometimes implies that the original linguistic interpretation of the rules are lost in the estimation step. To avoid such undesirable behavior it is necessary to impose certain restrictions on the MF parameters, and then solve the constrained minimization problem obtained.

For some model based control applications it is also extremely important that the applied models reflect certain non-structural system properties, e.g., a monotonically increasing steady-state gain curve and/or a non-oscillatory step response behavior. Whereas such features are in general difficult to guarantee when using neural networks or other flexible series expansions, these can be dealt with by employing the special fuzzy partition based model structure (4.19). Experiments on real-world data – in this case a tank system – as well as other applications (Lindskog, 1996) have demonstrated the feasibility and the usefulness of this approach.

To this end, let us finally point to some extensions and open problems related to the fuzzy identification framework discussed above.

1. Stability and various robustness issues are very important when the models are going to be used in control applications. How does the choice of MFs then affect stability? How and to what extent can the linguistic system knowledge be exploited for robust control design? Can we apply modern stability tools stemming from the robust control field (unstructured uncertainties, etc.)? For this problem, (Suykens et al., 1995) has

already suggested an interesting method based on a particular neural network model. The basic idea is to view the neural network as a nominal linear model with bounded nonlinear feedback perturbations, and then use a standard robust control design scheme. The obvious question here is whether a similar procedure can be devised for fuzzy models as well.

2. To ensure a monotonic steady-state gain curve we restricted the MFs to correspond to fuzzy partitions. What other and perhaps better MF configurations are able to preserve this knowledge? Also, what other kinds of non-structural properties can be captured within the fuzzy framework?

3. A water heating system (Koivisto, 1995) with a known increasing steady-state gain curve behavior is successfully modeled in (Lindskog, 1996) using the fuzzy model structure (4.19). Because of seasonal temperature variations and some other factors this model is only valid under certain operational conditions. To also handle long-term seasonal changes there is here a need for a fuzzy specific and monotonicity preserving recursive estimation algorithm.

4. The applications considered and mentioned above are rather small. However, it is our belief that the use of linguistic expert knowledge really pays off for more involved processes. To investigate this it is worth looking further into application fields where verbal knowledge dominates, as is the case, e.g., for many biomedical or biochemical systems.

References

Aguirre, L. A. and S. A. Billings (1995). Improved structure selection for nonlinear models based on term clustering. *International Journal of Control*, **62**(3), 569–587.

Åström, K. J. and B. Wittenmark (1995). *Adaptive Control,* 2nd ed. Electrical Engineering: Control Engineering. Addison-Wesley.

Babuška, R. and H. B. Verbruggen (1994). Applied fuzzy modeling. In: *Proceedings of the IFAC Symposium on Artificial Intelligence in Real Time Control*. Valencia, Spain, pp. 61–66.

Björk, Å. (1996). *Numerical Methods for Least Squares Problems*. SIAM.

Bohlin, T. (1991). *Interactive System Identification: Prospects and Pitfalls*. Communications and Control Engineering. Springer.

Breiman, L. (1993). Hinging hyperplanes for regression, classification, and function approximation. *IEEE Trans. Information Theory*, **39**(3), May, 999–1013.

Breiman, L., J. H. Friedman, R. A. Olshen, and C. J. Stone (1984). *Classification and Regression Trees*. The Wadsworth Statistics/Probability Series. Wadsworth and Brooks.

Brown, M. and C. Harris (1994). *Neurofuzzy Adaptive Modelling and Control*. Systems and Control Engineering. Prentice Hall International.

Chen, C. H., ed., (1996). *Fuzzy Logic and Neural Network Handbook*. McGraw-Hill.

Chen, S. and S. A. Billings (1992). Neural networks for nonlinear dynamic system modelling and identification. *International Journal of Control*, **56**(2), 319–346.

Dennis, J. E. and R. B. Schnabel (1983). *Numerical Methods for Unconstrained Optimization and Nonlinear Equations*. Prentice Hall.

Draper, N. and H. Smith (1981). *Applied Regression Analysis*, 2nd ed. John Wiley and Sons.

Driankov, D., H. Hellendoorn, and M. Reinfrank (1993). *An Introduction to Fuzzy Control*. Springer. 2nd ed. 1996.

Dubois, D. and H. Prade (1992). Fuzzy sets in approximate reasoning, part 1. *Fuzzy Sets and Systems*, **40**(1), 65–74.

Fletcher, R. (1987). *Practical Methods of Optimization*. John Wiley and Sons.

Forssell, U. and P. Lindskog (1997). Combining semi-physical and neural network modeling: an example of its usefulness. Submitted to the 11th IFAC Symposium on System Identification (SYSID'97) to be held in Fukuoka, Japan, July 1997.

Golub, G. H. and C. F. Van Loan (1989). *Matrix Computations*, 2nd ed. Johns Hopkins University Press.

Hangos, K. M., ed., (1995). *International Journal of Adaptive Control and Signal Processing: Special Issue on Grey Box Modelling*, Vol. **9**(6), November/December. John Wiley and Sons.

Haykin, S. (1994). *Neural Networks: A Comprehensive Foundation*. Macmillan.

Higgins, C. M. and R. M. Goodman (1994). Fuzzy rule-based networks for control. *IEEE Trans. Fuzzy Systems*, **2**(1), February, 82–88.

Ishigami, H., T. Fukuda, T. Shibata, and F. Arai (1995). Structure optimization of fuzzy neural network by genetic algorithm. *Fuzzy Sets and Systems*, **71**(3), May, 257–264.

Johansen, T. A. (1994). *Operating Regime Based Process Modeling and Identification*. Phd thesis 94:109-W, Division of Engineering Cybernetics, University of Trondheim, Trondheim, Norway, November.

Johansen, T. A. and B. A. Foss (1994). Identification of non-linear system structure and parameters using regime decomposition. In: *Preprints of the 10th IFAC Symposium on System Identification*, (M. Blanke and T. Söderström, eds.), Vol. 1, July. Copenhagen, Denmark, pp. 131–136.

Kaymak, U. and R. Babuška (1995). Compatible cluster merging for fuzzy modeling. In: *Proceedings FUZZ-IEEE/IFES'95*. Yokohama, Japan, pp. 897–904.

Kirkpatrick, S., C. D. Gelatt, and M. P. Vecchi (1983). Optimization by simulated annealing. *Science*, **220**(4598), May, 671–680.

Koivisto, H. (1995). *A Practical Approach to Model Based Neural Network Control*. Phd thesis 170, Tampere University of Technology, Tampere, Finland, December.

Kosko, B. (1992). Fuzzy Systems as Universal Approximators. In: *Proceedings of the 1st IEEE International Conference on Fuzzy Systems*. San Diego, CA, pp. 1153–1162.

Kung, S. Y. (1993). *Digital Neural Networks*. Prentice Hall.

Lee, C. C. (1990). Fuzzy logic in control systems: fuzzy logic controller – parts I and II. *IEEE Trans. Systems, Man, and Cybernetics*, **SMC-20**(2), March/April, 404–435.

Lin, C.-T. and C. S. G. Lee (1996). *Neural Fuzzy Systems: A Neuro-Fuzzy Synergism to Intelligent Systems*. Prentice Hall.

Lindskog, P. (1996). *Methods, Algorithms and Tools for System Identification Based on Prior Knowledge*. Phd thesis 436, Department of Electrical Engineering, Linköping University, Linköping, Sweden, May.

Lindskog, P. and L. Ljung (1995). Tools for semiphysical modelling. *International Journal of Adaptive Control and Signal Processing*, **9**(6), November-December, 509–523.

Ljung, L. (1987). *System Identification: Theory for the User*. Prentice Hall.

Ljung, L., J. Sjöberg, and H. Hjalmarsson (1996). On neural network model structures in system identification. In: *Identification, Adaptation, Learning: The Sci-*

ence of Learning Models from Data, (S. Bittanti and G. Picci, eds.), Vol. 153 of *Nato ASI Series F: Computer and Systems Sciences*, pp. 366–399. Springer.

Mamdani, E. H. and S. Assilian (1975). An experiment in linguistic synthesis with a fuzzy logic controller. *International Journal of Man-Machine Studies*, **7**(1), 1–13.

Marks II, R. J., ed., (1994). *Fuzzy Logic Technology and Applications*. IEEE Technology Update. IEEE Technical Activities Board.

Poggio, T. and F. Girosi (1990). Networks for approximation and learning. *Proceedings of the IEEE*, **78**(9), 1481–1497.

Pucar, P. and J. Sjöberg (1995a). On the hinge finding algorithm for hinging hyperplanes – revised version. Technical Report LiTH-ISY-R-1804, Department of Electrical Engineering, Linköping University, Linköping, Sweden. Available by anonymous ftp 130.236.24.1.

Pucar, P. and J. Sjöberg (1995b). Parameterization and conditioning of hinging hyperplane models. Technical Report LiTH-ISY-R-1809, Department of Electrical Engineering, Linköping University, Linköping, Sweden. Available by anonymous ftp 130.236.24.1.

Roger Jang, J.-S. and N. Gulley (1995). *Fuzzy Logic Toolbox*. The MathWorks, Inc., Cochituate Place, Natick, MA.

Roger Jang, J.-S. and C.-T. Sun (1995). Neuro-fuzzy modeling and control. *Proceedings of the IEEE*, **83**(3), March, 378–406.

Scales, L. E. (1985). *Introduction to Non-linear Optimization*. Computer Science Series. Macmillan.

Sjöberg, J., Q. Zhang, L. Ljung, A. Benveniste, B. Delyon, P.-Y. Glorennec, H. Hjalmarsson, and A. Juditsky (1995). Nonlinear black-box modeling in system identification: a unified overview. *Automatica*, **31**(12), December, 1691–1724.

Söderström, T. and P. Stoica (1989). *System Identification*. Prentice Hall International.

Strömberg, J.-E., F. Gustafsson, and L. Ljung (1990). Trees as black-box model structures for dynamical systems. Technical Report LiTH-ISY-I-1122, Department of Electrical Engineering, Linköping University, Linköping, Sweden.

Sugeno, M. and G. T. Kang (1988). Structure identification of fuzzy model. *Fuzzy Sets and Systems*, **28**(1), 15–33.

Sugeno, M. and T. Yasukawa (1993). A fuzzy-logic-based approach to qualitative modeling. *IEEE Trans. Fuzzy Systems*, **1**(1), February, 7–31.

Sun, C.-T. (1994). Rule-base structure identification in an adaptive-network-based fuzzy inference system. *IEEE Trans. Fuzzy Systems*, **2**(1), February, 64–73.

Suykens, J. A. K., B. L. R. De Moor, and J. Vandewalle (1995). Nonlinear system identification using neural state space models, applicable to robust control design. *International Journal of Control*, **62**(1), July, 129–152.

Takagi, T. and M. Sugeno (1985). Fuzzy identification of systems and its applications to modeling and control. *IEEE Trans. Systems, Man, and Cybernetics*, **SMC-15**(1), January/February, 116–132.

The MathWorks, Inc. (1992). *MATLAB: High-Performance Numeric Computation and Visualization Software*. The MathWorks, Inc., Cochituate Place, Natick, MA.

Wang, L.-X. (1992). Fuzzy systems are universal approximators. In: *Proceedings of the 1st IEEE International Conference on Fuzzy Systems*. San Diego, CA, pp. 1163–1170.

Wang, L.-X. (1994). *Adaptive Fuzzy Systems and Control: Design and Stability Analysis*. Prentice Hall.

Wang, L.-X. (1995). Design and analysis of fuzzy identifiers of nonlinear dynamic systems. *IEEE Trans. Automatic Control*, **AC-40**(1), January, 11–23.

Wang, L.-X. and J. M. Mendel (1992). Fuzzy basis functions, universal approximation, and orthogonal least-squares learning. *IEEE Trans. Neural Networks*, **3**(5), September, 807–814.

Watson, G. (1969). Smooth regression analysis. *Sankhya, Series*, **A**(26), 359–372.

Yoshinari, Y., W. Pedrycz, and K. Hiroto (1993). Construction of fuzzy models through clustering techniques. *Fuzzy Sets and Systems*, **54**(2), march, 157–165.

Zadeh, L. A. (1965). Fuzzy sets. *Information and Control*, **8**, 338–353.

Zhang, Q. and A. Benveniste (1992). Wavelet networks. *IEEE Trans. Neural Networks*, **3**(6), 889–898.

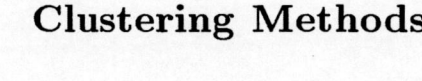

Clustering Methods

Constructing Fuzzy Models
by Product Space Clustering

R. Babuška and H.B. Verbruggen

Delft University of Technology, P.O. Box 5031, NL-2600 GA Delft, The Netherlands

1. Introduction

There are several different approaches to modeling of complex nonlinear systems. The main distinction can be made between global and local methods. Global methods describe the system under study using nonlinear functional relationships between the system's variables. Examples are nonlinear state space models or input-output black-box models such as the popular NARX (Nonlinear AutoRegressive with eXogenous input) structure used often in connection with neural or wavelet networks. Local approaches, on the other hand, attempt to cope with complexity and nonlinearity of systems by decomposing the modeling problem into a number of simpler, in most cases, linear sub-problems (Johansen and Foss, 1993; Banerjee et al., 1995). These methods are conceptually simple and intuitively appealing, as they are close to the way human solve problems. Local models are usually more easily interpretable than complicated global models.

Modeling techniques based on fuzzy sets can be seen as local modeling methods, as they use partitioning of the process domains into a number of fuzzy regions. For each region in the input space, a rule is defined that specifies the output of the model. The rules can be seen as local submodels of the system. The exact nature of these submodels, as well as the way they are combined, depend on the particular type of rules and on the inference mechanism involved. A main distinction can be made between locally constant submodels expressed as the Mamdani-type (linguistic) or relational models and locally linear submodels represented by the Takagi–Sugeno (TS) rules. Fuzzy models can be constructed basically in two ways:

1. As fuzzy expert systems, using human knowledge, in a manner similar to the design of knowledge-based fuzzy controllers (Driankov et al., 1993).
2. Using numerical data and suitable identification techniques.

For developing models of complex real systems, both approaches may be combined. In this chapter, attention is focused on identification techniques for building fuzzy models from data. Acquisition of a rule-based fuzzy model from data requires the identification of the input and output variables, of the antecedent and consequent structure, of the membership functions, and other parameters associated with the particular model structure. Some of these identification tasks can be formulated as nonlinear optimization problems, for which a number of techniques have been proposed, such as neural learning methods (Jang, 1992; de Oliveira, 1993; Glorennec, 1994), orthogonal least squares (Wang, 1994), inductive learning (Ross, 1995), evidential reasoning (Baldwin et al., 1995), or fuzzy clustering (Yoshinari et al., 1993; Zhao et al., 1994; Babuška and Verbruggen, 1994; Kaymak and Babuška, 1995).

This chapter focuses on approximation of nonlinear systems by local linear models, using a class of fuzzy clustering methods. It is argued that with a sufficiently informative identification data set, this approach does not require any prior knowledge on the partitioning of the domains, the number of rules needed, etc. Moreover, the obtained models are locally interpretable and often correspond well with the local behavior of the identified process. The form and the overlap of the membership functions, extracted from the obtained fuzzy partition, provide information about the character of the system's nonlinearity. Hence, the information obtained can be used not only to derive prediction models but also to analyze the underlying system and to validate the model. Moreover, smooth systems can be identified in the same framework as systems characterized by abrupt changes in their behavior.

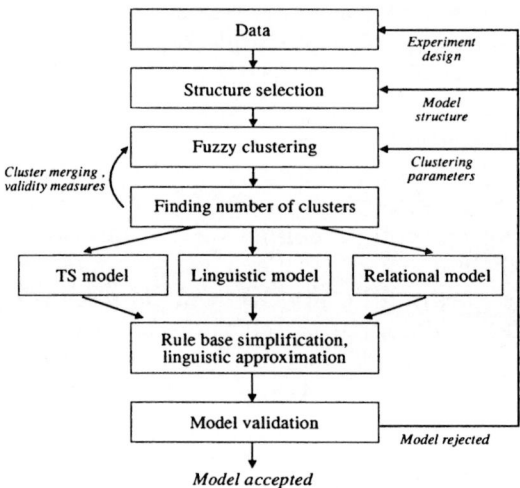

Fig. 1.1. Overview of the identification approach based on fuzzy clustering

Figure 1.1 outlines the main steps of the identification procedure: data collection, structure selection, clustering and the choice of the number of clusters, derivation of the fuzzy model, simplification and validation of the model. As indicated in the figure, different types of fuzzy models can be derived from the fuzzy partition obtained by clustering. This chapter describes in detail only the methods for deriving TS models, the remaining approaches are briefly mentioned. For more details, refer to the cited literature.

The remaining part of this chapter describes these steps in detail and presents the corresponding algorithms. Section 2 reviews the Takagi–Sugeno, singleton and Mamdani fuzzy models, in order to establish the notation and terminology needed in the subsequent sections. Section 3 describes the different structures of dynamic models, to which the identification method can be applied. Section 4 presents the details of the identification method based on fuzzy clustering, and Section 5 describes the derivation of the consequent and antecedent parts of the Takagi–Sugeno fuzzy models. Section 6 presents a simulation example of a pH neutralization process. In Section 7 practical aspects of the presented technique are discussed. The appendix presents a MATLAB implementation of the clustering algorithm.

2. Overview of Fuzzy Models

This section reviews the fuzzy model structures used, and introduces the notation and terminology used in this chapter. Basic knowledge about modeling, fuzzy sets and fuzzy logic is assumed.

2.1 Affine TS Model

The affine *Takagi–Sugeno* (TS) fuzzy model (Takagi and Sugeno, 1985) consists of rules R_i with the following structure

$$R_i: \quad \textbf{If} \quad \underbrace{\textbf{x is } A_i}_{antecedent} \quad \textbf{then} \quad \underbrace{y_i = \textbf{a}_i^T \textbf{x} + b_i}_{consequent}, \quad i = 1, 2, \ldots, K \tag{2.1}$$

where $\textbf{x} \in \mathcal{X} \subset \mathbb{R}^p$ is a crisp input vector, A_i is a (multidimensional) fuzzy set: $\mu_{A_i}(\textbf{x}): \mathcal{X} \to [0, 1]$, $y_i \in \mathbb{R}$ is the (scalar) output of the i-th rule, $\textbf{a}_i \in \mathbb{R}^p$ is a parameter vector and b_i is a scalar offset. The index $i = 1, 2, \ldots, K$ denotes that a given variable is related to the i-th rule, and K is the number of rules in the rule base. Model (2.1) can represent multiple-input, multiple-output (MISO) static and dynamic systems, as explained in more detail in Sect. 3. Given the outputs of the individual consequents y_i, the global output y of the TS model (2.1) is computed using the weighted (fuzzy) mean formula

$$y = \frac{\sum_{i=1}^{K} \beta_i(\textbf{x}) y_i}{\sum_{i=1}^{K} \beta_i(\textbf{x})}. \tag{2.2}$$

Here $\beta_i(x)$ denotes the *degree of fulfillment* of the i-th rule's antecedent, computed simply as the membership degree of $\textbf{x}$ into the fuzzy set A_i: $\beta_i = \mu_{A_i}(\textbf{x})$. As it may be difficult to implement and interpret multidimensional fuzzy sets, the antecedent proposition in (2.1) is usually expressed as a combination of simple propositions with one-dimensional fuzzy sets defined for the individual components of $\textbf{x}$. The degree of fulfillment of the antecedent is computed by combining the membership degrees of the individual propositions using conjunctions, disjunctions, the complement and also linguistic hedges. Most common is the *conjunctive* form of the antecedent, given by

$$\text{If } x_1 \text{ is } A_{i,1} \text{ and } \ldots \text{ and } x_p \text{ is } A_{i,p} \text{ then } y_i = \mathbf{a}_i^T \mathbf{x} + b_i \qquad (2.3)$$

where the degree of fulfillment $\beta_i(\mathbf{x})$ is calculated as

$$\beta_i(\mathbf{x}) = \mu_{A_{i,1}}(x_1) \wedge \mu_{A_{i,2}}(x_2) \wedge \ldots \wedge \mu_{A_{i,p}}(x_p). \qquad (2.4)$$

The minimum operator ($\wedge$) can be replaced by other t-norms, such as the product, bold intersection, etc. (Zimmermann, 1991). Denoting the normalized degree of fulfillment $\gamma_i(\mathbf{x}) = \beta_i(\mathbf{x}) / \sum_{j=1}^{K} \beta_j(\mathbf{x})$, the global output of the TS model can be written as a global pseudo-linear model

$$y = (\sum_{i=1}^{K} \gamma_i(\mathbf{x}) \mathbf{a}_i^T) x + \sum_{i=1}^{K} \gamma_i(\mathbf{x}) b_i = \tilde{\mathbf{a}}^T(\mathbf{x}) x + \tilde{b}(\mathbf{x}) \qquad (2.5)$$

where $\tilde{\mathbf{a}}(\mathbf{x})$ and $\tilde{b}(\mathbf{x})$ are input-dependent parameters, computed as convex linear combinations of the constant parameters $\mathbf{a}_i$ and b_i

$$\tilde{\mathbf{a}}(\mathbf{x}) = \sum_{i=1}^{K} \gamma_i(\mathbf{x}) \mathbf{a}_i, \quad \tilde{b}(\mathbf{x}) = \sum_{i=1}^{K} \gamma_i(\mathbf{x}) b_i. \qquad (2.6)$$

A special case of the consequent function occurs when the offsets $b_i = 0$, $i = 1, \ldots, K$, and the model is called a *homogeneous* TS model

$$\text{If } \mathbf{x} \text{ is } A_i \text{ then } y_i = \mathbf{a}_i^T \mathbf{x}. \qquad (2.7)$$

This model can approximate only a certain class of systems, as all the consequent models contain the origin. On the other hand, the absence of the offset term facilitates controller design and stability analysis in a quasi-linear framework, using Lyapunov theory and linear matrix inequalities (Tanaka and Sugeno, 1992; Tanaka et al., 1996).

2.2 Mamdani and Singleton Fuzzy Model

The Mamdani or linguistic fuzzy model (Zadeh, 1973; Mamdani, 1977; Driankov et al., 1993) consists of rules where both the antecedent and the consequent are fuzzy propositions

$$R_i: \quad \text{If } \mathbf{x} \text{ is } A_i \text{ then } y \text{ is } B_i, \quad i = 1, 2, \ldots, K \qquad (2.8)$$

where $\mathbf{x} \in \mathbb{R}^p$ is the antecedent variable, and $y \in \mathbb{R}$ is the consequent variable. A_i and B_i represent *linguistic terms* (labels) defined by fuzzy sets $\mu_{A_i}(\mathbf{x}): \mathbb{R}^p \to [0, 1]$ and $\mu_{B_i}(y): \mathbb{R} \to [0, 1]$, respectively. When a defuzzified output is required, the consequent fuzzy sets can be replaced by singletons (fuzzy representation of real numbers). This model is called the *singleton model*

$$\text{If } \mathbf{x} \text{ is } A_i \text{ then } y = b_i. \qquad (2.9)$$

and can also be seen as a special case of the affine TS model, where the parameter vectors $\mathbf{a}_i = \mathbf{0}$, $i = 1, \ldots, K$. The singleton fuzzy models belong to the general class of function approximators, called *basis functions expansion*, taking the form $y = \sum_{i=1}^{K} B_i(\mathbf{x}) b_i$. In the case of the singleton model, the basis functions $B_i(\mathbf{x})$ are given by the (normalized) degrees of fulfillment of the rule antecedents, see, for instance, (2.4), and the constants b_i are the consequents. Other structures such as radial basis function networks, splines, etc., also belong to this class of systems, whose approximation properties are quite well understood. For instance, with triangular membership functions forming a partition,[1] the singleton model results in a multi-linear interpolation between the consequents (Brown and Harris, 1994; Jager, 1995).

The use of membership functions (validity functions, basis functions) for partitioning the operating space of the system into overlapping regions is not restricted to fuzzy modeling. A similar approach is applied for instance by Johansen (1994) or Banerjee et al. (1995) in the context of operating regime based modeling and control. Analogies can also be found in nonlinear statistical regression using splines (Boor, 1978; Friedman, 1991), phase-regression models (Seber and Wild, 1989), or radial basis functions (Jang and Sun, 1993).

3. Structure Selection for Modeling of Dynamic Systems

In fuzzy modeling, the problem of structure selection[2] can be divided into three sub-problems: 1) choice of the input and output variables of the model, 2) representation of the system's dynamics, and 3) determining the number of membership functions per variable. These three steps are discussed below.

Choice of the input and output variables of the model. Although most identification methods assume that the input and output variables of the process are known (Ljung, 1987), in reality, especially for multivariable and closed-loop systems, it is often not clear which variables should be considered as the model inputs. The presented approach does not provide any special tools for determining the relevant process variables. Well-known statistical techniques, such as correlation analysis, and of course prior knowledge should be used at this step.

Representation of the system's dynamics. The purpose of this step is to transform the identification of a dynamic system into a static regression problem. Fuzzy models can approximate a wide class of static nonlinear regression problems $y = f(\mathbf{x})$, where f is an unknown smooth function (Wang, 1992; Kosko, 1994; Zeng and Singh, 1994; Zeng and Singh, 1995). In general, dynamic systems can be modelled by static regression structures, using the concept of system's state. One can distinguish mainly between the state-

---

[1] The sum of membership degrees of a domain element into all the membership functions is one.

[2] Structure selection is also called structure identification in the literature, see for instance (Sugeno and Yasukawa, 1993; Johansen, 1994)

space models and the input–output models. Both forms are discussed in more details in this section.

Number of membership functions per variable. This choice also relates to the number of rules and hence to the approximation properties and the complexity of the fuzzy model. The approach based on fuzzy clustering provides some tools for determining the appropriate number of rules in the model, as described in more detail in Sect. 4.2. In the fuzzy modeling literature, the term structure identification pertains mainly to this step, i.e., to the identification of the number and parameters of the antecedent membership functions, see (Sugeno and Kang, 1988).

The remaining part of this section discusses in more details the choice of the representation of the system's dynamics.

3.1 State-Space Modeling

State-space models use a state-transition function, which maps the current state and the current input of the dynamic system into the change (derivative) of the state (continuous-time case), or into the state at the next sampling instant (discrete-time case). The state transition function is just a static mapping, which can be represented by a fuzzy model, such as the following discrete-time TS model:

$$\textbf{If } \xi(k) \textbf{ is } A_i \textbf{ and } \mathbf{u}(k) \textbf{ is } B_i \textbf{ then } \begin{cases} \xi_i(k+1) = A_i\xi(k) + B_i\mathbf{u}(k) \\ \mathbf{y}_i(k) = C_i\xi(k). \end{cases} \quad (3.1)$$

Here $\xi(\mathbf{k})$ is the system's state, $\mathbf{u}(k)$ is the input, and A_i, B_i, C_i are matrices of appropriate dimensions, associated with the i-th rule. The state space representation is suitable when the available prior knowledge allows to determine the structure of the system under study and to identify the state variables. The advantage is that the structure of the model may be related to the structure of the real system and consequently also the model rules and parameters are often physically relevant.

3.2 Input–Output Models

For building fuzzy models from data, generated by poorly understood dynamic systems, the input–output representation is often applied. Instead of using a physically relevant state vector, the state of the system is represented by a finite number of past inputs and outputs of the system. The most common structure is the NARX (Nonlinear AutoRegressive with eXogenous input) model, which can represent the observable and controllable modes of a large class of discrete-time nonlinear systems (Leonaritis and Billings, 1985; Chen and Billings, 1989). This model is used in most nonlinear identification methods such as neural networks (Sjöberg et al., 1994), radial basis functions (Chen et al., 1991), CMAC (Brown and Harris, 1994), and also fuzzy models

(Yager and Filev, 1994; Wang, 1994). The NARX model establishes a relation between the collection of past input–output data and the predicted output

$$\hat{y}(k+1) = F\left(y(k), \ldots, y(k-n_y+1), u(k), \ldots, u(k-n_u+1)\right) \qquad (3.2)$$

where k denotes discrete time samples, n_y and n_u are integers related to the system's order. In terms of rules, the model is given by

R_i: **If** $y(k)$ is $A_{i,1}$ **and** $y(k-1)$ is $A_{i,2}$ **and** ... **and** $y(k-n_y+1)$
    is A_{i,n_y} **and** $u(k)$ is $B_{i,1}$ **and** $u(k-1)$ is $B_{i,2}$ **and** ... **and**
    $u(k-n_u+1)$ is B_{i,n_u} **then**

$$\hat{y}(k+1) = \sum_{j=1}^{n_y} a_{i,j} y(k-j+1) + \sum_{j=1}^{n_u} b_{i,j} u(k-j+1) + c_i \qquad (3.3)$$

where $a_{i,j}$, $b_{i,j}$ and c_i are the consequent parameters. The NARX model can represent MISO systems directly and MIMO systems in a decomposed form as a set of coupled MISO models. It should be noted that the dimension of the regression problem in input–output modeling is often larger than with the state-space models, since the state of the system can usually be represented by a vector of a lower dimension than, for instance, in the NARX model given by (3.2).

Other common input–output models are the nonlinear output error (NOE), which includes the past model outputs (predictions) $\hat{y}(k), \ldots, \hat{y}(k-n_y)$ instead of the measured outputs of the process, as in (3.2), and the so-called innovations forms (e.g., NARMAX), including the past values of the prediction error in the regression vector. These structures are not used for clustering-based identification, as the regression vector cannot be constructed directly from the data and the corresponding regression problem must be solved in a recurrent way.

Once the type of the dynamic model is fixed, it remains to select the order of the system (the integers n_u, n_y in the NARX model) and the input delays. A set of candidate structures (combinations of model orders and delays) is typically chosen by the user, based on prior knowledge, assumptions about the system, purpose of modeling, desired accuracy, etc. These structures are then compared in terms of some suitable performance index, such as the mean square prediction error evaluated on a fresh data set to avoid fitting the noise (Ljung, 1987). In connection with fuzzy models also the regularity criterion was applied (Sugeno and Yasukawa, 1993), which is based on cross-validation of two data sets. This step can be done automatically by computer programs. As linear identification techniques are much simpler and numerically more robust than nonlinear methods, it is usually recommended to start with a linear model, and to determine the structure using a variety of available tools, including the well-known information criteria (Akaike, 1974; Rissanen, 1978). The structure of the best linear model is than used as a starting point for the nonlinear modeling. In the context of clustering, the cluster validity measures can be used to select the model order and the number of clusters within the given nonlinear structure, as shown later.

3.3 Hybrid Approaches

In many systems, such as chemical and biochemical processes, the modeling task can be divided into two subtasks: modeling of well-understood mechanisms based on, e.g., mass and energy balances (first-principle modeling) and approximation of partially known relationships such as specific reaction rates. The latter task often involves the use of empirical models, valid under certain assumption about the process. These models may be very inaccurate if the assumptions are not met or if the process is only partially known. In such a case, the first-principles model based on differential equations can be combined with a fuzzy model representing the unknown relationships based on heuristic knowledge, experience and data. As an example consider modeling a fed-batch stirred bioreactor described by the following equations derived from the mass balances (Psichogios and Ungar, 1992)

$$\frac{dX}{dt} = \eta(\cdot)X - \frac{F}{V}X, \quad \frac{dS}{dt} = -k_1\eta(\cdot)X + \frac{F}{V}[S_i - S], \quad \frac{dV}{dt} = F \qquad (3.4)$$

where X is the biomass concentration, S is the substrate concentration, V is the reactor's volume, F is the inlet flow rate, k_1 is the substrate to cell conversion coefficient and S_i is the inlet feed concentration. These mass balances provide a partial model. The kinetics of the process are represented by the specific growth rate $\eta(\cdot)$ which accounts for the conversion of the substrate to biomass and is typically a complex nonlinear function of the process variables. Many different models have been proposed to describe this function. Hence, choosing the right model for a given process may not be straightforward. An alternative approach is to approximate $\eta(\cdot)$ using process measurements and to incorporate the identified nonlinear relation into the known white-box model given by equations (3.4). The data for identification can be obtained for instance from batch experiments, for which $F = 0$ and the first equation in (3.4) reduces to $dX/dt = \eta(\cdot)X$ where $\eta(\cdot)$ appears explicitly. The advantage of fuzzy models in this setting is that they can serve as good numerical predictors and at the same time can provide a posteriori qualitative knowledge about the unknown relationship. An example of application of this approach to the modeling of the kinetics of enzymatic penicillin-G conversion is given by Babuška et al. (1996).

The first choice the user needs to make is to select the representation of the model, e.g., the input–output regression model, state-space, or hybrid model, as discussed above. As a general rule, as much available knowledge as possible should be incorporated at this step. For well-understood systems, the state-space modeling may be chosen. For partially known systems and systems with higher-order dynamics, the hybrid approach may be most suitable. Low-order or poorly understood processes, on the other hand, can be conveniently represented as input–output models.

4. Fuzzy Clustering

This section presents the identification techniques based on fuzzy clustering. By choosing the structure of the model, the identification problem is transformed into static nonlinear regression $y = F(\mathbf{x})$. In the context of regression and system modeling, the model input $\mathbf{x}$ is called the *regression vector* or simply the *regressor*, the output y is called the *regressand* or the *response variable*. The product space of the regressor and the regressand, $\mathcal{Z} = (\mathcal{X} \times \mathcal{Y}) \subset \mathbb{R}^n$ is called the *regression space*, where $n = p + 1$ is the dimension of this space. Recall, that p is the dimension of the regression vector $\mathbf{x}$. In this space, the equation $y = F(\mathbf{x})$ defines a hypersurface (subspace of the dimension $\mathbb{R}^p$), which is called the *regression surface*. Figure 4.1 gives examples of the regression surfaces of two first-order discrete-time systems; a Hammerstein system with an input dead-zone/saturation nonlinearity, and a system given by the equation $y(k+1) = y(k) \exp(-u(k))$.

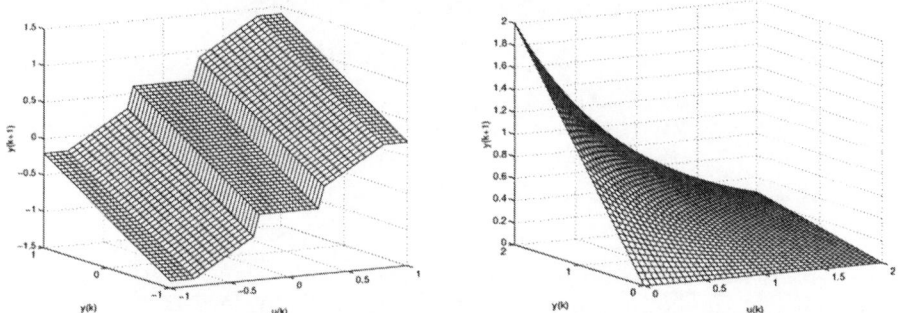

Fig. 4.1. Example of regression surfaces of two nonlinear dynamic systems. Left: system with dead-zone and saturation, Right: system $y(k+1) = y(k) \exp(u(k))$

Geometrically, the consequents of the affine TS model (2.1) can be seen as hyperplanes in the regression space. By means of the antecedent fuzzy sets, the regression space is partitioned into smaller regions, in which the regression surface can be locally approximated by these hyperplanes. The purpose of identification is to find the number, locations and parameters of the hyperplanes, such that the regression surface is accurately approximated. This is achieved by applying a class of fuzzy clustering methods called *subspace clustering algorithms*. These methods can detect clusters which lie in subspaces of the regression space. Note that there is a major difference between this method and other techniques using fuzzy c-means clustering. For instance, Pedrycz (1984) uses c-means clustering to find membership functions for each antecedent variable separately and then establishes a relational model between the identified membership functions. Sugeno and Yasukawa (1994), on the other hand, applied c-means clustering in the consequent space and the antecedent membership functions are induced from the consequent ones. These methods do not aim directly at the approximation of the system that generated the data. Moreover, the membership functions can only be found inside regions densely covered by the data. Berenji and Khedar (1993) applies fuzzy clustering in the product space, but uses the c-means algorithm, which induces spherical clusters and not hyperplanar ones. A similar approach to the one presented in this chapter has been used by Yoshinari et al. (1993) and Zhao et al. (1994). Subspace clustering has been also applied in image processing for straight line or contour extraction (Dave, 1992; Krishnapuram and Freg, 1992), which are applications conceptually similar to the approximation of nonlinear systems by local linear models.

The set (matrix) of data to be clustered, denoted by Z, is constructed by concatenating a matrix containing the regression vectors in its columns and a vector containing the regressands. As an example, consider a SISO system for which a set of N measurements is available: $S = \{(y(k), u(k)) | k = 1, 2, \ldots, N\}$. Postulating, for instance, a second order NARX structure, $y(k + 1) = F(y(k), y(k-1), u(k), u(k-1))$, the data set for clustering is constructed as:

$$Z = \begin{bmatrix} y(2) & y(3) & \ldots & y(N-1) \\ y(1) & y(2) & \ldots & y(N-2) \\ u(2) & u(3) & \ldots & u(N-1) \\ u(1) & u(2) & \ldots & u(N-2) \\ y(3) & y(4) & \ldots & y(N) \end{bmatrix}. \tag{4.1}$$

The first four rows contain the regressors and the last row the regressand. The vector in the k-th column of the matrix Z will be denoted by z_k. This vector contains the complete information about the system at time instant k, i.e., the system's state, the input and also the output at the next time instant $k + 1$.

The subspace clustering algorithms can be classified into the following three groups:

1. Algorithms using an adaptive distance measure, such as the Gustafson–Kessel algorithm (Gustafson and Kessel, 1979) or the fuzzy maximum likelihood estimation algorithm (Gath and Geva, 1989).
2. Algorithms based on hyperplanar prototypes, so-called fuzzy linear varieties and fuzzy c-elliptotypes (Bezdek, 1981; Bezdek et al., 1981a; Bezdek et al., 1981b).
3. Fuzzy c-regression algorithms, using prototypes defined by regression functions (Hathaway and Bezdek, 1993).

All three classes of algorithms can be seen as extensions of the c-means functional (Bezdek, 1981), minimizing the variance of the data from the cluster prototypes, and as such are in fact nonlinear optimization algorithms. The main difference among these algorithms is in the definition of the distance measure in combination with the prototypical structure of the cluster. For instance, the fuzzy c-lines algorithm measures the orthogonal distance of the data points from prototypes that are defined as hyperplanes, while the Gustafson–Kessel and maximum likelihood algorithms are based on proto-typical points, and the distance measure is adaptive to match the cluster shape. Because of these differences, each algorithm performs in a different way for the same problem (data set) and consequently, different algorithms may be suitable for different problems. Experience shows that the algorithms with adaptive distance measure, namely the Gustafson–Kessel (GK) algorithm, perform well with most practical problems. This algorithm is briefly reviewed in the next section.

4.1 The Gustafson–Kessel Algorithm

The vectors $\mathbf{z}_k$, $k = 1, 2, \ldots, N$, contained in the columns of the data matrix Z, will be partitioned into c clusters, represented by their prototypical vectors $\mathbf{v}_i = [v_{i,1}, \ldots, v_{i,n}]^T \in \mathbb{R}^n$, $i = 1, \ldots, c$. Denote $V \in \mathbb{R}^{n \times c}$ the matrix having $\mathbf{v}_i$ in its i-th column. This matrix is called the *prototype matrix*. The fuzzy partitioning of the data among the c clusters is represented as the *fuzzy partition matrix* $U \in \mathbb{R}^{c \times N}$ whose elements denoted $\mu_{i,k} \in [0,1]$ are the membership degree of the data vector $\mathbf{z}_k$ in the i-th cluster. A class of clustering algorithms search for the partition matrix and the cluster prototypes such that the following objective function is minimized

$$J(Z; V, U) = \sum_{i=1}^{c} \sum_{k=1}^{N} (\mu_{i,k})^m d^2(\mathbf{z}_k, \mathbf{v}_i), \tag{4.2}$$

subject to the following constraints

$$\sum_{i=1}^{c} \mu_{i,k} = 1, \quad k = 1, \ldots, N, \tag{4.3}$$

$$0 < \sum_{k=1}^{N} \mu_{i,k} < N, \quad i = 1, \ldots, c. \tag{4.4}$$

In (4.2), $m > 1$ is a parameter that controls the fuzziness of the clusters. With higher values of m the clusters overlap more, and as m approaches one from above, the partition becomes crisp ($\mu_{i,k} \in \{0, 1\}$). The usual setting with $m = 2$ is suitable for most applications. The function $d(\mathbf{z}_k, \mathbf{v}_i)$ is the distance of the data vector $\mathbf{z}_k$ from the cluster prototype $\mathbf{v}_i$. The constraint (4.3) avoids the trivial solution $U = 0$ and the constraint (4.4) guarantees that clusters are neither empty nor contain all the points to degree 1. The optimization problem defined by the functional (4.2) subject to the constraints (4.3) and (4.4) can be solved by different nonlinear optimization techniques. The most popular one is the so-called alternating optimization (Bezdek et al., 1987), which leads to an iterative scheme, known as the fuzzy c-means algorithm (Bezdek, 1981).

The shape of the clusters is determined by the particular distance measure $d(\mathbf{z}_k, \mathbf{v}_i)$ involved. Gustafson and Kessel (1979) extended the c-means algorithm for an inner-product metric norm

$$d^2(\mathbf{z}_k, \mathbf{v}_i) = (\mathbf{z}_k - \mathbf{v}_i)^T \mathbf{M}_i (\mathbf{z}_k - \mathbf{v}_i) \tag{4.5}$$

where $\mathbf{M}_i$ is a positive definite matrix adapted according to the actual shapes of the individual clusters, described approximately by the cluster covariance matrices $\mathbf{F}_i$

$$\mathbf{F}_i = \frac{\sum_{k=1}^{N}(\mu_{i,k})^m (\mathbf{z}_k - \mathbf{v}_i)(\mathbf{z}_k - \mathbf{v}_i)^T}{\sum_{k=1}^{N}(\mu_{i,k})^m}. \tag{4.6}$$

It can be shown that the distance inducing matrix $\mathbf{M}_i$ is calculated as the normalized inverse of the cluster covariance matrix

$$\mathbf{M}_i = \det(\mathbf{F}_i)^{\frac{1}{n}} \mathbf{F}_i^{-1}. \tag{4.7}$$

The normalization by the determinant of $\mathbf{F}_i$ is involved in order to constrain $\mathbf{M}_i$. Without this constraint, the objective function (4.2), which is linear with respect to $\mathbf{M}_i$, could be made as small as desired by making $\mathbf{M}_i$ less positive definite. In the iterative optimization scheme of the GK algorithm below, the superscript (l) denotes the value of a given variable at the l-th iteration.

Gustafson–Kessel fuzzy clustering algorithm

Given the data matrix Z, choose the number of clusters $1 < c < N$, the weighting exponent $m > 1$ and the termination tolerance $\epsilon > 0$. Initialize the fuzzy partition matrix $U^{(0)}$ randomly, such that it satisfies the conditions (4.3) and (4.4).

Repeat for $l = 1, 2, \ldots$

Step 1: Compute the cluster prototypes (means):

$$\mathbf{v}_i^{(l)} = \frac{\sum_{k=1}^{N}(\mu_{i,k}^{(l-1)})^m \mathbf{z}_k}{\sum_{k=1}^{N}(\mu_{i,k}^{(l-1)})^m}, \quad 1 \leq i \leq c.$$

Step 2: Compute the cluster covariance matrices:

$$\mathbf{F}_i = \frac{\sum_{k=1}^{N}(\mu_{i,k}^{(l-1)})^m (\mathbf{z}_k - \mathbf{v}_i^{(l)})(\mathbf{z}_k - \mathbf{v}_i^{(l)})^T}{\sum_{k=1}^{N}(\mu_{i,k}^{(l-1)})^m}, \quad 1 \leq i \leq c.$$

Step 3: Compute the distances:

$$d_{i,k}^2 = (\mathbf{z}_k - \mathbf{v}_i^{(l)})^T [\det(\mathbf{F}_i)^{\frac{1}{n}} \mathbf{F}_i^{-1}](\mathbf{z}_k - \mathbf{v}_i^{(l)}), \quad 1 \leq i \leq c, \ 1 \leq k \leq N.$$

Step 4: Update the fuzzy partition matrix:

$$\mu_{i,k}^{(l)} = \frac{1}{\sum_{j=1}^{c}(d_{i,k}/d_{j,k})^{2/(m-1)}}, \quad 1 \leq i \leq c, \ 1 \leq k \leq N.$$

if $d_{i,k} = 0$ for some $i = s$, set $\mu_{k,s} = 1$ and $\mu_{i,k} = 0$, $\forall i \neq s$.

until $\|U^{(l)} - U^{(l-1)}\| < \epsilon$.

This algorithm simply loops through the estimates of the cluster centers V, the covariance matrices F_i and the fuzzy partition matrix U, and terminates when $\|U^{(l)} - U^{(l-1)}\| < \epsilon$. Equivalently, the algorithm can be initialized with $V^{(0)}$ and terminate on $\|V^{(l)} - V^{(l-1)}\| < \epsilon$. Different results may be obtained for the same values of ϵ, since the termination criterion used in the above algorithm requires more parameters (membership degrees in the partition matrix) to become close to one another. The error norm in the termination criterion is usually chosen as $\max(|U^{(l)} - U^{(l-1)}|)$, and the termination criterion is set to $\epsilon = 0.001$, even though $\epsilon = 0.01$ works well in most cases. The convergence properties of the alternating optimization schemes are analyzed for instance in (Bezdek et al., 1987; Hathaway and Bezdek, 1991).

The GK algorithm computes the fuzzy partition matrix U, the cluster prototype matrix V and the covariance matrices F_i, $i = 1, \ldots, c$. In Sect. 5 we explain how to derive fuzzy models from these matrices. Before that, the important issue of determining the number of clusters is discussed.

4.2 Determining the Number of Clusters

The number of clusters must be specified before clustering. The more clusters, the finer approximation of the nonlinear system can be eventually obtained, but also more parameters have to be estimated and their variance is therefore higher due to overfitting. If no prior knowledge is available, which would suggest the number of clusters, automated procedures can be applied. Two approaches are presented in this section: validity measures and compatible cluster merging.

4.2.1 Validity Measures. Validity measures are criteria that assess the qualities of the clusters, such as the fuzzy hypervolume, the within-cluster distance, the fuzzy partition density, etc. Cluster validity measures have been extensively studied in the context of pattern recognition (Bezdek, 1981; Gath and Geva, 1989; Pal and Bezdek, 1995). The number of clusters is found by evaluating a given validity measure for a range of c, and selecting the number of clusters that minimizes (maximizes) the validity measure. Instead of a (local) maximum or minimum a "knee" point may be sought on a graph of the validity measure plotted against c, if the validity measure is monotonic. Gath and Geva (1989) proposed the fuzzy hypervolume and average partition density measures for the adaptive distance clustering algorithms.

The *fuzzy hypervolume* measure V_h is defined as

$$V_h = \sum_{i=1}^{c} [\det(F_i)]^{1/2} .$$

(4.8)

The *average partition density* D_A is given by

$$D_A = \frac{1}{c} \sum_{i=1}^{c} \frac{S_i}{[\det(F_i)]^{1/2}}$$

(4.9)

where S_i is computed using only those vectors z_k that lie within a hyperellipsoid whose radii are the standard deviations of the cluster features

$$S_i = \sum_k \mu_{i,k}, \ \forall k, \text{ such that } (z_k - v_i)^T F_i^{-1}(z_k - v_i) < 1. \tag{4.10}$$

Krishnapuram and Freg (1992) applied also the average within-cluster distance D_W

$$D_W = \frac{1}{c} \sum_{i=1}^{c} \frac{\sum_{k=1}^{N}(\mu_{i,k})^m d_{i,k}^2}{\sum_{k=1}^{N}(\mu_{i,k})^m}. \tag{4.11}$$

A good partition is indicated by small values of V_h and D_A, and by large values of D_W. A dedicated validity measure was proposed for the clustering identification method (Babuška and Verbruggen, 1995a) based on a flatness index of the clusters. Denote the eigenvalues of the cluster covariance matrix F_i, sorted in a descending order, $\lambda_{i,1} \geq \lambda_{i,2} \geq \cdots \geq \lambda_{i,n}$. When approximating the regression surface, the obtained hyperellipsoidal clusters are flat, i.e., one of the axes is much shorter than the others. Consequently, the smallest eigenvalue λ_{in} is significantly smaller than the remaining ones. In order to efficiently approximate the regression surface by hyperplanes, the clusters should be as flat (planar) as possible. The flatness index t_i of a cluster is defined as a ratio between the smallest and the largest eigenvalue, that is $t_i = \lambda_{in}/\lambda_{i1}$. An aggregate measure called the *average cluster flatness* is defined as

$$t_A = \frac{1}{c} \sum_{i=1}^{c} \frac{\lambda_{i,n}}{\lambda_{i,1}}. \tag{4.12}$$

Aiming simultaneously at a good accuracy of the obtained fuzzy model, one can combine the flatness index with some measure of the prediction error. Such a measure will prefer a few larger, flat clusters to a greater number of small ones, if both settings lead to the same prediction error. This approach conceptually resembles the use of information criteria in linear system identification (Akaike, 1974) and can be used also for selecting the structure of the model.

4.2.2 Compatible Cluster Merging. Cluster merging approaches start with a higher number of clusters than are expected for the particular problem. This initial number of clusters should be chosen sufficiently high, such that the nonlinearity of the regression hypersurface can be captured accurately enough and small clusters in the data set are not missed. However, the maximum number of clusters that may be chosen is limited by the dimension of the regression vector and by the number of measurements available. Few data points cannot be partitioned into many clusters because in such a case the clusters are not sufficiently determined and the covariance matrices cannot be computed.

The initial number of clusters is reduced by successively merging compatible clusters until some threshold is reached and no more clusters can be merged (Krishnapuram and Freg, 1992; Kaymak and Babuška, 1995). Two clusters i and j are considered compatible when they are approximately parallel, i.e., the dot-product of their normal vectors $|\boldsymbol{\Phi}_{i,n} \cdot \boldsymbol{\Phi}_{j,n}|$ is close to one, and when they are sufficiently close, i.e., the distance of their prototypical points $\|\mathbf{v}_i - \mathbf{v}_j\|$ is small. Figure 4.2 gives a geometrical illustrations of these conditions.

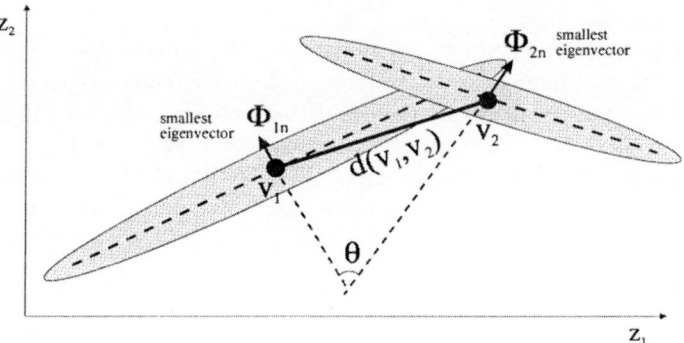

Fig. 4.2. Geometric representation of the cluster compatibility criteria

By evaluating these two criteria for all pairs of clusters, one obtains two matrices S_1 and S_2, whose elements indicate the degree of compatibility between the clusters in each pair. Using a fuzzy aggregation operator (Kaymak and Babuška, 1995) the two matrices are combined, allowing for a partial compensation between the two criteria. Finally, a decision is made on which clusters should be merged, using a threshold specified by the user. This threshold is related to the complexity/accuracy trade-off of the resulting model, and in practice, some experimentation is needed for finding the right value, especially for noisy data.

By merging the clusters, a new partition matrix is created with fewer rows. The membership degrees of the combined clusters are computed by adding up the membership degrees of the compatible clusters, i.e., the new partition matrix still fulfills the conditions (4.3) and (4.4). The clustering algorithm is initialized with this new partition matrix and run again. Because of this initialization, the convergence is now much faster, compared to a random initialization.

The benefits from using the cluster merging technique are twofold.

1. Since several clusters can be merged in one step, the number of clusters (and hence also the number of rules in the obtained model) can be optimized without testing all the cluster numbers from two to the specified

upper limit, as with the cluster validity measures. This may significantly reduce the computational effort.

2. By decreasing the number of clusters gradually, a better solution can be found since small regions with few data points can be captured using initially a higher number of clusters, see (Kaymak and Babuška, 1995) for details.

5. Deriving Takagi–Sugeno Fuzzy Models

Each cluster obtained by product space clustering of the identification data can be regarded as a local linear approximation of the regression hypersurface. The entire global model can be conveniently represented in the affine Takagi–Sugeno (TS) form (2.1). Each cluster is converted into one TS rule by estimating the consequent parameters and by deriving the antecedent membership functions from the fuzzy partition matrix. These two steps are described below.

5.1 Deriving Consequent Parameters

There are several methods to obtain the consequent parameters. Based on the geometrical interpretation of the TS model, the consequent parameters can be directly computed from the cluster prototypical points and the smallest eigenvectors of the cluster covariance matrices. As this method makes no distinction between the model inputs and outputs, it corresponds to the total least squares solution of the local linearization around the cluster center, as shown later.

A set of optimal parameters with respect to the model output can also be estimated from the identification data set by ordinary least squares. This approach can be formulated as the minimization of the total prediction error using the TS defuzzification formula (2.2), or as minimization of the prediction errors of the individual local models, solved as a set of independent weighted least squares problems. If the model should serve as a numerical predictor, the global least squares approach is usually preferred, as it gives the least prediction error. On the other hand, more accurate local models can be obtained by weighted ordinary or total least squares.

5.1.1 Deriving Rule Consequents from the Cluster Prototypes. The consequent parameters $\mathbf{a}_i$ and b_i of the affine TS model (2.1) can be derived from the geometrical structure of the clusters. The distance measure (4.5) defines each cluster as a hyperellipsoid whose shape is roughly described by the eigenstructure of the cluster covariance matrix F_i. The clusters, approximating the regression surface, lie in fact in p-dimensional linear subspaces of the regression space. In other words, in the obtained hyperellipsoids, the shortest axis is much shorter than the remaining ones, i.e., the smallest eigenvalue $\lambda_{i,n}$ is typically in the orders of magnitude smaller than the largest eigenvalue. The eigenvector $\Phi_{i,n}$ corresponding to the smallest eigenvalue of F_i determines the normal vector to a hyperplane spanned by the remaining eigenvectors of that cluster, as illustrated in Fig. 5.1 for the 3D case.

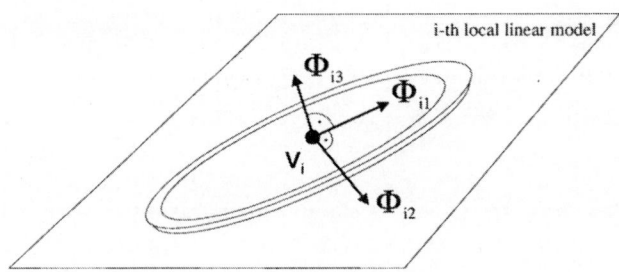

Fig. 5.1. The position and orientation of a local linear model obtained by clustering is determined by the cluster center $\mathbf{v}_i$ and by the eigenvectors $\Phi_{i,j}$.

For brevity, the smallest eigenvector of the i-th cluster will be denoted just Φ_i, omitting the subscript n. Recalling that $\mathbf{z} = [\mathbf{x}^T; y]^T$ is the data vector and $\mathbf{v}_i$ is the cluster's prototypical point, one can directly write the implicit normal form of the consequent's hyperplane:

$$\langle \Phi_i, (\mathbf{z} - \mathbf{v}_i) \rangle = 0. \tag{5.1}$$

The above expression states that the inner product of the normal vector Φ_i with any vector belonging to the hyperplane equals zero. It is convenient to divide the prototype $\mathbf{v}_i$ into a vector $\mathbf{v}_i^x$ corresponding to the regressor $\mathbf{x}$ and a scalar v_i^y corresponding to the regressand y, $\mathbf{v}_i = [(\mathbf{v}_i^x)^T; v_i^y]^T$. The smallest eigenvector Φ_i is partitioned in the same way, $\Phi_i = \left[(\Phi_i^x)^T; \phi_i^y \right]^T$. Expression (5.1) now can be written as

$$\left\langle [(\Phi_i^x)^T; \phi_i^y], ([\mathbf{x}^T; y]^T - [(\mathbf{v}_i^x)^T; v_i^y]^T) \right\rangle = 0 . \tag{5.2}$$

Carrying out the inner product leads to the following equality:

$$(\Phi_i^x)^T (\mathbf{x} - \mathbf{v}_i^x) + \phi_i^y (y - v_i^y) = 0 . \tag{5.3}$$

After a simple algebraic manipulation, an explicit equation for the hyperplane is obtained:

$$y = \underbrace{-\frac{1}{\phi_i^y} (\Phi_i^x)^T}_{\mathbf{a}_i^T} \mathbf{x} + \underbrace{\frac{1}{\phi_i^y} \Phi_i^T \mathbf{v}_i}_{b_i} . \tag{5.4}$$

By comparing the above expression with the affine TS rule (2.1), the equations for $\mathbf{a}_i$ and b_i directly follow:

$$\mathbf{a}_i = -\frac{1}{\phi_i^y} \Phi_i^x = -\frac{1}{\phi_{i,p+1}} [\phi_{i,1}, \phi_{i,2}, \ldots, \phi_{i,p}]^T , \tag{5.5}$$

$$b_i = \frac{1}{\phi_i^y} \Phi_i^T \mathbf{v}_i . \tag{5.6}$$

It can be shown that $\mathbf{a}_i$ given by (5.5) is a solution of a weighted total least squares (TLS) problem for the data detrended by subtracting the cluster means and weighted by the membership degrees. First, for each cluster, the data is transformed such that the cluster's center becomes the origin

$$\Delta \mathbf{x}_k^i = \mathbf{x}_k - v_i^x, \quad \Delta y_k^i = y_k - v_i^y. \tag{5.7}$$

By this transformation, the local model describing the i-th cluster becomes a truly linear (homogeneous) form, instead of the affine form. To express the varying relevance of the different data samples $(\Delta \mathbf{x}_k^i, \Delta y_k^i)$ to the i-th local linear model, the data is weighted by $w_{i,k} = \sqrt{(\mu_{i,k})^m}$:

$$\tilde{\mathbf{x}}_k^i = w_{i,k} \Delta \mathbf{x}_k^i, \quad \tilde{y}_k^i = w_{i,k} \Delta y_k^i. \tag{5.8}$$

It is now straightforward to show that (5.5) is the unique solution of $\tilde{\mathbf{y}} \approx \tilde{\mathbf{X}} \mathbf{a}_i$ in the total least squares sense, where the matrix $\tilde{\mathbf{X}} \in \mathbb{R}^{N \times p}$ has the vectors $\tilde{\mathbf{x}}_k^i$ are in its rows, and the column vector $\tilde{\mathbf{y}} \in \mathbb{R}^N$ contains the scalars $\tilde{y}_k^i$. The proof is based on Theorem 2.6 in (Huffel and Vandewalle, 1991), stating that the TLS solution is given by

$$\mathbf{a}_i = -\frac{1}{q_{p+1,p+1}} [q_{1,p+1}, q_{2,p+1}, \ldots, q_{p,p+1}]^T \tag{5.9}$$

where q_{p+1} is the last column of the matrix $\mathbf{Q}_2$ obtained by the singular value decomposition (SVD)

$$[\tilde{\mathbf{X}}; \tilde{\mathbf{y}}] = \mathbf{Q}_1 \Sigma \mathbf{Q}_2^T, \tag{5.10}$$

assuming that the singular values are ordered such that $\sigma_1 \geq \sigma_2 \geq \ldots \geq \sigma_{p+1} > 0$ and $\Sigma = \text{diag}(\sigma_1, \ldots, \sigma_{p+1})$. For any matrix A decomposed by SVD into $\mathbf{A} = \mathbf{Q}_1 \Sigma \mathbf{Q}_2^T$ the columns of $\mathbf{Q}_2$ are the unit eigenvectors of $\mathbf{A}^T \mathbf{A}$, see, e.g., (Strang, 1976), p. 142. In our case, $\mathbf{A} = [\tilde{\mathbf{X}}; \tilde{\mathbf{y}}]$ and

$$[\tilde{\mathbf{X}}; \tilde{\mathbf{y}}]^T [\tilde{\mathbf{X}}; \tilde{\mathbf{y}}] = \sum_{k=1}^{N} (\mu_{i,k})^m (\mathbf{z}_k - \mathbf{v}_i)(\mathbf{z}_k - \mathbf{v}_i)^T = \mathbf{F}_i \sum_{k=1}^{N} (\mu_{i,k})^m \tag{5.11}$$

where $\mathbf{F}_i$ is the cluster covariance matrix. The vector $\mathbf{q}_{p+1}$ is thus equal to the smallest eigenvector Φ_i of $\mathbf{F}_i$.

The offset parameters b_i can be derived as follows. In the incremental coordinates (5.7), the i-th local linear model is given by

$$\Delta y^i \quad = \quad \mathbf{a}_i^T \Delta \mathbf{x}, \tag{5.12}$$

$$y - v_i^y \quad = \quad \mathbf{a}_i^T (\mathbf{x} - \mathbf{v}_i^x), \tag{5.13}$$

$$y \quad = \quad \mathbf{a}_i^T \mathbf{x} + v_i^y - \mathbf{a}_i^T \mathbf{v}_i^x, \tag{5.14}$$

from which $b^i = v_i^y - \mathbf{a}_i^T \mathbf{v}_i^x$. By substituting for $\mathbf{a}_i$ from (5.5) the following expression is obtained:

$$
\begin{aligned}
b^i &= v_i^y + \frac{1}{\phi_i^y}(\Phi_i^x)^T \mathbf{v}_i^x \\
&= \frac{1}{\phi_i^y}(\phi_i^y v_i^y + (\Phi_i^x)^T \mathbf{v}_i^x) \\
&= \frac{1}{\phi_i^y}\Phi_i^T \mathbf{v}_i
\end{aligned}
\tag{5.15}
$$

which is equivalent to (5.6).

As shown above, the adaptive distance (GK) clustering algorithm can be regarded as a method that decomposes a global nonlinear regression problem to a set of local linear TLS problems. Instead of applying the weighted TLS, only those data samples that belong to the i-th cluster to a degree greater than a specified threshold (α-level) can be used to obtain the estimate for the TS consequent parameters. This is achieved by computing the cluster means and covariance matrices from the α-cut[3] of the fuzzy partition matrix U instead of the partition itself. The estimate of the parameters $\mathbf{a}_i$ and b_i are again given by (5.5).

5.1.2 Estimation of Consequent Parameters by Ordinary Least Squares – Local Approach.
This section describes the estimation of the consequent parameters $\mathbf{a}_i$, b_i by formulating a set of independent ordinary least squares (OLS) problems. Let X denote the matrix in $\mathbb{R}^{N \times p}$ having $\mathbf{x}_k$ in its k-th row, $\mathbf{y}$ denote the vector in $\mathbb{R}^N$ having y_k as its k-th component, and W_i denote a diagonal matrix in $\mathbb{R}^{N \times N}$ having $\mu_{i,k}$ as its k-th diagonal element. The consequent parameters of the rule belonging to the i-th cluster, $\mathbf{a}_i$ and b_i are concatenated into a single parameter vector $\theta_i = \left[\mathbf{a}_i^T; b_i\right]^T$. In order to estimate the offset b_i, a unitary column is appended to X, yielding the extended regressor matrix $X_e = [X; \mathbf{1}]$.

Assuming that each cluster represents a local linear model of the system, the consequent parameter vectors θ_i can be estimated independently by weighted least squares, where the membership degrees of the fuzzy partition express the relevance of the data pair $(\mathbf{x}_k, y_k)$ to the particular local model. If the columns of X_e are linearly independent and $\mu_{i,k} > 0$ for $1 \leq k \leq N$, then

$$
\theta_i = \left[X_e^T W_i X_e\right]^{-1} X_e^T W_i \mathbf{y}
\tag{5.16}
$$

is the solution of the least squares problem $X_e \theta \approx \mathbf{y}$ where the k-th data pair $(\mathbf{x}_k, y_k)$ is weighted by $\mu_{i,k}$. The parameters $\mathbf{a}_i$ and b_i are then given by

$$
\mathbf{a}_i = [\theta_1, \theta_2, \ldots, \theta_p], \quad b_i = [\theta_{p+1}].
\tag{5.17}
$$

If the columns of X_e are linearly dependent, techniques based on orthogonal factorization of X_e should be used. Also here, instead of applying the weighted least squares, only those data samples that belong to the i-th cluster to a

---

[3] The α-cut of a fuzzy set A is a crisp subset of the domain $\mathcal{X}$, whose elements all have membership degrees greater than or equal to α.

degree greater than a specified threshold α can be used to obtain the estimate for the consequent parameters of the i-th rule. If a sufficient number of data samples is available in each cluster, the α-cut approach gives less biased estimates, as only the data belonging to a high degree to the cluster can be considered. On the other hand, the variance of the estimate may increase, as fewer data samples are available.

5.1.3 Estimation of Consequent Parameters by Ordinary Least Squares – Global Approach.

The weighted least squares approach gives an optimal estimate of the parameters of the local models, but it does not provide an optimal TS model in terms of minimal prediction error. In order to obtain an optimal global predictor, the aggregation of the rules should be taken into account. When using the fuzzy mean defuzzification (2.2), which is a convex linear combination, a global least squares problem can be solved to obtain the consequent parameter estimates. The membership degrees $\beta_{i,k} = \mu_{A_i}(\mathbf{x}_k)$, representing the degrees of fulfillment of the i-th rule for each data point, can be obtained from the fuzzy partition matrix U. Recall that each row of U contains a pointwise definition of the membership function for the data in the product space $\mathcal{X} \times \mathcal{Y}$. In order to obtain the membership function A_i in the regressor space $\mathcal{X}$, the i-th row of U, denoted $U_{(i)}$ must be projected onto the regressor space

$$\beta_{i,k} = \text{proj}_{\mathbb{N}_p}^{\mathbb{N}_p^{+1}}(U_{(i)}) \tag{5.18}$$

where $\text{proj}(\cdot)$ is the point-wise projection operator (Kruse et al., 1994). The result of the projection step is that a set of data vectors with repeated regressors $\mathbf{x}_k$ are assigned the maximum membership degree from this set. In order to write (2.2) in a matrix form for all the data $(\mathbf{x}_k, y_k)$, $1 \leq k \leq N$, denote B_i a diagonal matrix in $\mathbb{R}^{N \times N}$ having the normalized membership degree γ_i as its k-th diagonal element. Finally, denote X' the matrix in $\mathbb{R}^{N \times cN}$ composed from matrices products of B_i and X_e as

$$X' = [(B_1 X_e); (B_2 X_e); \ldots; (B_c X_e)]^T . \tag{5.19}$$

Denote θ' the vector in $\mathbb{R}^{c(p+1)}$ given by

$$\theta' = \left[\theta_1^T; \theta_2^T; \ldots; \theta_c^T \right] \tag{5.20}$$

where $\theta_i = [a_i^T b_i]^T$ for $1 \leq i \leq c$. The resulting global least squares problem $X'[\theta'] \approx \mathbf{y}$ has the solution

$$\theta' = \left[(X')^T X' \right]^{-1} (X')^T \mathbf{y} . \tag{5.21}$$

From (5.20) the parameters $\mathbf{a}_i$ and b_i are obtained by

$$\mathbf{a}_i = [\theta'_{q+1}, \theta'_{q+2}, \ldots, \theta'_{q+p}]^T, \ b_i = [\theta_{q+p+1}], \ q = (i-1)(p-1). \tag{5.22}$$

5.2 Deriving Antecedent Membership Functions

The fuzzy partition matrix U projected onto the antecedent space defines the membership functions pointwise, for the available data. In order to obtain a prediction model or a model suitable for controller design, the antecedent membership functions need to be expressed in a form that allows one to compute the membership degrees for any input data. This can be achieved by using an inverse of the distance function of the clustering algorithm in the antecedent product space or by pointwise projections on the antecedent variables. The following sections describe both approaches in a greater detail.

5.2.1 Multidimensional Membership Functions.
With this method, the degrees of fulfillment of the rules are computed by evaluating the distance function $d(\cdot)$, see (4.5), only for the regressor $\mathbf{x}$ and the regressor part of the cluster prototype $\mathbf{v}_i^x$, using the corresponding partition of the cluster covariance matrix

$$\mathbf{F}^x = [f_{ij}], \quad 1 \le i, j \le p. \tag{5.23}$$

The inner product norm then measures the distance of the antecedent vector from the projection of the cluster center to the antecedent space. Then the inner product norm can be evaluated as

$$d(\mathbf{x}_k, \mathbf{v}_i^x) = (\mathbf{x} - \mathbf{v}_i^x)^T \mathbf{F}_i^x (\mathbf{x} - \mathbf{v}_i^x), \tag{5.24}$$

and transformed into the membership degree (degree of fulfillment), using some kind of inversion. One possible choice is to use the same formula as in the clustering algorithm

$$\beta_i(\mathbf{x}_k) = \frac{1}{\sum_{j=1}^{c} [d(\mathbf{x}_k, \mathbf{v}_i^x)/d(\mathbf{x}_k, \mathbf{v}_j^x)]^{2/(m-1)}}, \tag{5.25}$$

which takes into account all the rules and computes the degree of fulfillment of one rule relative to the other rules. The sum of the membership degrees also equals one as with clustering, hence $\gamma_i = \beta_i$. Another option is to apply the 'possibilistic' transformation

$$\beta_i(\mathbf{x}_k) = \frac{1}{1 + d(\mathbf{x}_k, \mathbf{v}_i^x)}, \tag{5.26}$$

which computes the membership degree to the i-th antecedent fuzzy set independently of the remaining rules, as it is usual in fuzzy reasoning systems.

The location of the membership functions obtained by both methods is the same, since both measure the distance relative to the cluster center. The shape of the membership functions, on the other hand, differs significantly. Expression (5.25) generates fuzzy sets similar to those obtained by clustering. The sets can be non-symmetrical, which is an advantage, but at the same can be non-convex, which is a drawback. Another disadvantage of this approach is that (5.25) may assign relatively high membership degrees to data points that are 'outside' the clustered data set. The reason is the normalization constraint, which forces the sum of the membership degrees to be equal to one. Consequently, the method will give reliable results only when the identification data set covers most of the antecedent domain. Membership functions generated by (5.26) are always convex, as the distance function has its global minimum in $\mathbf{v}_i^x$ and increases monotonically in all directions from $\mathbf{v}_i^x$. For the same reason, the sets are also symmetrical around the cluster centers. Data points outside the identification data set are assigned low membership degrees in all clusters, since they are far from all cluster centers and the constraint (5.25) is not present.

An advantage of multivariable membership functions is the simplicity and speed of computations involved in evaluating the degrees of fulfillment of the rules. A drawback is a limited interpretability of the rule locations and less numerical accuracy of the model, as the expressions (5.25) and (5.26) may not give an accurate approximation of the actual cluster shape.

5.2.2 Generating Membership Functions by Projection. The principle of this method is to project the multidimensional fuzzy sets defined pointwise in the rows of the partition matrix U onto the individual antecedent variables of the rules. These variables can be the original regression variables in which case the projection is just an orthogonal projection of the data. Transformed antecedent variables also can be obtained by means of eigenvector projection, using the first p eigenvectors of the cluster covariance matrices. This eigenvector projection is useful for clusters which are opaque to the axis of the regression space, and cannot be accurately represented by axis-orthogonal projection. In both cases, the projected (pointwise defined) membership functions are approximated by some suitable parametric functions, as illustrated in Fig. 5.2.

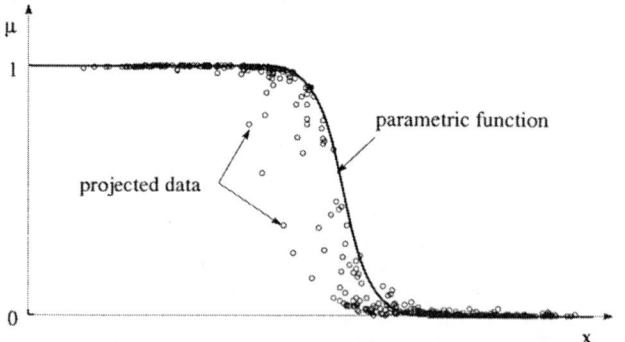

Fig. 5.2. Approximation of the projected data by a parametric membership function

Axis Orthogonal Projection. To obtain the TS rules in the conjunctive form (2.3), the multivariate fuzzy sets A_i are projected onto the regressor components $r_1, r_2, \ldots, r_p$, to obtain the univariate fuzzy sets $A_{i,j}$:

$$\mu_{A_{i,j}} = \text{proj}_j^{;\,\mathbb{N}_{p+1}} U_{(i)} \tag{5.27}$$

where $U_{(i)}$ denotes the i-th row of the fuzzy partition matrix. In computing the degree of fulfillment of the i-th rule, the original cluster in the antecedent product space is 'reconstructed' by the intersection operator (2.4). Notice that this reconstruction is, in general, not exact. The clusters may be rotated in the space such that they are not parallel to the axes or due to the distribution of the data can take various shapes which cannot be generated by intersecting one-dimensional fuzzy sets. Hence, by projection some information is always lost, resulting in a *decomposition error*. On the other hand, the axis orthogonal projection is useful, as it provides the possibility to interpret the model in a qualitative (linguistic) way, using the membership functions defined directly for the regressors.

Eigenvector Projection. The decomposition error can be reduced by means of eigenvector projection. The eigenvectors carry information about the orientation of the clusters in the regression space and form an orthonormal basis. The clustered data can be projected onto this basis by

$$X'_i = XH_i, \tag{5.28}$$

where $H_i \in \mathbb{R}^{p \times p}$ is the projection matrix which has in its columns the (unitary) eigenvectors of F_i^x. In general, each cluster will have its own projection matrix. Consequently, each rule will have different antecedent variables, computed as linear combinations of the original antecedent variables (regressors). This transformation is closely related to principal component analysis and can be regarded as a kind of data preprocessing for each rule, as illustrated in Fig. 5.3.

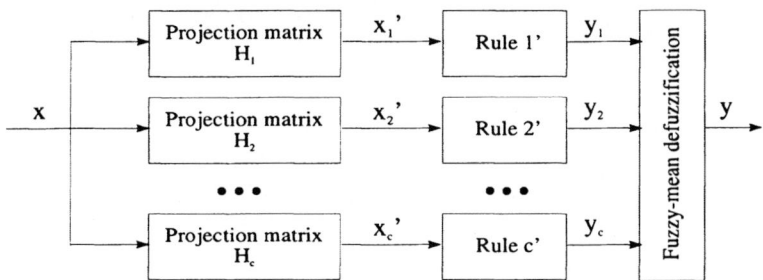

Fig. 5.3. Transformation of antecedent variables when using a different projection matrix for each rule

A significant drawback of eigenvector projection is that in general it hampers the linguistic interpretation of the rule antecedents, as the transformed variables may not be relevant to the system dynamics. On the other hand, it can lead to a reduction of the antecedent dimension and gaining information about the nature of the system's nonlinearity in the cases, when all, or some, of the eigenvectors are equal to one another. In such a case, a global projection matrix H is obtained which suggest a new model structure with transformed antecedent variables. These aspects are still a topic of current research.

5.3 Rule Base Simplification

The automated approaches for acquisition of fuzzy models from data, such as product space fuzzy clustering or parameter adaptation (neuro-fuzzy) techniques, lead to a certain degree of redundancy in the obtained model. With fuzzy clustering techniques, redundancy occurs when the clusters are projected onto the individual antecedent variables, as illustrated in Fig. 5.4. The projection of clusters C_1, C_2 and C_3 onto antecedent variables x_1 and x_2 results in similarity between A_1 and A_2; furthermore B_3 is similar to the universal set ($\mu = 1$) as it covers the entire domain of x_2.

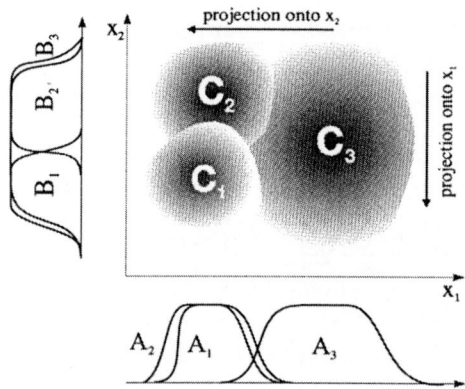

Fig. 5.4. Redundancy in terms of similar membership functions as a cluster projection

Simplification of the rule base is achieved by merging similar fuzzy sets, and by removing fuzzy sets similar to the universal set (these fuzzy sets do not contribute to the rule base). A large number of similarity measures are known from the literature. Based on an extensive research study (Setnes, 1995), the following similarity measure was selected for rule base simplification

$$S(A, B) = \frac{\mid \min(A, B) \mid}{\mid \max(A, B) \mid} = \frac{\sum_{j=1}^{m}(\mu_A(x_j) \wedge \mu_B(x_j))}{\sum_{j=1}^{m}(\mu_A(x_j) \vee \mu_B(x_j))} \qquad (5.29)$$

where $\mid . \mid$ denotes cardinality of a set. Membership functions $\mu_A(x_j)$ and $\mu_B(x_j)$ are defined on a discretized domain $\mathcal{D} = \{x_q \mid q = 1, 2, ..., n_d\}$, where n_d is the number of discrete values.

The simplification algorithm merges similar fuzzy sets iteratively using two thresholds: $\eta \in (0, 1)$ for merging fuzzy sets that are similar to one another, and $\eta_r \in (0, 1)$ for removing fuzzy sets similar to the universal set.

In each iteration, the similarity between all fuzzy sets for each antecedent variable is considered. The pair of membership functions with the highest similarity $s \geq \eta$ are merged. The rule base is updated by substituting the new fuzzy set for the ones merged. The algorithm repeatedly evaluates the similarities in the updated rule base, until there are no more fuzzy sets for which $s \geq \eta$. Finally, fuzzy sets similar to the universal set are removed from the antecedents of the rules in which they occur. The algorithm only merges one pair of fuzzy sets per iteration and merging in different dimensions is independent of each other. Merging two fuzzy sets $A_{l,q}$ and $A_{m,q}$ is accomplished by taking the support of the new fuzzy set A as the support of $A_{l,q} \cup A_{m,q}$. This guarantees preservation of the coverage of the antecedent space. The kernel of A is given by averaging the kernels of $A_{l,q}$ and $A_{m,q}$, making a trade-off between the two rules l and m. The algorithm can be summarized as follows

Rule base Simplification Algorithm

Given a TS fuzzy rule base obtained by clustering, select thresholds $\eta_r, \eta \in (0, 1)$.

Repeat:

Step 1: Select the two most similar fuzzy sets in the rule base. Calculate $s_{i,j,k} = S(A_{i,j}, A_{k,j})$, $j = 1, 2, \ldots, p$, $i, k = 1, 2, \ldots, K$. Select $A_{l,q}$ and $A_{m,q}$, such that $s_{l,m,q} = \max_{i,j,k,\ i \neq k} \{s_{i,j,k}\}$.

Step 2: Merge the two most similar fuzzy sets and update the rule base. If $S(A_{l,q}, A_{m,q}) \geq \eta$ merge $A_{l,q}$ and $A_{m,q}$ to create a new fuzzy set A and replace $A_{l,q} = A$ and $A_{m,q} = A$.

Until: no more fuzzy sets have similarity $s_{i,j,k} \geq \eta$.

Step 3: Remove fuzzy sets similar to the universal set. For each fuzzy set $A_{i,j}$ calculate $S(A_{i,j}, U)$, where $\mu_U = 1$, $\forall x_j$. If $S(A_{i,j}, U) \geq \eta_r$ remove $A_{i,j}$ from the antecedent of the i-th rule.

Merging similar fuzzy sets may result in rules with equal antecedent parts. These rules can be combined into a single rule, allowing for the reduction of the number of rules as well (Setnes, 1995).

Simplification and interpretation of the rule base can also be achieved by means of linguistic approximation where a similarity measure is used to compare the fuzzy sets obtained from data to some reference fuzzy sets and their modifications by linguistic hedges (Sugeno and Yasukawa, 1993).

5.4 Summary of the Identification Procedure

This section summarizes the described identification method and discusses some practical aspects concerning experiment design and model validation.

Step 1: *Design identification experiments and collect a set of representative process measurements.* This is an important initial step for any identification method, since it determines the information content of the identification data set. As opposed to linear techniques, pseudo-random binary excitation signals are not suitable for nonlinear identification in general and for fuzzy clustering in particular. Although the choice of the excitation signal may be problem-dependent, the input data should preferably excite the system in the entire range of the considered variables both in amplitude and frequency. For instance, a multi-sinusoidal signal, which consists of several sine waves of different amplitudes and frequencies, is usually preferred to pseudo-random (binary) signals containing only a few amplitude levels. Data recorded during the routine process operation can be used as well, provided they fulfil the above requirements. If the identification data set does not cover the entire operating range, additional rules can be supplied by the user, see also Step 6. No special preprocessing of the data is needed, the clustering method can deal with noisy data. The variance of the parameter estimates and also of the number of rules, position of membership functions, etc., will naturally increase with increasing levels of noise.

Step 2: *Choose the model structure.* The relevant system variables must be selected for the model inputs and when identifying dynamic systems, the structure and the order of the dynamic model must be chosen. After the appropriate structure is defined, the matrix Z is constructed from the identification data, see Sect. 3.

Step 3: *Cluster the data.* The initial number of clusters c, and the clustering parameters m and ϵ must be defined before clustering. The choice of c has been discussed in Sect. 4.2. For the remaining parameters, $m = 2$ and $\epsilon = 0.01$ are good settings for most problems. Note that for large data sets, the clustering procedure may be computationally demanding.

Step 4: *Determine the number of clusters.* Using cluster validity measures, the compatible cluster merging technique or a combination of both, an appropriate numbers of clusters can be found. This step obviously involves several repetitions of Step 3 with different c and a different initial partition matrix U.

Step 5: *Generate the rules.* By projecting the clusters on the antecedent variables, the membership functions are obtained. The consequent parameters are extracted from the eigenstructure of the cluster covariance matrix or estimated by least squares, as described in Sect. 5.

Step 6: *Validate the model.* The validation of fuzzy models has several facets, namely a standard validation through numerical simulations and comparisons with the process data, analysis of the linear consequent models (stability, step responses, gains, time-constants, non-minimum phase behavior, etc.) and analysis of the coverage of the input space by the rules. For an incomplete rule base, additional rules can be provided based on prior knowledge or local linearization of first-principle models. The antecedents of these

rules can be created from unused combinations of the membership functions
of the initial model.

As mentioned in Sect. 1, also other types of fuzzy models than the TS
model can be extracted from the fuzzy partition. Babuška and Verbruggen
(1995b) present a method for building singleton and linguistic models. An
extension to fuzzy relational models is presented in (Babuška and Verbruggen,
1995a). These approaches are based on the observation that a piecewise linear
TS model can also be approximately represented as a singleton model, see
Sect. 2.2. To obtain linear interpolation between the constant consequents b_i,
the antecedent fuzzy sets must be defined by triangular membership functions
such that $\sum_{i=1}^{K} \mu_i = 1$, i.e., they are in fact piecewise linear basis splines of
order 2 (Brown and Harris, 1994) and the product intersection operator must
be used. The cores (kernels) of the membership functions (cf. knots of the
basis functions) are placed such that they coincide with the intersection points
of the adjacent membership functions of the affine TS model, and additional
sets are added at the extreme points of the domain. For further details see
the above references.

6. Example: pH Neutralization

This simulation example illustrates the individual steps of the identification
technique. An affine TS model is identified for a pH neutralization process
from a set of input–output measurements, using fuzzy clustering. This highly
nonlinear process is well-known for posing difficulties to most black-box identification techniques. It is shown that by means of fuzzy clustering, an accurate prediction model is obtained which is also locally interpretable.

A neutralization tank with three influent streams (acid, buffer and base)
and one effluent stream is considered. The identification and validation data
sets are obtained by simulating the model of Hall and Seborg (1989) for
random changes of the influent base stream flow-rate Q. The influent buffer
stream and the influent acid stream are kept constant. The output is the pH
in the tank. The identification data set, containing $N = 480$ samples with
sampling time 15 s, is shown in Fig. 6.1.

The process is represented as a first-order discrete-time NARX model

$$\mathrm{pH}(k + 1) = f(\mathrm{pH}(k), Q(k)) \tag{6.1}$$

where k denotes the sampling instant and f is an unknown relationship approximated by the TS fuzzy model. Based on the prior knowledge, this structure is considered appropriate for approximation of the pH–neutralization
dynamics. The data matrix Z is constructed from the identification data set
as

$$Z = \begin{bmatrix} \mathrm{pH}(1) & \mathrm{pH}(2) & \ldots & \mathrm{pH}(N-1) \\ Q(1) & Q(2) & \ldots & Q(N-1) \\ \mathrm{pH}(2) & \mathrm{pH}(3) & \ldots & \mathrm{pH}(N) \end{bmatrix}, \tag{6.2}$$

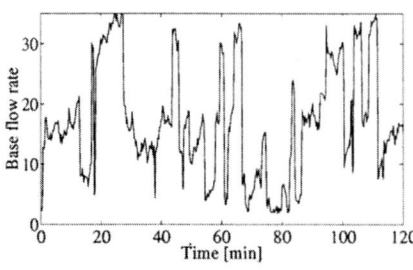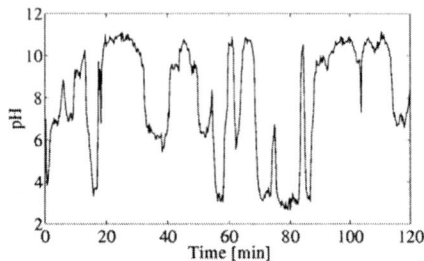

Fig. 6.1. Identification data

and it is clustered by the GK algorithm. To determine the number of clusters, the validity measures from Sect. 4.2.1 were applied. The upper limit on the number of cluster was set to $c = 10$, assuming that nonlinearity of the process can be sufficiently well approximated by 10 local linear models.

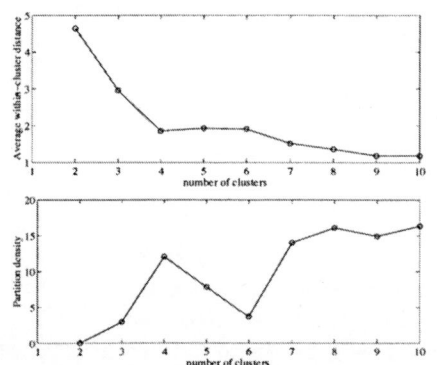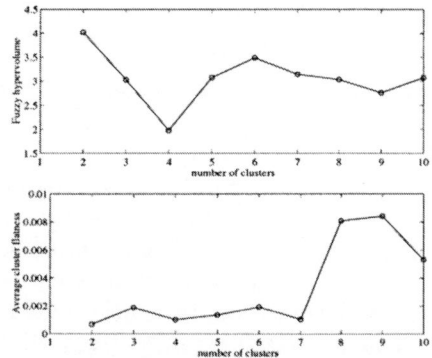

Fig. 6.2. Cluster validity measures

Figure 6.2 shows the result obtained with validity measures for $c = 2, 3, \ldots, 10$. One can see that the average within-cluster distance, the fuzzy hypervolume, and average cluster flatness all attain its first local minimum at $c = 4$. The partition density attains its first local maximum for the same value. The membership functions for each of the regressors are obtained by fitting the cluster projections by parametric exponential functions as shown in Fig. 6.3.

The consequent parameters were obtained from the cluster covariance matrices and the cluster means, using the equations (5.5) and (5.6). The resulting rule base is given by

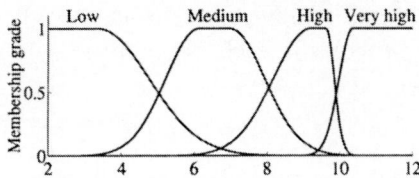 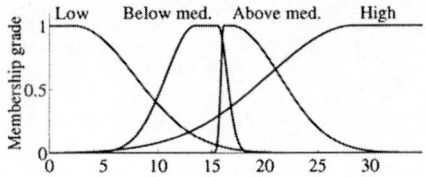

Fig. 6.3. Membership functions for the pH (left) and for the base flow-rate (right)

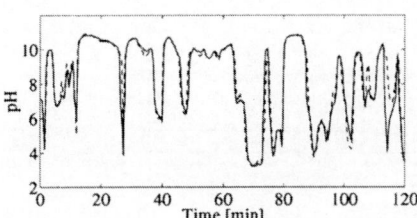 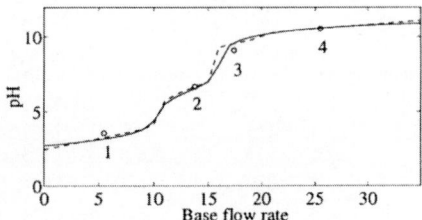

Fig. 6.4. Left: model validation on a fresh data set. Right: steady-state responses (titration curves) of the true process and the model. Solid line is the process, dashed-dotted line the model. The circles represent the centers of the obtained clusters.

1) **If** pH(k) is LOW **and** $Q(k)$ is LOW
    then pH($k + 1$) $= 0.56$pH(k) $+ 0.07Q(k) + 1.09$.
2) **If** pH(k) is MEDIUM **and** $Q(k)$ is BELOW MEDIUM
    then pH($k + 1$) $= 0.60$pH(k) $+ 0.10Q(k) + 1.27$.
3) **If** pH(k) is HIGH **and** $Q(k)$ is ABOVE MEDIUM
    then pH($k + 1$) $= 0.40$pH(k) $+ 0.08Q(k) + 4.37$.
4) **If** pH(k) is VERY HIGH **and** $Q(k)$ is HIGH
    then pH($k + 1$) $= 0.81$pH(k) $+ 0.01Q(k) + 1.77$.

Figure 6.4 shows the validation of the obtained fuzzy model using a fresh data set and the comparison of the steady state response, the so-called titration curve, of the 'true' system and of the model. Both curves are in good agreement and the results obtained here from numerical data only are comparable to those reported in (Johansen, 1994), where several different sources of information were combined (e.g., smoothness assumption, data ranges, simplified mass-balance equations, etc.).

In addition to accurate approximation of the system, the TS fuzzy model gives certain insights into the nonlinear dynamics of the system, as it is a set of locally valid linear ARX models. The validity regions for these models are defined by the antecedent membership functions. It can be observed how the parameters of the local models (the gains and the time constants) vary from region to region. By converting the consequent models to continuous-time domain, the following gains K_i and time constants τ_i are obtained:

$$\mathbf{K} = \{0.15, \ 0.25, \ 0.13, \ 0.06\}, \qquad \tau = \{25.68, \ 29.23, \ 16.49, \ 69.47\}.$$

By plotting the step responses of the individual models, one can get an idea about the behavior of the model (and the process) in each particular region, see Fig. 6.5.

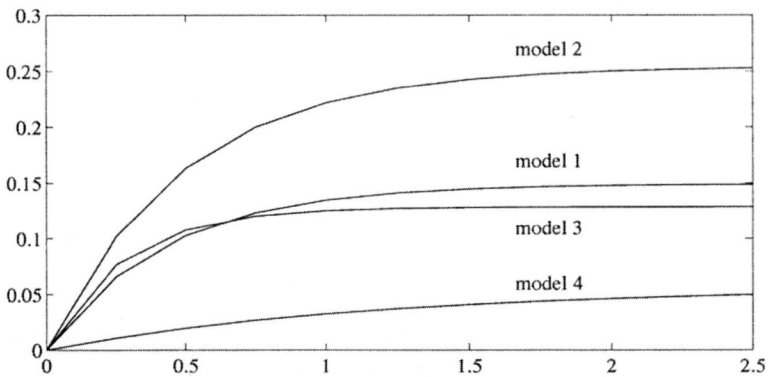

Fig. 6.5. Step responses of the consequent models

It can be seen that the consequent model of rule 3, for instance, has a smaller time constant and higher gain than the remaining models, which is in agreement with the region it represents, as given by the membership functions and also by the position of the cluster center on the titration curve, given in Fig. 6.4. Model 4, on the other hand, represents the saturated part of the titration curve and has a low gain and a low time constant, as one can see from the time plots in Fig. 6.1.

By examining the rule base and the membership functions, one can see that certain regions of the pH×Q space are not covered by any rule, i.e., the rule base is not complete. This is because very few or no data points are available in the identification data set for these regions. This is seen as an advantage of the clustering based identification technique, as it gives the user information about the reliability of the obtained model in different regions of the model variables. This information can be used to design new identification experiments, even though to our knowledge, no constructive method is available for the design of the excitation signals, especially for higher-order systems. The completeness of the rule base can be checked automatically when evaluating the degrees of fulfillment of the rules for different values of the antecedent variables.

Additional rules can be provided by the user or automatically, by combining the unused membership functions in the antecedents. The consequent part can be defined by the user, based, e.g., on an approximate mechanistic model (Johansen, 1994), or can be computed by interpolating the neighboring rules. For the pH model, an additional rule can be created, for instance, to cover the domain of MEDIUM pH and LOW Q:

If pH(k) **is** MEDIUM **and** $Q(k)$ **is** LOW
then pH$(k + 1) = 0.48$pH$(k) + 0.07Q(k) + 2.73$.

The consequent parameters in this rule were obtained by taking the average of two neighboring rules (i.e., rules with two largest values of the fulfillment degree) in the original rule base. Note that this addition of new rules also can be done on-line, when low activation of the current rules occurs.

7. Practical Considerations and Concluding Remarks

Methods for generating fuzzy models from data by product space clustering were described. This section summarizes the advantages and drawbacks of the presented approach, focusing mainly on its practical aspects.

7.1 Use of Prior Knowledge

Prior knowledge about the system is needed to postulate the structure of the (dynamic) model, see Sect. 3. Techniques based on validity measures and compatible cluster merging then can be applied to automatically determine the number of clusters in the data set. The number of clusters determines the number of rules in the fuzzy model generated from the partition. By combining the validity measures with prediction error criteria, a compromise can be done between the model complexity and the requirements on its performance, without the need of using the usual trial and error approach.

A major advantage of the clustering identification approach is that the obtained models are locally interpretable and often correspond well with the local behavior of the identified process. The form and the overlap of the membership functions, extracted from the obtained fuzzy partition, provide information about the character of the system's nonlinearity. By examining these results, the user can gain insights in the behavior of the process and also qualitatively validate the obtained model (Babuška et al., 1996). This contrasts with many pure black-box techniques, such as neural networks, where the validation relies mainly on numerical simulation. Prior knowledge also can be incorporated in the obtained model, in the form of additional rules, based on approximate mechanistic models, step response identification, or heuristics and experience.

7.2 Model Complexity

In many cases, product space clustering generates less complex models than methods based on optimization of the antecedent membership functions separately for each variable (adaptive spline methods (Brown and Harris, 1994), the Takagi–Sugeno approach (Takagi and Sugeno, 1985; Sugeno and Tanaka, 1991), table look-up schemes (Wang, 1994), etc.). The reason is, that the rules are created directly in the product space of the model variables and take into account the form of the system's nonlinearity and coverage of the space by identification data. Hence, more rules are obtained for regions characterized by complex nonlinear behavior, while (almost) linear regions are described by fewer rules or just one rule. In this way, the information about the system's nonlinearity is reflected in the obtained membership functions. Apart from acquisition of fuzzy models, clustering can be used as a data analysis and preprocessing step for other techniques, such a training of neural networks (Braake, 1997). No rules are generated in the regions that do not contain any data. With regard to generalization capabilities, this can be seen as a drawback, as the model gives no output for these regions. On the other hand, one can see it as an advantage, since analysis of the rule base completeness helps validate the model and design new identification experiments.

Redundancy may occur in the obtained model in terms of similar membership functions, describing the same region in the domain of a particular variable. This is a result od cluster projection, as shown in Sect. 5.3. Techniques based on similarity measures are applied to simplify the initial fuzzy models acquired from data.

7.3 Robustness of the Identification Method

Clustering requires sufficiently informative data. Piecewise constant or binary signals, often used for identification of linear systems, are not suitable. The technique also cannot use data sets with linearly dependent variables, as the clustering algorithm runs into numerical problems. Practical experience shows, that in this sense, the technique is somewhat more sensitive to poor data than other approaches. On the other hand, with a good data set, the user can gain valuable insights into the process behavior.

Large data sets in combination with high-dimensional regression problems may drastically increase the computational demands of the clustering algorithm. The technique has proven to be practically useful for medium-complexity identification problems, i.e., data sets of up to several thousand data points and low-order dynamical systems with few input variables. Large identification problems need to be decomposed (if possible), or solved using hybrid approaches, see Sect. 3.3. An advantage of the clustering approach is that is can be applied to well-behaved, smooth nonlinear systems, as well as to systems, whose behavior changes abruptly, as a result of 'hard' nonlinearities, such as dead-zones, hysteresis, stiction, etc. The character of the process nonlinearity is reflected in the form and overlap of the obtained membership functions.

It should be noted that the application of the presented clustering identification technique requires certain skills and experience of the user. Rather than a black-box, fully automated procedure, the approach should be seen as an interactive method facilitating an active participation of the user in a computer-assisted modeling session. For this purpose also special software tools are needed.

7.4 Real-World Applications

A number of practical applications have shown that the presented identification procedure can be applied to various problems, including modeling and controller design for physical processes such as pressure control (Braake et al., 1994; Babuška et al., 1996), hybrid modeling of biochemical processes (Babuška et al., 1996; Kaymak, 1994), or biomedical applications (Verhaegen and Babuška, 1996). Most of these applications are medium-complexity systems, with relatively simple dynamics, but strongly nonlinear behavior. In all cases, the possibility to interpret and qualitatively validate the model proved to be an important advantage of the approach. For efficient use of the presented algorithms, a MATLAB toolbox has been developed (Babuška, 1995).

Acknowledgements

Thanks are due to Dr. Tor Arne Johansen who provided us with MATLAB code of the pH neutralization model, and to our colleague João da Costa Sousa for proof-reading the manuscript.

A. Appendix: The Gustafson–Kessel Algorithm – MATLAB Implementation

This appendix presents a MATLAB function implementing the GK clustering algorithm. The main purpose is to give the reader a basic idea, rather than provide a 'friendly' program. The code is written as simply as possible and is not optimized either for speed or for memory usage. To become really usable, the function needs to be extended in a number of ways, including, for instance, automatic generation of the initial partition matrix, given the number of clusters, etc. Interested readers may contact the authors for MATLAB functions implementing the identification procedures described in this chapter.

```
function [U,V,F] = gk(Z,U0,m,tol)
% Clustering with fuzzy covariance matrix (Gustafson-Kessel algorithm)
%
% [U,V,F] = GK(Z,U0,m,tol)
%-------------------------------------------------------------------
% Input:   Z    ... N by n data matrix
%          U0   ... initial fuzzy partition matrix
```

```
%          m    ... fuzziness exponent (m > 1)
%          tol  ... termination tolerance
%-------------------------------------------------------------------
% Output: U    ... fuzzy partition matrix
%         V    ... cluster means (centers)
%         F    ... cluster covariance matrices concatenated in 1 matrix
%                   F = [F1;F2;...Fc], where c is the number of clusters
%-------------------------------------------------------------------

% (c) Robert Babuska, 1996

%---------------- prepare matrices ---------------------------------

[mz,nz] = size(Z);                       % data matrix size
c = size(U0,2);                          % number of clusters
mZ1 = ones(mz,1);                        % auxiliary variable
nZ1 = ones(nz,1);                        % auxiliary variable
V1c = ones(1,c);                         % auxiliary variable
U = zeros(mz,c);                         % partition matrix
d = U;                                   % distance matrix
F = zeros(c*nz,nz);                      % covariance matrix

%---------------- iterate ------------------------------------------

while  max(max(U0-U)) > tol              % iteration loop
  U = U0; Um = U.^m; sumU = sum(Um);     % auxil. variables
  V = (Um'*Z)./(nZ1*sumU)';              % calculate centers
  for j = 1 : c,                         % for all clusters
    ZV = Z - mZ1*V(j,:);                 % auxiliary variable
    f = nZ1*Um(:,j)'.*ZV'*ZV/sumU(j);    % calc. cov. matrix
    d(:,j)=sum((ZV*(det(f)^(1/nz)*inv(f)).*ZV)')'; % calc. distances
  end;
  d = (d+1e-10).^(-1/(m-1));
  U0 = (d ./ (sum(d')'*V1c));            % update part matrix
end

%---------------- update final F and U -----------------------------

U = U0; Um = U.^m; sumU = nZ1*sum(Um);
for j = 1 : c,
    ZV = Z - mZ1*V(j,:);
    F((j-1)*nz+(1:nz),:) = nZ1*Um(:,j)'.*ZV'*ZV/sumU(1,j);
end;

%---------------- end of function ----------------------------------
```

References

Akaike, H. (1974). A new look at the statistical model identification. *IEEE Trans. Automatic Control*, **19**(6), 716–723.

Babuška, R. (1995). Fuzzy modeling and cluster analysis toolbox for MATLAB. In: *Proc. Third European Congress on Intelligent Techniques and Soft Computing EUFIT'95*. Aachen, Germany, pp. 1479–1483.

Babuška, R., H.A.B. te Braake, A.J. Krijgsman, and H.B. Verbruggen (1996). Comparison of intelligent control schemes for real-time pressure control. *Control Engineering Practice*, **4**(11), 1585–1592.

Babuška, R. and H.B. Verbruggen (1994). Applied fuzzy modeling. In: *Proc. IFAC Symposium on Artificial Intelligence in Real Time Control*. Valencia, Spain, pp. 61–66.

Babuška, R. and H.B. Verbruggen (1995a). New approach to constructing fuzzy relational models from data. In: *Proc. Third European Congress on Intelligent Techniques and Soft Computing EUFIT'95*. Aachen, Germany, pp. 583–587.

Babuška, R. and H.B. Verbruggen (1995b). A new identification method for linguistic fuzzy models. In: *Proc. FUZZ-IEEE/IFES'95*. Yokohama, Japan, pp. 905–912.

Babuška, R., H.B. Verbruggen, and H.J.L. van Can (1996). Fuzzy modeling of enzymatic Penicillin–G conversion. In: *Preprints 13th IFAC World Congress*, Vol. N, July. San Francisco, CA, pp. 479–484.

Baldwin, J.F., T.P. Martin, and B.W. Pilsworth (1995). *FRIL – Fuzzy and Evidential Reasoning in Artificial Intelligence*. Research Studies Press, Taunton.

Banerjee, A., Y. Arkun, R. Pearson, and B. Ogunnaike (1995). H_∞ control of nonlinear processes using multiple linear models. In: *Proc. European Control Conference*. Rome, Italy, pp. 2671–2676.

Berenji, H.R. and P.S. Khedar (1993). Clustering in product space for fuzzy inference. In: *Proc. of Second International Conference on Fuzzy Systems*. San Francisco, CA, pp. 1402–1407.

Bezdek, J.C. (1981). *Pattern Recognition with Fuzzy Objective Function*. Plenum Press, New York.

Bezdek, J.C., C. Coray, R. Gunderson, and J. Watson (1981a). Detection and characterization of cluster substructure, I. linear structure: Fuzzy c-lines. *SIAM J. Appl. Math.*, **40**(2), 339–357.

Bezdek, J.C., C. Coray, R. Gunderson, and J. Watson (1981b). Detection and characterization of cluster substructure, II. fuzzy c-varieties and convex combinations thereof. *SIAM J. Appl. Math.*, **40**(2), 358–372.

Bezdek, J.C., R. Hathaway, R.E. Howard, C.A. Wilson, and M.P. Windham (1987). Local convergence analysis of a grouped variable version of coordinate descent. *Journal of Optimization Theory and Applications*, **54**(3), 471–477.

Boor, Carl de (1978). *A Practical Guide to Splines*. Springer, New York.

Braake, H.A.B. te (1997). Two step approach in training of regulated activation weights neural network (RAWN). Accepted for publication in *Engineering Applications of AI*.

Braake, H.A.B. te, R. Babuška, and H.J.L. van Can (1994). Fuzzy and neural models in predictive control. *Journal A*, **35**(3), Oct, 44–51.

Brown, M. and C. Harris (1994). *Neurofuzzy Adaptive Modelling and Control*. Prentice Hall, New York.

Chen, S. and S.A. Billings (1989). Representation of nonlinear systems: The NARMAX model. *International Journal of Control*, **49**, 1013–1032.

Chen, S., C.F.N. Cowan, and P.M. Grant (1991). Orthogonal least squares learning algorithm for radial basis function networks. *IEEE Trans. Neural Networks*, **2**(2), March, 302–309.

Dave, R.N. (1992). Boundary detection through fuzzy clustering. In: *IEEE Int. Conf. on Fuzzy Systems*. San Diego, CA, pp. 127–134.

Driankov, D., H. Hellendoorn, and M. Reinfrank (1993). *An Introduction to Fuzzy Control*. Springer, Berlin, 2nd ed. 1996.

Friedman, J.H. (1991). Multivariate adaptive regression splines. *Annals of Statistics*, **19**(1), 1–141.

Gath, I. and A.B. Geva (1989). Unsupervised optimal fuzzy clustering. *IEEE Trans. Pattern Analysis and Machine Intelligence*, **7**, 773–781.

Glorennec, P.Y. (1994). Learning algorithms for neuro-fuzzy networks. In: *Fuzzy Control Systems*, (Kandel and Langholz, eds.), pp. 3–18. CRC Press, Boca-Raton, Fl.

Gustafson, D.E. and W.C. Kessel (1979). Fuzzy clustering with a fuzzy covariance matrix. In: *Proc. IEEE CDC*. San Diego, CA, pp. 761–766.

Hall, R.C. and D.E. Seborg (1989). Modelling and self-tuning control of a multivariable pH neutralization process. part i: Modelling and multiloop control. In: *Proc. American Control Conference*, Vol. 2. Pittsburgh, PA, pp. 1822–1827.

Hathaway, R. and J.C. Bezdek (1991). Grouped coordinate minimization using Newton's method for inexact minimization in one vector coordinate. *Journal of Optimization Theory and Applications*, **71**(3), 503–516.

Hathaway, R.J. and J.C. Bezdek (1993). Switching regression models and fuzzy clustering. *IEEE Trans. Fuzzy Systems*, **1**(3), 195–204.

Huffel, S. van and J. Vandewalle (1991). *The Total Least Squares Problem; Computational Aspects and Analysis*. Frontiers in Applied Mathematics, SIAM, Philadelphia, PA.

Jager, R. (1995). *Fuzzy Logic in Control*. PhD dissertation, Delft University of Technology, Delft, The Netherlands.

Jang, J.S.R. (1992). Self-learning fuzzy controllers based on temporal back propagation. *IEEE Trans. Neural Networks*, **3**(5), 714–723.

Jang, J.-S.R. and C.-T. Sun (1993). Functional equivalence between radial basis function networks and fuzzy inference systems. *IEEE Trans. on Neural Networks*, **4**(1), January, 156–159.

Johansen, T.A. (1994). *Operating Regime Based Process Modelling and Identification*. PhD dissertation, The Norwegian Institute of Technology, University of Trondheim, Trondheim, Norway.

Johansen, T.A. and B.A. Foss (1993). Constructing NARMAX models using ARMAX models. *Int. J. Control*, **58**, 1125–1153.

Kaymak, U. (1994). Application of fuzzy methodologies to a washing process. Chartered designer thesis, Delft University of Technology, Delft, The Netherlands.

Kaymak, U. and R. Babuška (1995). Compatible cluster merging for fuzzy modeling. In: *Proc. FUZZ-IEEE/IFES'95*. Yokohama, Japan, pp. 897–904.

Kosko, B. (1994). Fuzzy systems as universal approximators. *IEEE Trans. Computers*, **43**, 1329–1333.

Krishnapuram, R. and Chin-Pin Freg (1992). Fitting an unknown number of lines and planes to image data through compatible cluster merging. *Pattern Recognition*, **25**(4), 385–400.

Kruse, R., J. Gebhardt, and F. Klawonn (1994). *Foundations of Fuzzy Systems*. John Wiley and Sons, Chichester.

Leonaritis, I.J. and S.A. Billings (1985). Input-output parametric models for nonlinear systems. *International Journal of Control*, **41**, 303–344.

Ljung, L. (1987). *System Identification, Theory for the User*. Prentice-Hall, NJ.

Mamdani, E.H. (1977). Application of fuzzy logic to approximate reasoning using linguistic systems. *Fuzzy Sets and Systems*, **26**, 1182–1191.

Oliveira, J. Valente de (1993). Neuron inspired rules for fuzzy relational structures. *Fuzzy Sets and Systems*, **57**(1), July, 41–55.

Pal, N.R. and J.C. Bezdek (1995). On cluster validity for the fuzzy c-means model. *IEEE Trans. Fuzzy Systems*, **3**(3), 370–379.

Pedrycz, W. (1984). An identification algorithm in fuzzy relational systems. *Fuzzy Sets and Systems*, **13**, 153–167.

Psichogios, D.C. and L.H. Ungar (1992). A hybrid neural network – first principles approach to process modeling. *AIChE J.*, **38**, 1499–1511.

Rissanen, J. (1978). Modeling by shortest data description. *Automatica*, **14**, 465–471.

Ross, Timothy J. (1995). *Fuzzy Logic with Engineering Applications*. McGraw-Hill, New York.

Seber, G.A.F. and C.J. Wild (1989). *Nonlinear Regression*. John Wiley & Sons, New York.

Setnes, M. (1995). *Fuzzy Rule Base Simplification Using Similarity Measures*. M.Sc. thesis, Delft University of Technology, Delft, The Netherlands.

Sjöberg, J., H. Hjalmarsson, and L. Ljung (1994). Neural networks in system identification. In: *Proc. SYSID'94*, Vol. 2. pp. 49–72.

Strang, G. (1976). *Linear Algebra and Its Applications*. Academic Press, New York.

Sugeno, M. and G.T. Kang (1988). Structure identification of fuzzy model. *Fuzzy Sets and Systems*, **28**, 15–33.

Sugeno, M. and K. Tanaka (1991). Successive identification of a fuzzy model and its application to prediction of a complex system. *Fuzzy Sets and Systems*, **42**, 315–334.

Sugeno, M. and T. Yasukawa (1993). A fuzzy-logic-based approach to qualitative modeling. *IEEE Trans. Fuzzy Systems*, **1**, 7–31.

Sugeno, Michio and Takahiro Yasukawa (1994). Qualitative modelling based on numerical data and knowledge data and its application to control. In: *Computational Intelligence: Imitating Life*, (Jacek M. Zurada, Robert J. Marks II, and Charles J. Robinson, eds.), pp. 304–315. IEEE Press, Piscataway, NJ.

Takagi, T. and M. Sugeno (1985). Fuzzy identification of systems and its application to modeling and control. *IEEE Trans. Systems, Man and Cybernetics*, **15**(1), 116–132.

Tanaka, K., T. Ikeda, and H.O. Wang (1996). Robust stabilization of a class of uncertain nonlinear systems via fuzzy control: Quadratic stability, h_∞ control theory and linear matrix inequalities. *IEEE Trans. on Fuzzy Systems*, **4**(1), 1–13.

Tanaka, K. and M. Sugeno (1992). Stability analysis and design of fuzzy control systems. *Fuzzy Sets and Systems*, **45**(2), 135–156.

Verhaegen, M. and R. Babuška (1996). Estimation of respiratory mechanical parameters by means of fuzzy clustering. In preparation.

Wang, L.-X. (1992). Fuzzy systems are universal approximators. In: *Proc. IEEE Int. Conf. on Fuzzy Systems 1992*. San Diego, USA, pp. 1163–1170.

Wang, L.-X. (1994). *Adaptive Fuzzy Systems and Control, Design and Stability Analysis*. Prentice Hall, NJ.

Yager, R.R. and D.P. Filev (1994). *Essentials of Fuzzy Modeling and Control*. John Wiley, New York.

Yoshinari, Y., W. Pedrycz, and K. Hirota (1993). Construction of fuzzy models through clustering techniques. *Fuzzy Sets and Systems*, **54**, 157–165.

Zadeh, L.A. (1973). Outline of a new approach to the analysis of complex systems and decision processes. *IEEE Trans. Systems, Man, and Cybernetics*, **1**, 28–44.

Zeng, X.J. and M.G. Singh (1994). Approximation theory of fuzzy systems – SISO case. *IEEE Trans. Fuzzy Systems*, **2**, 162–176.

Zeng, X.J. and M.G. Singh (1995). Approximation theory of fuzzy systems – MIMO case. *IEEE Trans. Fuzzy Systems*, **3**(2), 219–235.

Zhao, J., V. Wertz, and R. Gorez (1994). A fuzzy clustering method for the identification of fuzzy models for dynamical systems. In: *9th IEEE International Symposium on Intelligent Control*. Columbus, Ohio, USA.

Zimmermann, H.-J. (1991). *Fuzzy Set Theory and its Application,* 2nd ed. Kluwer, Boston, MA.

Identification of Takagi–Sugeno Fuzzy Models via Clustering and Hough Transform

Min-Kee Park[1], Seung-Hwan Ji[2], Eun-Tai Kim[2], and Mignon Park[2]

[1] Seoul National Polytechnic University, 172 Kongneung-dong, Nowon-gu, Seoul, 139–743, Korea
[2] Yonsei University, 134 Shinchon-dong, Seodaemun-gu, Seoul, 120–749, Korea

1. Introduction

In this chapter we consider the identification of a Takagi–Sugeno fuzzy model (TS fuzzy model) [2]. This type of fuzzy model is especially useful in the area of fuzzy model-based control [10]. The TS fuzzy model is a nonlinear system model represented by fuzzy rules of the type

$$\mathrm{R}^i \,:\, \textbf{If } x_1 \textbf{ is } A_1^i \textbf{ and} \ldots \textbf{and } x_m \textbf{ is } A_m^i$$
$$\textbf{then } y^i = a_0^i + a_1^i x_1 + \cdots + a_m^i x_m \qquad (1.1)$$

where R^i $(i = 1, 2, \ldots, n)$ denotes the i-th fuzzy rule, x_j $(j = 1, 2, \ldots, m)$ are input variables and y^i is an output. Furthermore, a_j^i are the parameters contained in the consequent (then-part) of the i-th fuzzy rule, and $A_1^i, A_2^i, \ldots, A_m^i$ are the linguistic values taken by the input variables in the antecedent (if-part) of the i-th rule. The meaning of these linguistic values is defined by corresponding membership functions. As shown in (1.1) and (1.2), this fuzzy model describes a nonlinear input-output relation. The overall output of a TS fuzzy model is given as

$$\hat{y} = \frac{\sum\limits_{i=1}^{c} \beta^i y^i}{\sum\limits_{i=1}^{c} \beta^i}, \qquad \beta^i = \prod_{j=1}^{m} \mu_{A_j^i}(x_j) \qquad (1.2)$$

where $\mu_{A_j^i}$ is the membership function defining the meaning of the linguistic value A_j^i.

The identification of a TS fuzzy model, using input-output data, is divided into two steps: structure identification and parameter estimation. The former consists of antecedent structure identification and consequent structure identification. The latter consists of antecedent and consequent parameter estimation where the consequent parameters are the coefficients of the linear expressions in the consequent of a fuzzy rule.

In this chapter, we describe an identification method for TS fuzzy models that is much simpler than the original identification method proposed by Takagi and Sugeno. The identification method described here consists of two steps: coarse tuning and fine tuning. In coarse tuning, consequent and antecedent parameters are roughly estimated by clustering. In fine tuning, the antecedent and consequent paremeters are estimated more precisely by a gradient descent algorithm. These two identification steps are shown in Fig. 2.1.

The algorithm described in this chapter bears some similarity to that of Babuška and Verbruggen (BV) also presented in this volume. Namely, the two identification methods are similar in the following two aspects:

1. The fuzzy model to be identified is an affine TS fuzzy model.
2. The consequent parameters are estimated before the estimation of the antecedent parameters (as opposed to the original TS identification method where the already determined antecedent parameters affect the estimation of the consequent parameters).

However, the two identification methods have also a number of differences:

1. In BV, consequent parameters are estimated first by clustering, and then the antecedent parameters are obtained. In our approach, after coarse tuning is completed, both the consequent and antecedent parameters are fine tuned.
2. We propose two consequent parameter estimation methods for coarse tuning: the HPC-MEANS algorithm and a Hough-transform-based identification strategy. HPC-MEANS is based on clustering and is similar to the BV consequent parameter estimation scheme (GK algorithm). The important differences are the following: (i) In HPC-MEANS, the prototypical representative of each cluster is a hyperplane and the cluster is in the shape of this hyperplane. In the GK algorithm, the prototypical representative of each cluster is a point and the cluster is in the shape of a deformed ellipsoid, which can be approximated by hyperplanar patches. (ii) In HPC-MEANS, a distance measure is simply how far away a sample data is from the hyperplane cluster. In the GK algorithm an adaptive distance measure is used. (iii) The Hough-transform-based identification strategy is different from the one presented in BV in that: in BV cluster merging starts with a sufficiently high number of clusters while in our approach, the initial number of clusters is automatically specified by a number of linear segments obtained by the Hough transform; in BV a clustering method is used to merge clusters, whereas we use clustering to split and merge clusters taking into account the continuity of the input-output data.
3. In our approach, the antecedent membership functions are assumed to be Gaussian functions and their parameters are roughly estimated by using the means and variances of the clustered data in course tuning. In fine tuning, the membership function parameters are adjusted precisely by the gradient descent algorithm in order to minimize the modeling

performance index (mean squared error). In BV, antecedent membership functions are estimated via the use of cluster projections without the use of any performance index.

2. The Identification Method

In this section, we describe the identification of the consequents of a TS fuzzy model independently of and prior to the identification of the antecedents of a TS fuzzy model. First, we present a method to identify consequents by a pattern clustering algorithm, HPC-MEANS (HyperPlane based C-MEANS), and then a Hough-transform-based identification strategy is proposed as an alternative to HPC-MEANS.

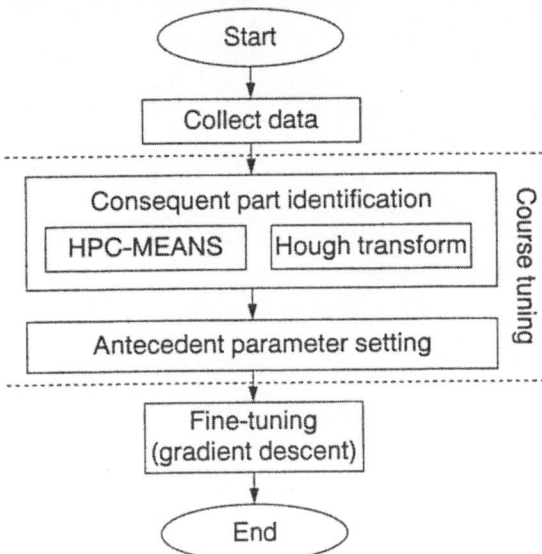

Fig. 2.1. The identification algorithm

2.1 Identification of Consequents

2.1.1 HPC-MEANS Algorithm. There have been several studies that have applied C-MEANS or FCM (Fuzzy C-MEANS) for the purpose of the identification of fuzzy models, but these two algorithms were used to cluster either the input or output space rather than to cluster both of these spaces. The reason for this is that what one needs in the identification of a fuzzy model is hyperplane-shaped clusters while the C-MEANS algorithm provides hypersphere-shaped clusters. Figure 2.2 shows hypersphere-shaped clusters obtained by C-MEANS with certain points as cluster representatives. Figure 2.3 shows hyperplane-shaped clusters with hyperplanes as cluster representatives. The latter cannot be obtained by conventional C-MEANS.

In this section, we present the HPC-MEANS (HyperPlan-based C-MEANS) algorithm, which provides hyperplane-shaped clusters and is thus relevant to fuzzy model identification. The algorithm can be compared to C-MEANS as follows:

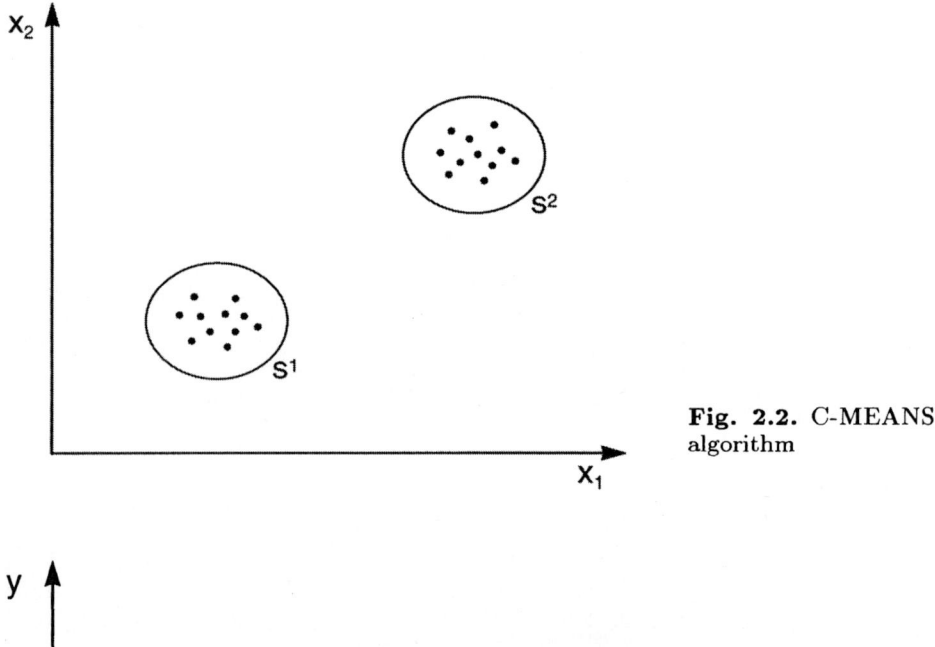

Fig. 2.2. C-MEANS algorithm

Fig. 2.3. HPC-MEANS algorithm

1. While C-MEANS obtains clusters having points as their prototypical representatives, HPC-MEANS obtains clusters with hyperplanes as their prototypical representatives.
2. While C-MEANS obtains the representative of each cluster by calculating the mean of the data associated with each cluster, HPC-MEANS does this by RLS (Recursive Least Squares).

On the other hand, HPC-MEANS can be compared to the clustering GK algorithm used by BV in the following way:

1. HPC-MEANS obtains hyperplane-shaped clusters with hyperplanes as their prototypical representatives. The GK algorithm obtains deformed ellipsoid-shaped clusters with points as their prototypical representatives.
2. HPC-MEANS uses a distance measure defined by how far away a sample data is from a hyperplane cluster. The GK algorithm uses an adaptive distance measure.

Before we go on to the stepwise description of HPC-MEANS, let us describe how the prototypical hyperplane of each cluster is obtained. For example, consider m inputs and a single output. Then the prototypical representative of the i-th cluster is then given as

$$y^i = a_0^i + a_1^i x_1 + a_2^i x_2 + \cdots + a_m^i x_m. \tag{2.1}$$

The above expression can be rewritten in the form of the following vector equation

$$y^i = \mathbf{X}^T \cdot \mathbf{P}^i \tag{2.2}$$

where $\mathbf{X} = \begin{bmatrix} 1 & x_1 & x_2 & \ldots & x_m \end{bmatrix}^T$ and $\mathbf{P}^i = \begin{bmatrix} a_0^i & a_1^i & a_2^i & \ldots & a_m^i \end{bmatrix}^T$. Suppose that there are N_i sets of sample data denoted by $(\mathbf{X}_1, y_1)$, $(\mathbf{X}_2, y_2)$, $(\mathbf{X}_3, y_3)$, ..., $(\mathbf{X}_{N_i}, y_{N_i})$ for the i-th cluster. The hyperplane representing the i-th cluster, $\mathbf{P}^i$, is obtained by the following recursive equations

$$\mathbf{P}_{j+1}^i = \mathbf{P}_j^i + \mathbf{K}_j \left[y_{j+1} - \mathbf{X}_{j+1}^T \mathbf{P}_j^i \right] \tag{2.3}$$

$$\mathbf{K}_j = \mathbf{S}_{j+1} \mathbf{X}_{j+1} = \frac{\mathbf{S}_j \mathbf{X}_{j+1}}{1 + \mathbf{X}_{j+1}^T \mathbf{S}_j \mathbf{X}_{j+1}} \tag{2.4}$$

$$\mathbf{S}_{j+1} = \left[1 - \mathbf{K}_j \mathbf{X}_{j+1}^T \right] \mathbf{S}_j \tag{2.5}$$

where $j = 1, 2, \cdots, N_i$ and $\mathbf{P}^i = \mathbf{P}_{N_i}^i$, and the initial values of the recursive equations are determined as

$$\mathbf{P}_0^i = 0, \qquad \mathbf{S}_0 = \alpha \mathbf{I} \tag{2.6}$$

where α is a sufficiently large number and $\mathbf{I}$ is an identity matrix. Then, the hyperplane-shaped fit for N_i in the i-th cluster is

$$\mathbf{P}^i = \mathbf{P}_{N_i}{}^i. \tag{2.7}$$

HPC-MEANS Algorithm. Assume that n sets of sample data $(\mathbf{X}_l, y_l)$, $1 \leq l \leq n$, are given and that a TS fuzzy model with c fuzzy rules is needed. The initial number of clusters c is assumed to be known. Using a performance index, an appropriate number is found iteratively (i.e., $N_1 + N_2 + \cdots + N_c = n$). This is almost the same strategy as the one employed in BV, or in the original TS identification method.

Step 1: *Assume c hyperplanes as initial cluster representatives:*

$$y^1 = \mathbf{X}^T \cdot \mathbf{P}^1(1)$$
$$y^2 = \mathbf{X}^T \cdot \mathbf{P}^2(1)$$
$$\cdots$$
$$y^c = \mathbf{X}^T \cdot \mathbf{P}^c(1) \tag{2.8}$$

where (1) denotes the first iteration.

Step 2: *Search for the nearest cluster.* At the k-th iteration, assign each set of sample data $(\mathbf{X}_\ell, y_\ell)$, $1 \leq \ell \leq n$ to the nearest cluster with $y^j = \mathbf{X}^T \cdot \mathbf{P}^j(k)$ as the representative. In other words, $(\mathbf{X}_\ell, y_\ell) \in S^j(k)$, if $|y_\ell - \mathbf{X}_\ell^T \mathbf{P}^j(k)| \leq |y_\ell - \mathbf{X}_\ell^T \mathbf{P}^i(k)|$, for all $i = 1, 2, \cdots, c$ where $S^j(k)$ is a cluster with the hyperplane $y^j = \mathbf{X}^T \cdot \mathbf{P}^j(k)$ as its representative.

Step 3: *Calculate total clustering error $D(k)$ for the k-th iteration:*

$$D(k) = \sum_{j=1}^{c} \frac{1}{N_j} \sum_{(\mathbf{X}_l, y_l) \in S^j(k)} |y_l - \mathbf{X}_l^T \mathbf{P}^j(k)| \tag{2.9}$$

where N_j is the number of sample data belonging to $S^j(k)$. Now, if $(D(k) - D(k-1))/D(k) \leq \delta$, the algorithm stops, or it proceeds to Step 4.

Step 4: *Calculate new cluster representatives, $y^j = \mathbf{X}^T \mathbf{P}^j(k+1)$, at the $(k+1)$-th iteration, for $S^1(k), S^2(k), \cdots, S^c(k)$ obtained in Step 2 by RLS given in (2.3)–(2.5).*

Step 5: *Go to Step 2 and let $k = k + 1$.*

Now simulation results obtained by the application of HPC-MEANS will be shown. For simplicity, let the input-output space be two-dimensional with a single input and single output. In this case, a consequent, a hyperplane for each cluster, is a first order linear expression given by $y^j = a_0^j + a_1^j x$.

Let the number of fuzzy rules, or the number of clusters formed by HPC-MEANS, be three. Figure 2.4 is two dimensional input-output data for a target system and Figs. 2.5–2.9 show clusters obtained by HPC-MEANS.

2.1.2 The Hough Transform Method. Alternatively to HPC-MEANS, we suggest an identification strategy based on the Hough transform and pattern clustering. Generally speaking, pattern-clustering is known to have problems with specifying the initial number of clusters. Also, for the identification

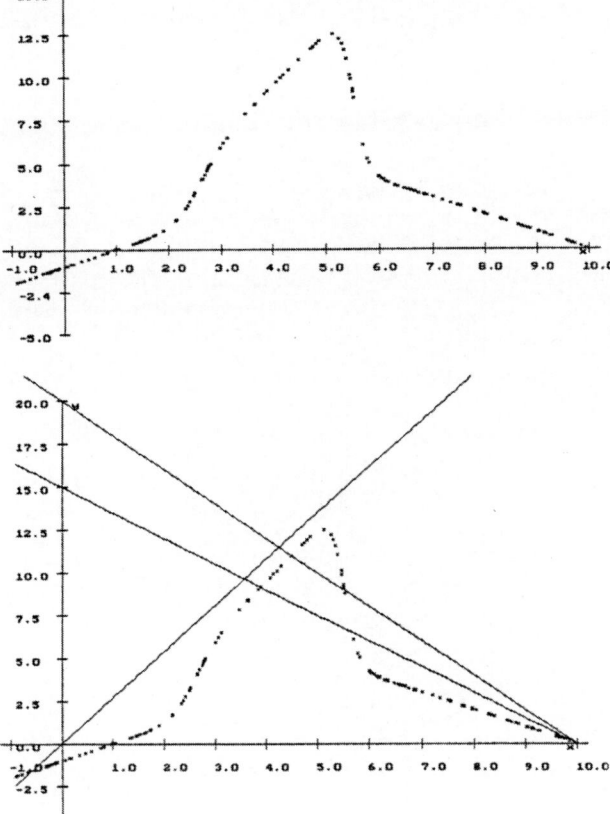

Fig. 2.4. The input-output data for the given SISO system

Fig. 2.5. The construction of clusters by HPC-MEANS (initial condition, $k = 1$)

of the consequents of the fuzzy rules we need to obtain linear expressions by taking into consideration the continuity of the input-output data. Furthermore, the HPC-MEANS algorithm has difficulty in setting initial values for consequent parameters. Therefore, we identify consequents by the Hough transform method, which is used to find a set of linear expressions. In order to identify antecedents, the input space should be partitioned to determine which fuzzy rule corresponds to which set of consequent parameters. The input space is partitioned by a clustering method in order to take into account the continuity of input-output data included in the linear expressions from the consequents. Finally, we suggest a gradient descent algorithm to fine-tune the antecedent and consequent parameters. The steps of this identification method are shown in Fig. 2.10.

98   Min-Kee Park et al.

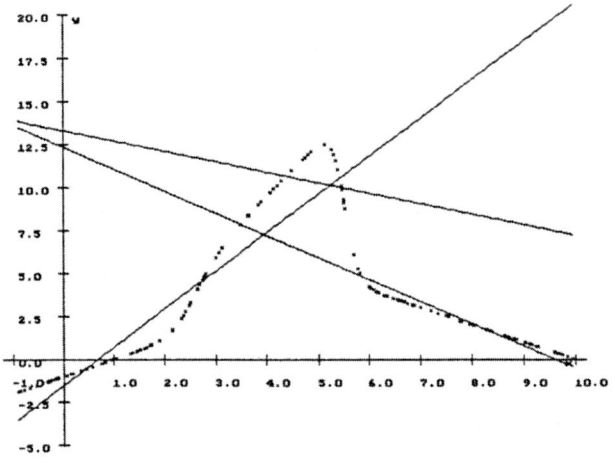

Fig. 2.6. The construction of clusters by HPC-MEANS ($k = 3$)

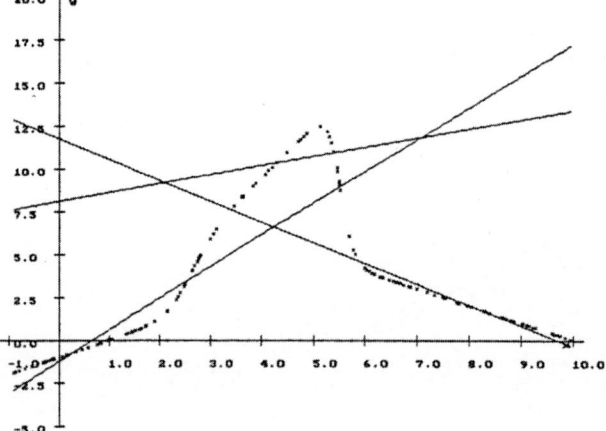

Fig. 2.7. The construction of clusters by HPC-MEANS ($k = 5$)

To determine the number of clusters, BV presents two approaches: use of validity measures and compatible cluster merging. Our approach is related to compatible cluster merging rather than the use of validity measures. In BV, cluster merging starts with a sufficiently high number of clusters, such that the nonlinearity of the regression hypersurface can be captured accurately. Finally, a decision is made on which clusters should be merged, using a threshold specified by the user. In our approach, the initial number of clusters is automatically specified by the number of linear segments obtained via the Hough transform method. Applying the Hough transform is very reasonable because it is a method to find out most likely hyperplanes from the input-output data. However, by using this approach two distant separate data points may be mistaken as belonging to the same cluster because hyperplanes are widely spread apart. Therefore, adaptive sample set

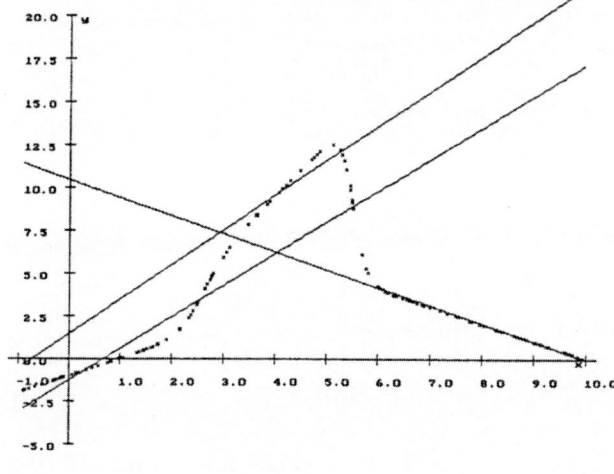

Fig. 2.8. The construction of clusters by HPC-MEANS ($k = 7$)

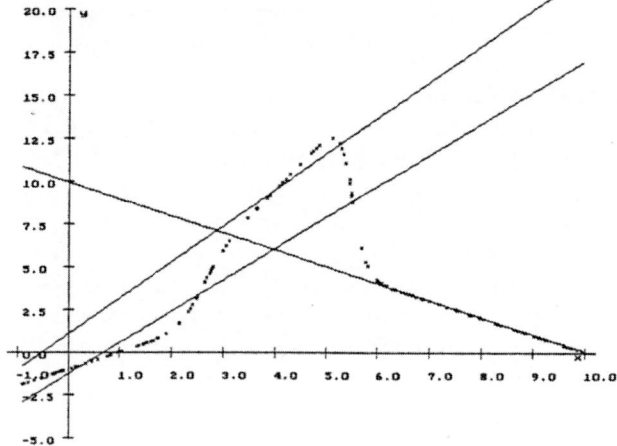

Fig. 2.9. The construction of clusters by HPC-MEANS ($k = 9$)

construction is used to split and merge clusters by taking into account the continuity of the input-output data. The number of clusters is determined using a membership boundary specified by the user.

In order to find out piecewise linear expressions from the input-output data, we employ the Hough transform method. In case of a SISO system, the input and output can be mapped into one point on a 2-dimensional plane. The Hough transform method can be used to find linear segments and thus identify the parameters of the corresponding fuzzy model.

The Hough Transform Algorithm. Consider a point (x_i, y_i), which represents an input-output pair and the equation of a line in a slope-intercept form, $y_i = ax_i + b$. There is an infinite number of lines that pass through (x_i, y_i), but they all satisfy the equation $y_i = ax_i + b$ for varying values of a and b. However, if we rewrite this equation as $b = -x_i a + y_i$, and consider the

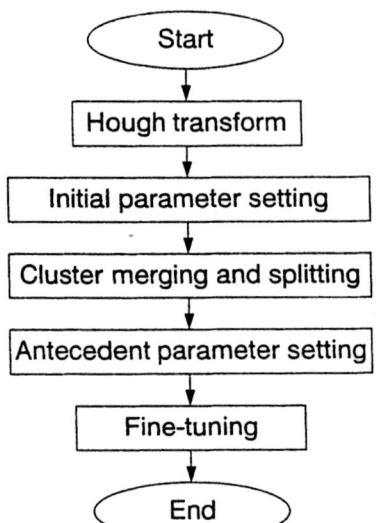

Fig. 2.10. Fuzzy model identification using the Hough transform method

ab plane (also called a parameter space), then we have the equation of a single line for a fixed pair (x_i, y_i). This is shown in the top-right part of Fig. 2.11. Furthermore, the second point (x_j, y_j) will also have a line in the parameter space associated with it, and this line will intersect the line associated with (x_i, y_i) at (a, b), where a is the slope and b the intercept of the line containing both (x_i, y_i) and (x_j, y_j) in the xy plane. In fact, all points on this line will have lines in the parameter space that intersect at (a, b). Because the slope and intercept can approach infinity as the line approaches a vertical position, the normal representation of a line used in the formation of accumulator cells in the $\rho\theta$ plane is

$$\rho = x\cos\theta + y\sin\theta \tag{2.10}$$

where ρ and θ are used to represent a straight line instead of a and b. This is shown in the lower right part of Fig. 2.11.

Although the Hough transform is usually applied to 2-dimensional data (especially in image processing), it is applicable to any function of the form $g(\mathbf{x}, \mathbf{c}) = 0$, where $\mathbf{x}$ is a vector of coordinates and $\mathbf{c}$ is a vector of coefficients. Therefore, the Hough transform method can be employed in the case of a multi-input and multi-output system.

In case of a double-input and single-output system, an input-output sample data should be 3-dimensional. The Hough transform equation for 3-dimensional cases is as follows:

$$\rho = (x\cos\theta + y\sin\theta)\sin\phi + z\cos\phi \tag{2.11}$$

where x and y are the inputs, z is the output, and ρ, θ, and ϕ are parameters in the Hough transform plane.

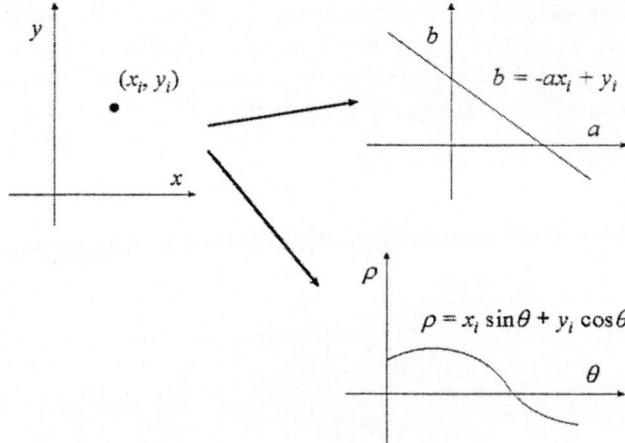

Fig. 2.11. 2D Hough transform of a point (x_i, y_i)

The results of the Hough transform are accumulated in the cells (accumulator cells) in the $\rho\theta$ plane and the cells with relatively high value are selected as candidates for the linear parts of the fuzzy model.

The Hough transform method can be represented in the form of the following agorithm:

Step 1: Divide the parameter space into small cells and construct an array of these cells. Initialize each cell to zero.

Step 2: Take one data sample and give arbitrary initial values to all parameters except one. For example, in (2.10), take a sample (x_i, y_i) and let $\theta = 0$.

Step 3: Calculate the left parameter value using the sample and the other parameter values. For example, in (2.10), calculate ρ with (x_i, y_i) and θ. This step is executed untill all the cells related to the point of the current data sample are covered.

Step 4: Find which cell the parameters belong to and update the cell by incrementing its value by one.

Step 5: Change the parameter values except for the one obtained in Step 3 so that an adjacent cell can be chosen for the next iteration.

Step 6: If all the cells are updated go to Step 7, otherwise go to Step 3.

Step 7: If all the samples have been processed through the loop, stop, otherwise go to Step 2.

Figure 2.12 shows the linear segments obtained by the Hough transform method from the input-output data for the SISO (single-input and single-output) system shown in Fig. 2.4. The parameters of each linear segment are used as the consequent parameters of the fuzzy rules from a TS fuzzy model.

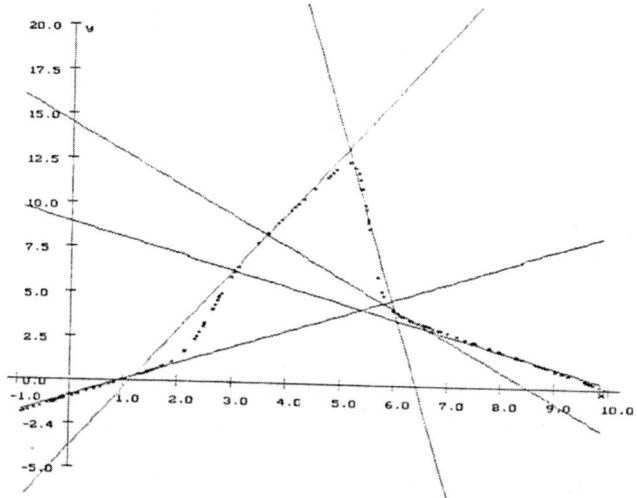

Fig. 2.12. The linear segments obtained by the Hough transform

Input Space Partitioning. The determination of the number of fuzzy rules in a TS fuzzy model requires that the input space should be partitioned. This partitioning can be achieved by using the linear segments from the Hough transform. If the linear segments are selected by setting a fixed threshold, several cells can be selected within a very small area in the parameter space. This can cause an undesirably large number of lines corresponding to a group of input-output data that has a linear characteristic. In order to avoid this problem, the cells at the local maximum points of the accumulator cells (in the Hough transform space, that is, in the parameter space) are chosen as settings for the initial consequent parameters.

The input space is partitioned in two steps: first, assign each sample to its nearest linear segment, and then determine the area occupied by the samples belonging to each such segment. The input space is partitioned according to the sample areas obtained. Once the input space is partitioned, each area is used to obtain a fuzzy rule. However, in partitioning the input space and determining the fuzzy model structure, the continuity of the input-output data should be taken into account, as well as the linearities hidden in this data. The sample groups which are included in one linear segment, but separated from other sample groups, should form different partitions or clusters as shown in Fig. 2.13.

The above is achieved by using an adaptive clustering method. The sample closest to a linear segment is to be included in the initial cluster of this

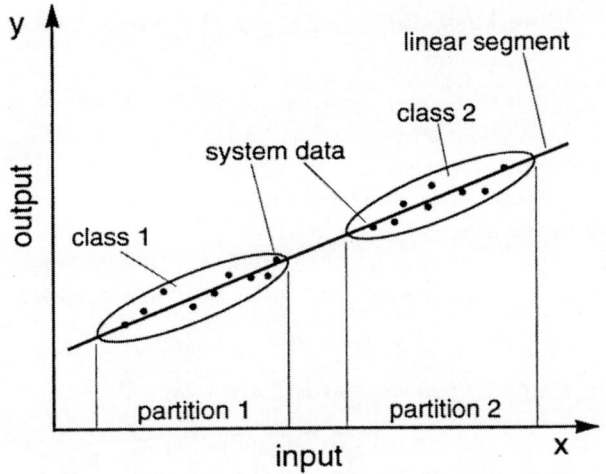

Fig. 2.13. Input space partitioning by clustering

segment. Thus, each segment forms an initial set of clusters. When all the input-output samples are assigned to the initial clusters, the initial cluster centers and deviations are calculated. In the case when the number of clusters and partitions is unknown, an adaptive sample set construction is one of the most common approaches. The clustering approach used in this chapter is as follows [23]:

Step 1: *Take the first sample as representative of the first cluster:*

$$\mathbf{z}_1 = \mathbf{x}_1 \tag{2.12}$$

where $\mathbf{z}_1$ is the first cluster center.

Step 2: *Take the next sample and compute its distance to all the existing clusters.*

1. Assign $\mathbf{x}$ to $\mathbf{z}_i$ if $d_i(\mathbf{x}, \mathbf{z}_i) \leq \theta\tau$, $0 \leq \theta \leq 1$ where τ is the membership boundary for a specified cluster. Its value is set by the user.
2. Do not assign $\mathbf{x}$ to $\mathbf{z}_i$ if $d_i(\mathbf{x}, \mathbf{z}_i) > \tau$.
3. No decision will be made on $\mathbf{x}$ if $\mathbf{x}$ falls in the intermediate region for $\mathbf{z}_i$.

Step 3: *Each time a new $\mathbf{x}$ is assigned to $\mathbf{z}_i$, compute $\mathbf{z}_i(t+1)$ and $\mathbf{C}(t+1)$* according to the expressions

$$\begin{aligned}
\mathbf{z}_i(t+1) &= (1/(t+1))[t\mathbf{z}_i(t) + \mathbf{x}]\mathbf{C}_i(t+1) \\
&= (1/(t+1))[t\mathbf{C}(t) + (\mathbf{x} - \mathbf{z}_i(t+1))^2]
\end{aligned}$$

where t is the number of pattern samples already assigned to $\mathbf{z}_i$ and $\mathbf{x}$ is the $(t+1)$-th such sample. $\mathbf{z}_i(t)$ and $\mathbf{C}(t)$, the variance, were already computed from the t samples. Form a new cluster $\mathbf{z}_j$ if $d(\mathbf{x}, \mathbf{z}_i) > \tau$.

Step 4: *Repeat Step 2 and Step 3 until all pattern samples have been assigned.* There might be some reassignment of $\mathbf{x}$'s when all $\mathbf{x}$'s are considered again.

This is because the means and variances have been adjusted with each $\mathbf{x}$ assigned to $\mathbf{z}_i$.

Step 5: *Extend to a larger number of pattern samples.* After the training is considered complete (this means that $\mathbf{x}$ no longer changes its class belonging, or some number of $\mathbf{x}$'s remain unassigned each time), we can perform clustering on a larger number of pattern samples. No indeterminate region will exist this time. All $\mathbf{x}$'s falling in the indeterminate region may be assigned to the nearest class according to the minimum distance rule. All those $\mathbf{x}$'s can be considered unclassified if their respective distances to all cluster centers are greater than τ.

The benefits of using the Hough transform method are twofold:

1. The initial number of clusters is automatically specified by the number of linear segments obtained by the Hough transform.
2. Since the Hough transform can automatically find out most likely hyperplanes from the input-output data, it can solve the problem of setting the initial values of the linear expressions in order to find out the linear hyperplanes in the c-varieties method and HPC-MEANS approach.

2.2 Antecedent Parameter Setting

After the input space is partitioned using the clustering method, the centers and variances of the clusters are used to form the antecedent parameters of the TS fuzzy model. Here, all membership functions are Gaussian functions (bell-type) defined as

$$\mu_{A_j^i}(x) = \exp\left\{-\left(\frac{x - p_{j1}^i}{p_{j2}^i}\right)^2\right\} \tag{2.13}$$

where p_{j1}^i is the center of the cluster corresponding to the j-th linguistic value in the i-th fuzzy rule and $p_{j2}^i/\sqrt{2}$ is the standard deviation of the cluster.

2.3 Fine Tuning

After the initial identification of the structure and parameters of the TS fuzzy model the parameters are adjusted so that the error between the model output and the available input-output data is decreased. We suggest Theorem 1 and Theorem 2, derived from gradient descent method, as an appropriate basis for the algorithms used to fine tune the parameters of the TS fuzzy model. Theorem 1 serves as the basis for the fine tuning algorithm for the antecedent parameters, Theorem 2 for the fine tuning of the consequent parameters.

Theorem 1 If the fuzzy model is represented by (2.14) and its output is obtained by (2.15):

> $\mathbf{R}^i$: **If** x_1 is $A_1^i(p_{11}^i, p_{12}^i)$ **and** ... **and** x_m is $A_m^i(p_{m1}^i, p_{m2}^i)$
> **then** $y^i = a_0^i + a_1^i x_1 + \cdots + a_m^i x_m.$ (2.14)

$$\hat{y} = \frac{\sum_{i=1}^{c} \beta^i y^i}{\sum_{i=1}^{c} \beta^i}, \quad \beta^i = \prod_{j=1}^{m} \mu_{A_j^i}(x_j) \tag{2.15}$$

where p_{jk}^i is the k-th ($k = 1, 2$) parameter of the membership function defining the meaning of the linguistic value $A_j{}^i$, then the antecedent parameters of the TS fuzzy model can be fine tuned as follows

$$p_{jk}^i(t+1) = p_{jk}^i(t) + \Delta p_{jk}^i \tag{2.16}$$

$$\Delta p_{jk}^i = \gamma(y_{\mathrm{des}} - \hat{y})(y^i - \hat{y}) \frac{1}{\sum_{i=1}^{m} \beta^i} \frac{\partial \beta^i}{\partial p_{jk}^i} \tag{2.17}$$

where γ denotes the learning rate, y_{des} denotes the desired output and $\hat{y}$ denotes the output obtained from a TS fuzzy model.

Proof. Given sample data $(\mathbf{x}, y_{\mathrm{des}})$, we define the error e as the difference between y_{des} and $\hat{y}$:

$$e = y_{\mathrm{des}} - \hat{y} = y_{\mathrm{des}} - \frac{\sum_{i=1}^{m} \beta^i y^i}{\sum_{i=1}^{m} \beta^i}. \tag{2.18}$$

The antecedent parameters should be adjusted to reduce the squared error e^2. Therefore, from the gradient descent algorithm we obtain

$$
\begin{aligned}
\Delta p_{jk}^i &= -\gamma \frac{\partial}{\partial p_{jk}^i}\left(\frac{e^2}{2}\right) = -\gamma e \frac{\partial e}{\partial p_{jk}^i} \\[2mm]
&= \gamma(y_{\mathrm{des}} - \hat{y}) \frac{\frac{\partial}{\partial p_{jk}^i}(\sum_{i=1}^{c} \beta^i y^i) \times \sum_{i=1}^{c} \beta^i - \sum_{i=1}^{c} \beta^i y^i \times \frac{\partial}{\partial p_{jk}^i}(\sum_{i=1}^{c} \beta^i)}{(\sum_{i=1}^{c} \beta^i)^2} \\[2mm]
&= \gamma(y_{\mathrm{des}} - \hat{y}) \frac{\frac{\partial \beta^i}{\partial p_{jk}^i} y^i \times \sum_{i=1}^{c} \beta^i - \sum_{i=1}^{c} \beta^i y^i \times \frac{\partial \beta^i}{\partial p_{jk}^i}}{(\sum_{i=1}^{c} \beta^i)^2}
\end{aligned}
$$

$$= \gamma(y_{\text{des}} - \hat{y}) \frac{\partial \beta^i}{\partial p^i_{jk}} \left(\frac{y^i}{\sum\limits_{i=1}^{c} \beta^i} - \frac{\sum\limits_{i=1}^{c} \beta^i y^i}{\sum\limits_{i=1}^{c} \beta^i} \frac{1}{\sum\limits_{i=1}^{c} \beta^i} \right)$$

$$= \gamma(y_{\text{des}} - \hat{y}) \frac{\partial \beta^i}{\partial p^i_{jk}} \frac{1}{\sum\limits_{i=1}^{c} \beta^i} (y^i - \hat{y})$$

$$= \gamma(y_{\text{des}} - \hat{y})(y^i - \hat{y}) \frac{1}{\sum\limits_{i=1}^{c} \beta^i} \frac{\partial \beta^i}{\partial p^i_{jk}}. \tag{2.19}$$

$\square$

The calculation of (2.17) depends on the type of membership function used. If the Gaussian function (2.13) is employed as a membership function, the learning method for the antecedent parameters is given as

$$\frac{\partial \beta^i}{\partial p^i_{j1}} = \frac{\partial A^i_j}{\partial p^i_{j1}} = \frac{2}{p^i_{j2}} \frac{x_j - p^i_{j1}}{p^i_{j2}} \exp\left\{ -\left(\frac{x_j - p^i_{j1}}{p^i_{j2}} \right)^2 \right\} \tag{2.20}$$

$$\frac{\partial \beta^i}{\partial p^i_{j2}} = \frac{\partial A^i_j}{\partial p^i_{j2}} = \frac{2}{p^i_{j2}} \left(\frac{x_j - p^i_{j1}}{p^i_{j2}} \right)^2 \exp\left\{ -\left(\frac{x_j - p^i_{j1}}{p^i_{j2}} \right)^2 \right\}$$

$$= \frac{x_j - p^i_{j1}}{p^i_{j2}} \frac{\partial \beta^i}{\partial p^i_{j1}}. \tag{2.21}$$

Theorem 2 If the TS fuzzy model is represented by (2.14) and its output is obtained by (2.15), then the consequent parameters can be fine tuned as follows

$$a^i_j(t+1) = a^i_j(t) + \Delta a^i_j \tag{2.22}$$

$$\Delta a^i_j = \gamma(y_{\text{des}} - \hat{y}) \frac{1}{\sum\limits_{i=1}^{c} \beta^i} \beta^i x_j \tag{2.23}$$

where γ denotes learning rate, y_{des} denotes desired output and $\hat{y}$ denotes output to be obtained from the fuzzy control.

Proof. The consequent parameters can be adjusted to reduce the squared error e^2 in the same way as this was done in the case of fine tuning the antecedent parameters by a gradient descent algorithm. That is

$$\Delta a^i_j = -\gamma \frac{\partial}{\partial a^i_j} \left(\frac{e^2}{2} \right) = -\gamma e \frac{\partial e}{\partial a^i_j}$$

$$= -\gamma(y_{\text{des}} - \hat{y})\frac{\partial}{\partial a_j^i}(y_{\text{des}} - \hat{y})$$

$$= \gamma(y_{\text{des}} - \hat{y})\frac{1}{\sum_{i=1}^{c}\beta^i}\beta^i x_j. \qquad (2.24)$$

$\square$

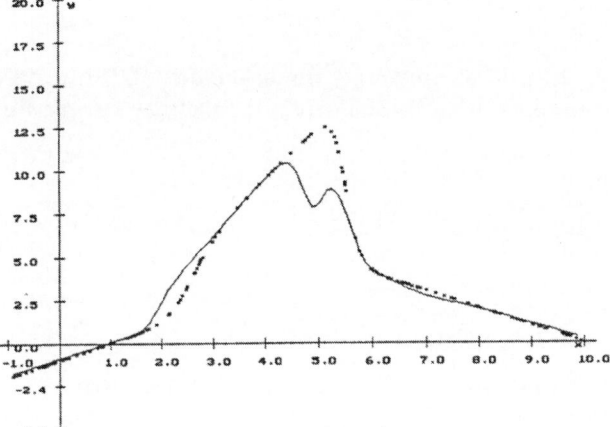

Fig. 2.14. The input/ output of the fuzzy model before fine tuning

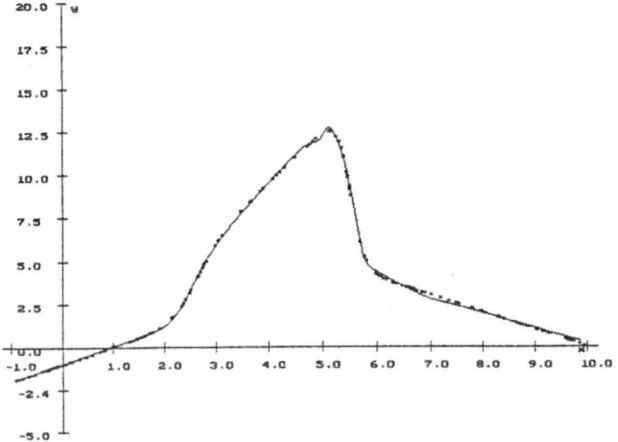

Fig. 2.15. The input/ output of the fuzzy model after fine tuning

Figure 2.14 illustrates the input and output of the TS fuzzy model constructed after initial parameter setting by means of the Hough transform and clustering method with the input-output data for the SISO system shown in Fig. 2.4. This figure shows that there is some error between the original system and the coarse TS fuzzy model. To minimize the error and adjust the TS fuzzy model parameters, fine tuning is performed by the gradient descent method (2.16)–(2.17) and (2.22)–(2.23). During the learning procedure, the antecedent and consequent parameters converge and the error decreases. Figure 2.15 is the input/output graph of the TS fuzzy model tuned by the gradient descent algorithm.

3. Example 1

The nonlinear system below, with two input and single output (DISO), described in the form of fuzzy rules, is used as the original fuzzy system to be modeled.

$$
\begin{aligned}
R_1 : \quad &\textbf{If } x_1 \text{ is } A_1^1(1.5, 1.0) \textbf{ and } x_2 \text{ is } A_2^1(1.5, 1.0) \\
&\textbf{then } y = -1.0 - 1.0x_1 + 4.0x_2 \\[2mm]
R_2 : \quad &\textbf{If } x_1 \text{ is } A_1^2(4.5, 1.0) \textbf{ and } x_2 \text{ is } A_2^2(1.5, 1.0) \\
&\textbf{then } y = -3.0 - 4.0x_1 + 5.0x_2
\end{aligned}
\tag{3.1}
$$

The input-output data are obtained from (3.1). They are shown in Table 3.1 and used in the identification. Figure 3.1 is a graphical representation of the data obtained from the given nonlinear fuzzy system.

After 500 iterations, the antecedent and consequent parameters converge with learning error rate γ of 0.00002. The resulting TS fuzzy model is as follows

$$
\begin{aligned}
R_1 : \quad &\textbf{If } x_1 \text{ is } A_1^1(1.506, 1.046) \textbf{ and } x_2 \text{ is } A_2^1(1.513, 1.073) \\
&\textbf{then } y = -0.995 - 1.010x_1 + 4.001x_2 \\[2mm]
R_2 : \quad &\textbf{If } x_1 \text{ is } A_1^1(4.015, 1.015) \textbf{ and } x_2 \text{ is } A_2^1(1.498, 1.078) \\
&\textbf{then } y = -3.313 - 3.942x_1 + 5.009x_2
\end{aligned}
\tag{3.2}
$$

The identification results are illustrated in Fig. 3.2. Figure 3.3 shows the error between the outputs of the identified TS fuzzy model and those of the original fuzzy system. MSE (mean squared error) decreases gradually and converges to a very small value during fine tuning.

$$
\text{MSE} = \bar{e^2} = \frac{1}{m} \sum_{i=1}^{m} \{y(i) - \hat{y}(i)\}^2
\tag{3.3}
$$

where $y(i)$ is the system data and $\hat{y}_i$ is the model output. The behavior of the MSE is shown in Fig. 3.4 and the MSE of the resulting TS fuzzy model, after 500 iterations of fine tuning, is 0.0098.

Table 3.1. The input-output data used in the identification

No.	x_1	x_2	y_{des}	$\hat{y}$	e^2
1	0.0000	0.0000	-1.0000	-0.9950	0.00003
2	0.0000	0.4375	0.7500	0.7554	0.00003
3	0.0000	0.8750	2.5000	2.5059	0.00003
4	0.0000	1.3125	4.2500	4.2563	0.00004
5	0.0000	1.7500	6.0000	6.0068	0.00005
6	0.0000	2.1875	7.7500	7.7572	0.00005
7	0.0000	2.6250	9.5000	9.5076	0.00006
8	0.0000	3.0625	11.2500	11.2581	0.00007
9	0.0000	3.5000	13.0000	13.0085	0.00007
10	0.0000	3.9375	14.7500	14.7589	0.00008
11	0.0000	4.3750	16.5000	16.5094	0.00009
12	0.0000	4.8125	18.2500	18.2598	0.00010
13	0.0000	5.2500	20.0000	20.0103	0.00011
14	0.0000	5.6875	21.7500	21.7607	0.00011
15	0.0000	6.1250	23.5000	23.5111	0.00012
16	0.0000	6.5625	25.2500	25.2616	0.00013
17	0.4375	0.0000	-1.4375	-1.4369	0.00000
18	0.4375	0.4375	0.3125	0.3136	0.00000
19	0.4375	0.8750	2.0625	2.0640	0.00000
20	0.4375	1.3125	3.8125	3.8144	0.00000
21	0.4375	1.7500	5.5625	5.5649	0.00001
22	0.4375	2.1875	7.3125	7.3153	0.00001
23	0.4375	2.6250	9.0625	9.0657	0.00001
24	0.4375	3.0625	10.8125	10.8162	0.00001
25	0.4375	3.5000	12.5625	12.5666	0.00002
26	0.4375	3.9375	14.3125	14.3171	0.00002
27	0.4375	4.3750	16.0625	16.0675	0.00002
28	0.4375	4.8125	17.8125	17.8179	0.00003
29	0.4375	5.2500	19.5625	19.5684	0.00003
30	0.4375	5.6875	21.3125	21.3188	0.00004
31	0.4375	6.1250	23.0625	23.0693	0.00005
32	0.4375	6.5625	24.8125	24.8197	0.00005
33	0.8750	0.0000	-1.8750	-1.8788	0.00001
34	0.8750	0.4375	-0.1250	-0.1283	0.00001
35	0.8750	0.8750	1.6250	1.6221	0.00001
36	0.8750	1.3125	3.3750	3.3726	0.00001
37	0.8750	1.7500	5.1250	5.1230	0.00000
38	0.8750	2.1875	6.8750	6.8734	0.00000
39	0.8750	2.6250	8.6250	8.6239	0.00000
40	0.8750	3.0625	10.3750	10.3743	0.00000
41	0.8750	3.5000	12.1250	12.1247	0.00000
42	0.8750	3.9375	13.8750	13.8752	0.00000
43	0.8750	4.3750	15.6250	15.6256	0.00000
44	0.8750	4.8125	17.3750	17.3761	0.00000
45	0.8750	5.2500	19.1250	19.1265	0.00000
46	0.8750	5.6875	20.8750	20.8769	0.00000
47	0.8750	6.1250	22.6250	22.6274	0.00001
48	0.8750	6.5625	24.3750	24.3778	0.00001
49	1.3125	0.0000	-2.3127	-2.3209	0.00007
50	1.3125	0.4375	-0.5627	-0.5705	0.00006
51	1.3125	0.8750	1.1873	1.1800	0.00005
52	1.3125	1.3125	2.9373	2.9305	0.00005
53	1.3125	1.7500	4.6873	4.6809	0.00004
54	1.3125	2.1875	6.4373	6.4314	0.00004
55	1.3125	2.6250	8.1874	8.1818	0.00003
56	1.3125	3.0625	9.9374	9.9323	0.00003
57	1.3125	3.5000	11.6874	11.6828	0.00002
⋮	⋮	⋮	⋮	⋮	⋮
254	6.5625	5.6875	-0.8125	-0.3927	0.17624
255	6.5625	6.1250	1.3750	1.7987	0.17956
256	6.5625	6.5625	3.5625	3.9902	0.18292

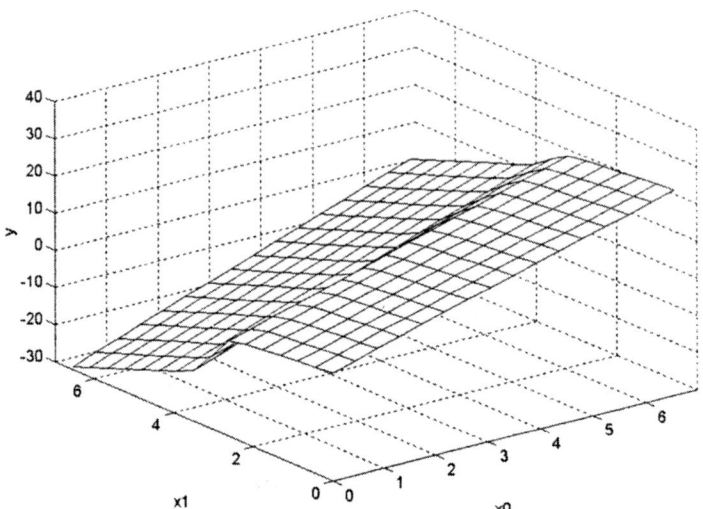

Fig. 3.1. The input/output of the given fuzzy system

4. Example 2

In this example, we apply the proposed fuzzy modeling scheme to the force control of an n-link robot manipulator. In the conventional approach, the contact force exerted on the environment is modeled as a linear spring. In this example, we use instead a nonlinear relationship between contact force and manipulator position and represent it in the form of a TS fuzzy model. By controlling the manipulator's position and specifying its relationship to the contact force, we can ensure that the manipulator is able to maneuver in a constrained environment while maintaining appropriate contact force. The manipulator's dynamic equation can be written as

$$
\begin{aligned}
\mathbf{T} &= \mathbf{M}(\mathbf{q})\ddot{\mathbf{q}} + \mathbf{V}(\mathbf{q},\dot{\mathbf{q}}) + \mathbf{G}(\mathbf{q}) + \mathbf{F}(\dot{\mathbf{q}}) + \mathbf{J}^T(\mathbf{q})\mathbf{f} \\
&= \mathbf{M}(\mathbf{q})\ddot{\mathbf{q}} + \mathbf{V}_m(\mathbf{q},\dot{\mathbf{q}})\dot{\mathbf{q}} + \mathbf{G}(\mathbf{q}) + \mathbf{F}(\dot{\mathbf{q}}) + \mathbf{J}^T(\mathbf{q})\mathbf{f}
\end{aligned} \tag{4.1}
$$

where $\mathbf{q}$ is an n-dimensional vector of joint angles and $\mathbf{T}$ is the n-dimensional vector of generalized forces. $\mathbf{M}(\mathbf{q})$ denotes the inertia matrix, $\mathbf{V}(\mathbf{q},\dot{\mathbf{q}})$ the coriolis/centripetal vector, $\mathbf{G}(\mathbf{q})$ the gravity vector, and $\mathbf{F}(\dot{\mathbf{q}})$ a friction term. $\mathbf{f}$ is the $n \times 1$ vector of contact forces and torques in task space and $\mathbf{J}(\mathbf{q})$ is an $n \times n$ task Jacobian matrix denoting the relationship between the joint space coordinate system and task space coordinate system.

Equation (4.1) represents the robot manipulator dynamics in a form which includes the environmental interaction forces. A force controller for the n-link robot manipulator can be formulated as follows.

The force exerted on the environment is defined as

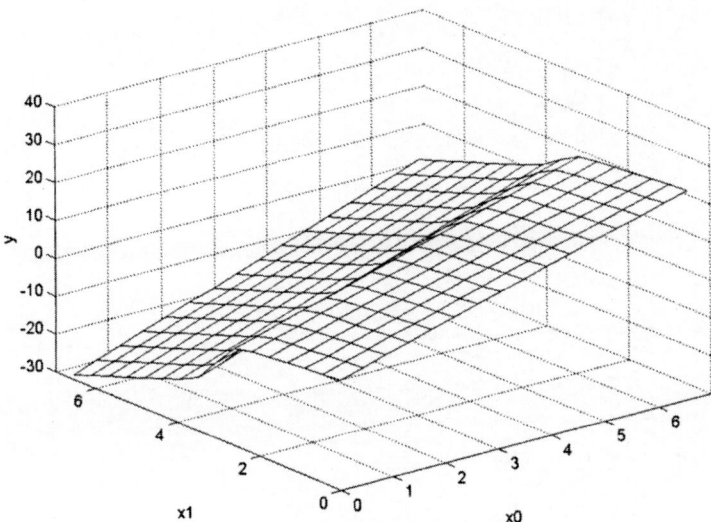

Fig. 3.2. The resulting TS fuzzy model after fine tuning

$$\mathbf{f} = \mathbf{g}(\mathbf{x}) \tag{4.2}$$

where $\mathbf{f} = [f_1, f_2, \cdots, f_n]^T$ and $\mathbf{x} = [x_1, x_2, \cdots, x_n]^T$. The $n \times 1$-vector function $\mathbf{g}(\mathbf{x}) = [g_1(x_1), g_2(x_2), \cdots, g_n(x_n)]^T$ represents the relationship between the interaction forces and manipulator position and is actually a nonlinear function represented by the following fuzzy model:

$$R_k^i : \textbf{If } x_k \textbf{ is } A_k^i \textbf{ then } g_k^i = a_{k0}^i x_k + a_{k1}^i \quad (\text{for } i = 1, 2, \ldots, c) \tag{4.3}$$

$$f_k = g_k(x_k) = \frac{\sum\limits_{i=1}^{c} \mu A_k^i(x_k) g_k^i}{\sum\limits_{i=1}^{c} \mu A_k^i(x_k)} \quad (\text{for } k = 1, 2, \ldots, n) \tag{4.4}$$

where R_k^i represents the i-th fuzzy rule for k-axis output $f_k = g_k(x_k)$, x_k is an input variable representing k-axis displacement, g_k^i is the output from the i-th fuzzy rule, a_{k0}^i, a_{k1}^i are the consequent parameters of the TS fuzzy model, and A_k^i are the linguistic values from the antecedents.

The multi-dimensional force controller is the PD-type controller

$$\mathbf{T} = \mathbf{J}^T(\mathbf{q})(-\mathbf{K}_v \dot{\mathbf{x}} + \mathbf{K}_p \tilde{\mathbf{x}}) + \mathbf{G}(\mathbf{q}) + \mathbf{F}(\dot{\mathbf{q}}) \tag{4.5}$$

where $\mathbf{K}_v$ and $\mathbf{K}_p$ are $n \times n$ diagonal, constant, positive definite matrices, and the task space tracking error is defined as

$$\tilde{\mathbf{x}} = \mathbf{x}_d - \mathbf{x} \tag{4.6}$$

where $\mathbf{x}_d$ is an $n \times 1$ vector that is used to denote the desired constant end-effector position.

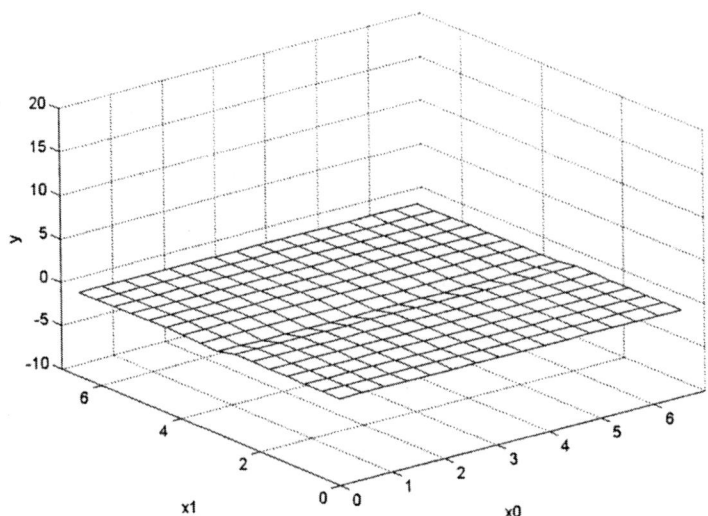

Fig. 3.3. The error between the original system and the TS fuzzy model

The substitution of (4.2) and (4.5) in (4.1) yields the closed-loop dynamics

$$\mathbf{M}(\mathbf{q})\ddot{\mathbf{q}} + \mathbf{V}_m(\mathbf{q}, \dot{\mathbf{q}})\dot{\mathbf{q}} = \mathbf{J}^T(\mathbf{q})\left[-\mathbf{K}_v \dot{\mathbf{x}} + \mathbf{K}_p \tilde{\mathbf{x}} - \mathbf{g}(\mathbf{x})\right]. \qquad (4.7)$$

In steady state manipulator position we have that $\dot{\mathbf{q}} = 0$ and hence $\ddot{\mathbf{q}} = 0$. By substituting $\dot{\mathbf{q}} = 0$ and $\ddot{\mathbf{q}} = 0$ in (4.7) we obtain

$$\lim_{t \to \infty}\left[\mathbf{K}_p \tilde{\mathbf{x}} - \mathbf{g}(\mathbf{x})\right] = 0. \qquad (4.8)$$

Therefore, the steady state manipulator position satisfies the equation

$$\mathbf{x} = \mathbf{x}_d - \mathbf{K}_p^{-1}\mathbf{g}(\mathbf{x}). \qquad (4.9)$$

To obtain the steady state force exerted on the environment, we substitute (4.9) into (4.2) to yield

$$g_i\left(x_{di} - \frac{f_{di}}{K_{pi}}\right) = f_{di}. \qquad (4.10)$$

In the above, f_d is the steady state force exerted on the environment and the subscript i is used to represent the i-th axis component of $\mathbf{x}_d$, $\mathbf{K}_p$, and $\mathbf{f}_d$. Because $\mathbf{K}_p$ is diagonal, (4.9) can be converted into (4.10). Using (4.10), we can obtain a desired position from a desired force and the desired force is created by commanding a desired position that is slightly inside the contact force.

We implement a force controller for the robot manipulator system given in Fig. 4.1 using the fuzzy model. The task space coordinate system is given by the directions u and v since we wish to move the end-effector along the

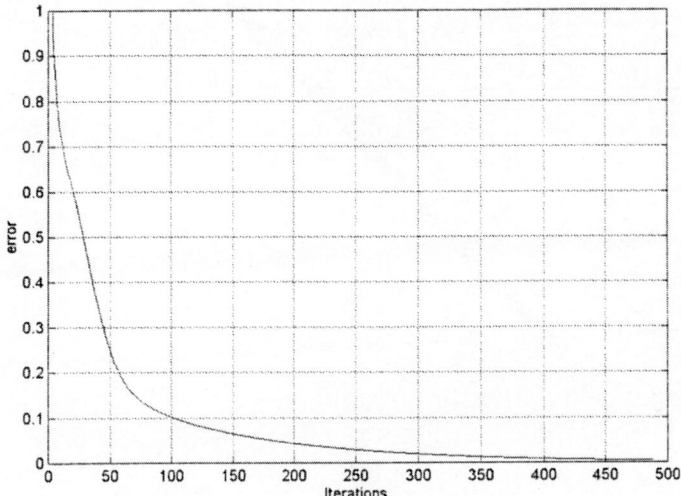

Fig. 3.4. MSE during fine tuning

surface in the direction v while applying a force normal to the surface along the direction u. The control objective is to move the end-effector to a desired final position of $v_d = 2$ m while exerting a final desired normal force of $f_{d1} = 1000$ unit force (uf). We neglect the surface friction and joint friction. The robot link masses are assumed to be unity and the initial end-effector position is given by

$$v(0) = 0 \text{ m}, \quad u(0) = 0 \text{ m}. \tag{4.11}$$

$\mathbf{T}$, $\mathbf{q}$, $\mathbf{G}$, $\mathbf{M}$, $\mathbf{J}$, and $\mathbf{x}$ are defined as

$$\mathbf{T} = \begin{bmatrix} T_1 \\ T_2 \end{bmatrix}, \ \mathbf{q} = \begin{bmatrix} q_1 \\ q_2 \end{bmatrix}, \ \mathbf{G} = \begin{bmatrix} 0 \\ (m_1 + m_2)g \end{bmatrix}$$
$$\mathbf{M} = \begin{bmatrix} m_1 & 0 \\ 0 & m_1 + m_2 \end{bmatrix}, \ \mathbf{J} = \frac{1}{\sqrt{2}} \begin{bmatrix} 1 & -1 \\ 1 & 1 \end{bmatrix}, \ \mathbf{x} = \begin{bmatrix} u \\ v \end{bmatrix}. \tag{4.12}$$

To accomplish the control objective, the force controller is given by

$$\mathbf{T} = \mathbf{J}^T(\mathbf{q})(-\mathbf{K}_v \dot{\mathbf{x}} + \mathbf{K}_p \tilde{\mathbf{x}}) + \hat{\mathbf{G}}(\mathbf{q}) \tag{4.13}$$

where $\mathbf{x} = \begin{bmatrix} u \\ v \end{bmatrix}$ and $\tilde{\mathbf{x}} = \begin{bmatrix} u_d - u \\ v_d - v \end{bmatrix}$. The gain matrices K_v and K_p have been taken as $\mathbf{K}_v = K_v \mathbf{I}$ and $\mathbf{K}_p = K_p \mathbf{I}$. For the purpose of simulation, we have selected $K_p = 1000$ and $K_v = 80$.
Assume that the normal force satisfies the relationship

$$f_1 = g(u) \tag{4.14}$$

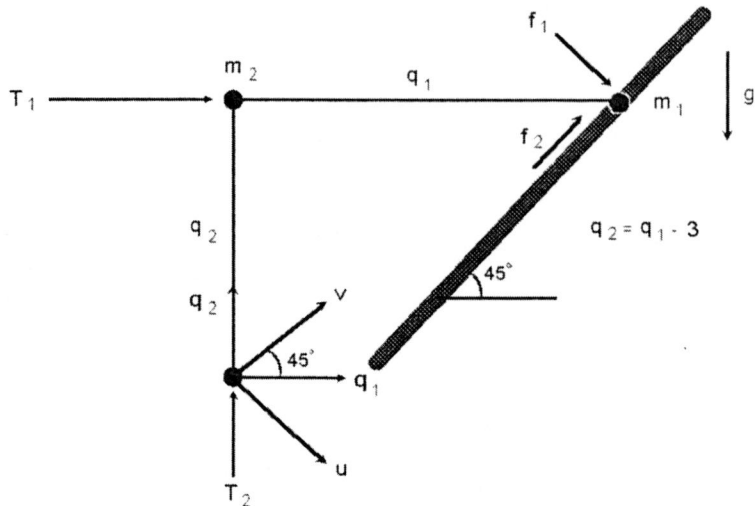

Fig. 4.1. Manipulator moving along slanted surface

where u is the normal distance to the surface and $g(u)$ is the relation between force and distance obtained from the TS fuzzy model. Figure 4.2 shows the relationship between interaction forces and manipulation position that is obtained from experiments. As shown in Fig. 4.2, the environment becomes progressively stiffer as interaction forces and manipulator position deflect and the contact force that is exerted on the environment by the link is not linear.

Figure 4.3 shows the linear segments obtained by the Hough transform method from the measured force-displacement data. Figure 4.4 is the input and output of the constructed fuzzy model after initial parameter setting by means of the Hough transform and clustering method applied to the force-displacement data shown in Fig. 4.2. This figure shows that there is some degree of error between the original system and the coarsely modeled system. To minimize error and adjust the system parameters, fine tuning is performed by the gradient descent method. During the learning procedure, the antecedent and consequent parameters converge and the error decreases. Figure 4.5 is the input/output graph of the TS fuzzy model fine tuned by the gradient descent algorithm. In Figs. 4.2–4.5 units of the x-axis and y-axis are mm and unit force (uf), respectively.

The identified TS fuzzy model is

$$
\begin{array}{lll}
R_1: & \text{If } x \text{ is } A^1(15.64, 3, 39), & \text{then } y = -4397.58 + 461.55x \\
R_2: & \text{If } x \text{ is } A^2(12.92, 1, 95), & \text{then } y = -1142.41 + 182.31x \\
R_3: & \text{If } x \text{ is } A^3(0.79, 3.37), & \text{then } y = -565.1 + 122.01x \\
R_4: & \text{If } x \text{ is } A^4(2.09, 5.82), & \text{then } y = -49.9 + 71.2x
\end{array} \tag{4.15}
$$

where $\mu_{A^i}(p_1^i, p_2^i) = \exp\{-(\frac{x - p_1^i}{p_2^i})^2\}$.

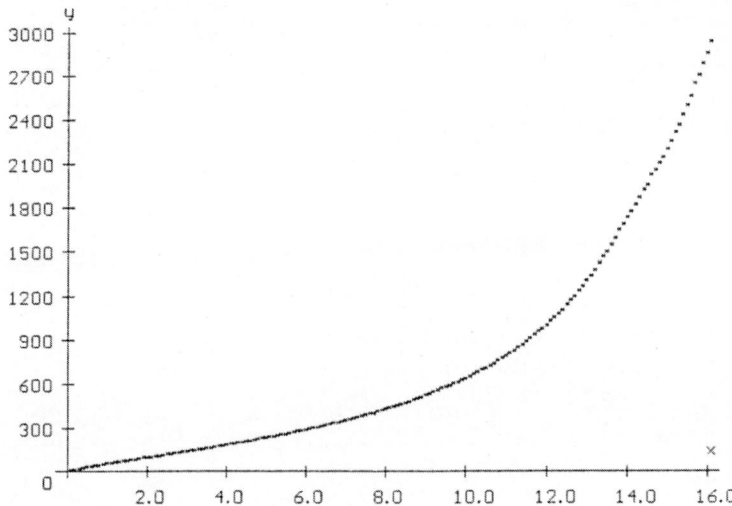

Fig. 4.2. Measured force/displacement data

To satisfy the control objective $f_{d1} = 1000$ uf, we utilize (4.10). Substituting the values of f_{d1} and K_p in (4.10) yields the desired normal distance to the surface $u_d = 12$ mm. The simulation results for the force controller are given in Fig. 4.6. It shows that the manipulator's force is controlled successfully using the TS fuzzy model of the environment.

5. Summary of the Identification Procedure

The proposed identification method can be summarized as follows:

Step 1: *Determination of the input variables.*

Step 2: *Identification of the consequent parameters.* Two strategies, HPC-MEANS and Hough transform and clustering, are used to cluster the available input-output data.

Step 3: *Determination of the number of clusters.* In HPC-MEANS, an appropriate number of clusters is determined using the performance index (MSE). In the Hough transform, the initial number of clusters is automatically specified by the number of linear segments obtained by the Hough transform. The final number of clusters is determined by merging and splitting the clusters using the adaptive sample set construction.

Step 4: *Setting of antecedent parameters.* Using the results of clustering, antecedent parts are determined and TS fuzzy model rules are constructed.

Step 5: *Fine tuning.* Antecedent and consequent parameters are fine tuned using the gradient descent method applied after coarse tuning.

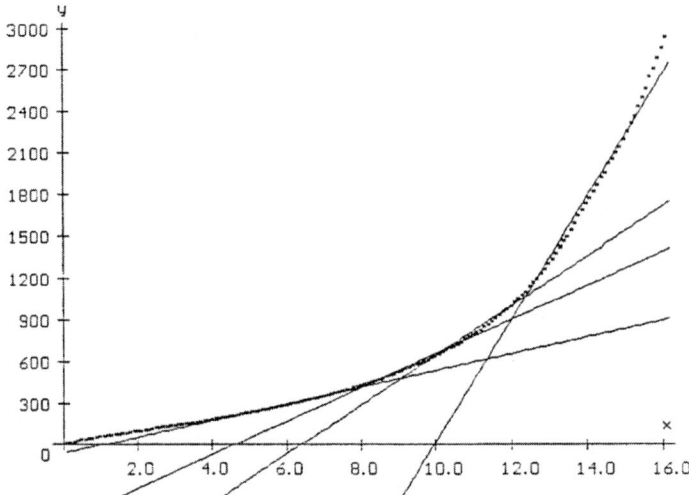

Fig. 4.3. Line segments obtained by Hough transform

6. Practical Considerations and Concluding Remarks

This section summarizes the advantages and drawbacks of the present identification method.

6.1 Use of Prior Knowledge

Prior knowledge about the input and output variables of the model is needed to select the structure of the TS fuzzy model. The Hough transform method then can be applied to specify automatically the initial number of clusters. The final number of cluster is determined using a membership boundary specified by the user. The major advantages and drawbacks of the proposed method are:

1. While general clustering methods have difficulty in setting the initial number of clusters, the Hough transform solves this problem by automatically specifying the number of linear segments.
2. Using the Hough transform method it is easy to identify the consequent structure and parameters without *a priori* knowledge since the method can automatically find most likely hyperplanes only from the input-output data.
3. The problem situation in which two distant data samples are mistakenly perceived to be in the same cluster, derived by HPC-MEANS and c-varieties methods, is avoided here because the adaptive sample set construction approach is used to split and merge clusters taking into account continuity of input-output data.

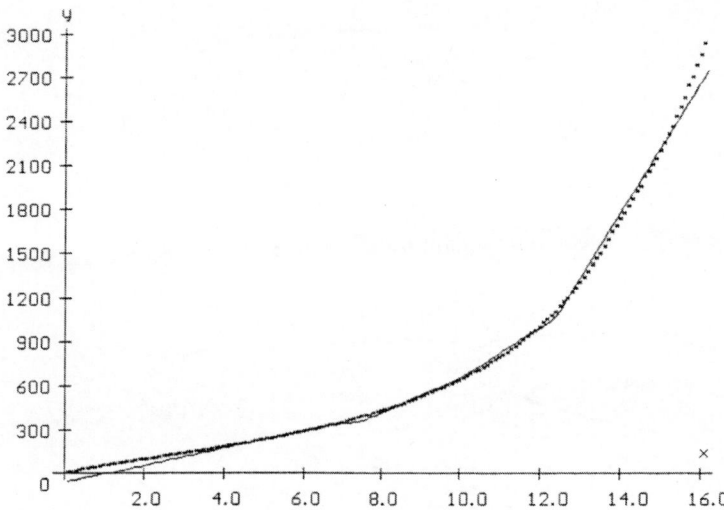

Fig. 4.4. TS fuzzy model before fine tuning

4. In the proposed algorithm, fine tuning is carried out after coarse tuning to minimize mean squared error. This is quite time consuming, but the result is a much better TS fuzzy model.

6.2 Model Complexity

The proposed method is much simpler when compared to the original Takagi and Sugeno approach to identification. The reason for this is that the consequent parameters are identified first and then antecedent parts are determined as opposed to the original Takagi and Sugeno method. However, the proposed method becomes complex at the point at which fine tuning begins in order to minimize the mean square error. All membership functions used in this paper are Gaussian functions and the identified TS fuzzy model provides output for regions that do not contain any data.

6.3 Robustness of the Identification Method

In order to identify piecewise linear consequents we employ the Hough transform method. This technique is less sensitive to noise and disturbances than other approaches because the Hough transform is a method for finding the most likely hyperplanes from input-output data. However, we do not take into consideration the correlations among the input data and rather address them independently, which results in an ineffective partitioning of input space.

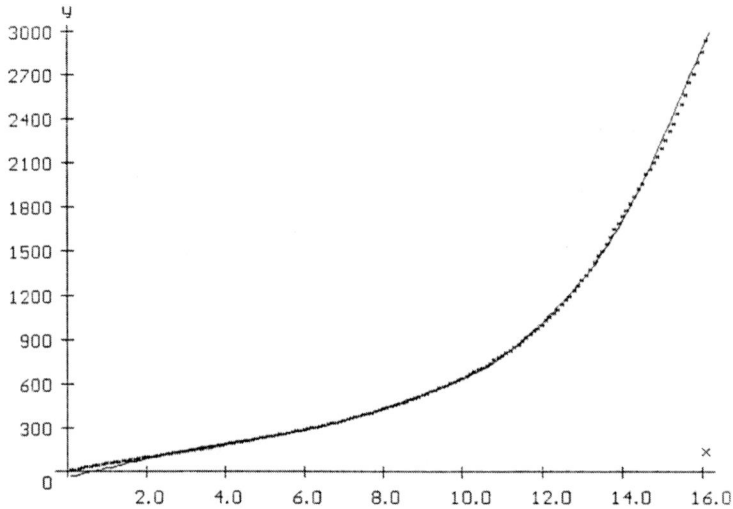

Fig. 4.5. TS fuzzy model after fine tuning

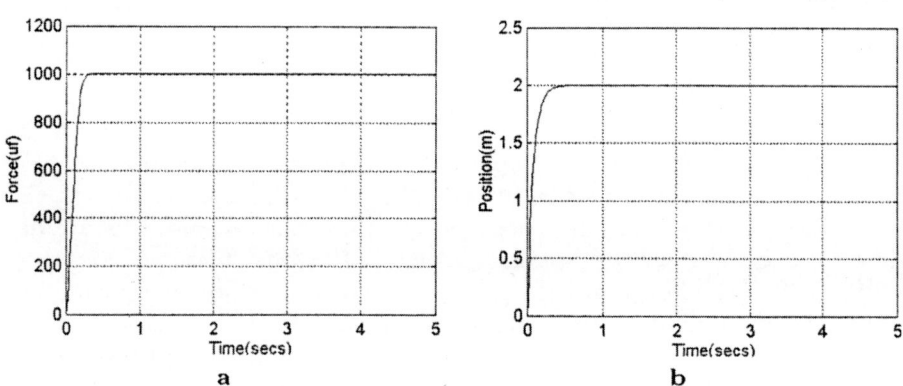

Fig. 4.6. Simulation for the force controller

6.4 Real-World Applications

The proposed identification method is applied to the modeling of Box and Jenkins furnace data and a camera calibration problem.

Acknowledgements

This work was supported by the MIC of Korea.

References

1. J. C. Bezdek, Fuzzy models – what are they, and why?, *IEEE Trans. Fuzzy Systems*, **1**(1)(1993)1–6.
2. T. Takagi and M. Sugeno, Fuzzy identification of systems and its applications to modeling and control, *IEEE Trans. Systems, Man and Cybernetics*, **15**(1)(1985)116–132.
3. M. Sugeno and G. T. Kang, Structure identification of fuzzy model, *Fuzzy Sets and Systems*, **28**(1988)15–33.
4. M. Sugeno and T. Yasukawa, Fuzzy-logic-based approach to qualitative modeling, *IEEE Trans. Fuzzy Systems*, **1**(1)(1993)7–31.
5. M. Sugeno and K. Tanaka, Successive identification of a fuzzy model and its applications to prediction of a complex system, *Fuzzy Sets and Systems*, **42**(1991)315–334.
6. J. C. Bezdek et al., Detection and characterization of cluster substructure I. Linear structure: fuzzy c-lines, *SIAM J. Appl. Math.*, **40**(2)(1981)339–357.
7. J. C. Bezdek et al., Detection and characterization of cluster substructure II. Fuzzy c-varieties and convex combinations thereof, *SIAM J. Appl. Math.*, **40**(2)(1981)358–372.
8. R. M. Tong, The evaluation of fuzzy models derived from experimental data, *Fuzzy Sets and Systems*, **4**(1980)1–12.
9. W. Pedrycz, An identification algorithm in fuzzy relational systems, *Fuzzy Sets and Systems*, **13**(1984)153–167.
10. K. Tanaka and M. Sugeno, Stability analysis and design of fuzzy control systems, *Fuzzy Sets and Systems*, **45**(1992)136–156.
11. Sam-Kit Sin and Rui J. P. de Figueiredo, Fuzzy system design through fuzzy clustering and optimal predefuzzification, *2nd IEEE Int. Conf. on Fuzzy Systems*, pp. 190–195, 1993.
12. Li-Xin Wang, *Adaptive fuzzy systems and control: design and stability analysis*, Prentice-Hall International, pp. 83–99.
13. D. H. Ballard, Generating the Hough transform to detect arbitrary shapes, *Pattern Recognition*, **13**(2)(1981)111–122.
14. K. Jajuga, Linear fuzzy regression, *Fuzzy Sets and Systems*, **20**(1986)343–354.
15. W. Pedrycz, An identification algorithm in fuzzy relational systems, *Fuzzy Sets and Systems*, **13**(1984)153–167.
16. R. M. Tong, The evaluation of fuzzy models derived from experimental data, *Fuzzy Sets and Systems*, **4**(1980)1–12.
17. S. Horikawa, T. Furuhashi and Y. Uchikawa, On fuzzy modeling using fuzzy neural networks with the back-propagation algorithm, *IEEE Trans. Neural Networks*, **3**(5)(1992)801–806.
18. Y. Lin and G. A. Cunningham III, A new approach to fuzzy-neural modeling, *IEEE Trans. Fuzzy Systems*, **3**(2)(1995)190–197.
19. C. W. Xu and Y. Z. Lu, Fuzzy model identification and self-learning for dynamic systems, *IEEE Trans. Systems, Man, and Cybernetics*, **17**(1987)683–689.
20. L. Wang and R. Langari, Building Sugeno-type models using fuzzy discretization and orthogonal parameter estimation techniques, *IEEE Trans. Fuzzy Systems*, **3**(4)(1995)454–458.
21. L. Ljung, *System identification: Theory for the user*, Prentice-Hall, 1987.
22. Rafael C. Gonzalez and Paul Wintz, *Digital image processing*, Addison Wesley, 2nd edition, 1987.
23. Sing-Tze Bow, *Pattern recognition and image preprocessing*, Dekker, 1992.
24. L. A. Zadeh, Fuzzy algorithms, *Information and Control*, **12**(2)(1968)94–102.

Rapid Prototyping of Fuzzy Models Based on Hierarchical Clustering

M. Delgado[1], M.A. Vila[1], and A.F. Gomez-Skarmeta[2]

[1] Dept. of Computer Science and Artificial Intelligence, E.T.S.I. Informatica, University of Granada, E-18071 Granada, Spain
[2] Dept. Informatics and Systems, University of Murcia, Murcia, Spain

1. Introduction

Throughout this chapter we consider the identification of a MISO (multi-input/single-output) system. The vector of system inputs $\mathbf{x}$ has p components, i.e., $\mathbf{x} = (x_1, \ldots, x_p) \in \mathcal{X} = (X_1 \times X_2 \times \ldots \times X_p) \subseteq \mathbb{R}^p$ and for the system output y we have that $y \in Y \subseteq \mathbb{R}$. The input-output data space is the product space $Z = (\mathcal{X} \times Y)$. Furthermore, a set of sample I/O pairs is denoted as $\Omega = \{((x_{t1}, x_{t2}, \ldots, x_{tp}), y_t), t = 1, \ldots, N\} = \{(x_t, y_t) \in Z, t = 1, \ldots, N\}$.

In the so-called *descriptive approach* to the identification of a fuzzy model the linguistic values for the inputs and outputs of the system under identification, together with their membership functions, belong to a predefined set of linguistic values, i.e., a predefined term set. Thus the set of all possible fuzzy rules is just the combination of all linguistic values, i.e., $R = L_1 \times L_2 \times \cdots L_p \times T$, where L_i is the predefined set of linguistic values for the i-th system input and T is the predefined set of linguistic values for the system output. The identification of a fuzzy model is concerned with the selection of these fuzzy rules that provide the best match with the input-output data. The methods in the descriptive approach try to construct a linguistic fuzzy model, that is a fuzzy model consisting of fuzzy if-then rules like

$$\textbf{If } x_1 \textbf{ is } A^1 \textbf{ and } \ldots \textbf{and } x_p \textbf{ is } A^p \textbf{ then } y \textbf{ is } B \qquad (1.1)$$

where A^i and B are linguistic values from the predefined term sets for the input and output variables. To achieve this goal, neural networks, associative memories, or cluster analysis is used. Here one can mention Pedrycz's work [Ped84] about the use of the standard fuzzy c-means clustering algorithm (FCM) [Bez81].

In the *approximative approach* to fuzzy model identification there are no predefined term sets and one arrives directly at membership functions and the fuzzy rules incorporating them via identification on the set of input-output data. Regarding the use of fuzzy clustering techniques we can mention here a number of identification methods belonging to the approximative approach [Tak85, Sug91, Sug93, Bab95a, Ber93, Sin93] that are aimed at the construction of affine Takagi–Sugeno (TS) fuzzy models or singleton fuzzy models.

In this chapter we present an approach to identification which is midway between the descriptive and the approximative approaches. Our working hypothesis is that each grouping of input-output data represents a local 'rule' (in a general sense) describing a local input/output relationship for the system under identification. To identify the set of such local rules, fuzzy clustering methods will be used, and the rough description provided by these will be considered as the initial fuzzy model. A more accurate fuzzy model can then be constructed from the initial fuzzy model by using tuning techniques like genetic algorithms or neural networks. Alternatively, the initial fuzzy model can be used to further obtain a Mamdani, a TS, or a singleton fuzzy model.

In this context we describe a *rapid prototyping approach* to the identification of a fuzzy model whose basic objective is to provide a prototype fuzzy model that can be further refined. For this purpose we use a fuzzy clustering algorithm, in particular the FCM algorithm [Bez81], for three main reasons:

1. A fuzzy partition of the Z space, in the form of fuzzy clusters, is obtained, which is an important characteristic of every fuzzy model.
2. Once the fuzzy clusters are obtained, the parameters of the fuzzy sets associated with these clusters depend only on the centers of the clusters and the distance of the input/output data to these centers, so we can define the membership functions in any domain, for example, directly in $\mathcal{X}$.
3. The amount of information required for the construction of the fuzzy model is reduced, and no a priori assumption about non-interactivity between the components of the input space is needed.

The computation of the overall system output is done via the use of Mizumoto's simplified approximate reasoning method [Miz89], mainly because of the simplicity of this method. Thus each consequent of a fuzzy rule is a singleton, i.e., a real number. Furthermore, we use the center-of-gravity defuzzification method.

To illustrate our working hypothesis suppose that the FCM algorithm is applied on Ω. This will group together those input-output (I/O) sample data in Z that are close to each other with respect to some similarity measure. Since we do not use any information at all about the structure of the I/O data (for example, possible linearity of the relationship between inputs and output), we use the FCM just to generate k fuzzy clusters in the product space Z with centers (or centroids) denoted as $c_{XY}^h = \left(c_X^h, c_Y^h\right)$ for $h = 1, \ldots, k$. The membership function of the fuzzy relation associated with the h-th cluster is then given as

$$\mu_{C_{XY}^h}(x_t, y_t) = \left\{ \frac{\sum_{l=1}^{k} \left\| (x_t, y_t) - (c_X^h, c_Y^h) \right\|^2}{\left\| (x_t, y_t) - (c_X^l, c_Y^l) \right\|^2} \right\}^{-\frac{1}{m-1}} \qquad (1.2)$$

where $m > 1$. Notice that $\mu_{C_{XY}^h}(x_t, y_t) = 1$ if $(x_t, y_t) = (c_X^h, c_Y^h)$ and that $0 \leq \mu_{C_{XY}^h}(x_t, y_t) < 1$ otherwise.

Thus we have identified a local relationship between inputs and output around the centroids which is captured by the above fuzzy relation. Furthermore, the latter can be used directly for computing the output, given certain inputs, according to the compositional rule of inference.

Section 2 describes the classical fuzzy c-means algorithm, as it will be used for the identification of fuzzy models. The first problem one must deal with when using a clustering technique is the estimation of the number of clusters present on the I/O data. Thus, in Sect. 3 we discuss the use of hierarchical clustering for the purpose of preprocessing the I/O data. Also, some 'goodness' measures to be used in hierarchical clustering when no previous information about the structure of I/O data is available are introduced. Sections 4 and 5 present different identification methods that use the information obtained from hierarchical clustering. These methods belong either to the descriptive or to the approximative approaches in the identification of fuzzy models. Section 6 contains an example of how to use the proposed identification methods within a two-step identification procedure. The first step is the construction of the fuzzy rules. Implicit in this step is the determination of the necessary number of fuzzy rules. The output of this step is a collection of fuzzy rules that can be seen as rough, initial version of the final fuzzy rule base. The second step tunes the initial fuzzy rules in order to provide the final fuzzy rule base. Since the construction of the fuzzy rules is the most difficult step to automate [Yag93a], we have concentrated our efforts on the development of rapid prototyping methods for the identification of fuzzy models.

2. The Fuzzy C-Means Algorithm

The fuzzy c-means algorithm was initially developed by Dunn and later generalized by Bezdek. The *sum of squared errors* performance index, or objective function, is the most widely used clustering criterion. Dunn extended the classical *within-groups sum of squared errors* (WGSS) objective function to a fuzzy version of it and the fuzzy c-means clustering algorithm minimizes this extended fuzzy objective function via an iterative method. Bezdek further generalized the fuzzy objective function proposed by Dunn by introducing the weighting exponent m, $1 \leq m < \infty$,

$$J_m(U, P : X) = \sum_{k=1}^{n} \sum_{i=1}^{c} (u_{ik})^m D_{ik,A}^2 \qquad (2.1)$$

where $U \in M_{\text{fcn}}$ is a c-fuzzy c-partition X, $P = v = (v_1, v_2, \ldots, v_c)$ are the c prototype vectors in $\mathbb{R}^p$, and A is any $(p \times p)$ symmetric positive definite matrix defined as

$$D_{ik,A} = \|x_k - v_i\|_A = \sqrt{(x_k - v_i)^T A (x_k - v_i)} \qquad (2.2)$$

where $D_{ik,A}$ is an inner product induced norm on $\mathbb{R}^p$, u_{ik} is referred to as the grade of membership of x_k to the cluster i. This grade of membership satisfies the following constraints:

$$
\begin{aligned}
&\text{i) } 0 \le u_{ik} \le 1, \quad 1 \le i \le c,\ 1 \le k \le n, \\
&\text{ii) } 0 < \sum_{k=1}^{n} u_{ik} < n, \quad 1 \le i \le c, \\
&\text{iii) } \sum_{i=1}^{c} u_{ik} = 1, \quad 1 \le k \le n.
\end{aligned}
\qquad (2.3)
$$

The FCM uses iterative opimization of the objective function, based on a weighted similarity measure between x_k and the cluster center v_i.

Fuzzy C-Means Algorithm

Step 1: Given a data set $X = \{x_1, x_2, \ldots, x_n\}$, select the number of clusters $2 \le c < N$, the maximum number of iterations T, the distance norm $\|\cdot\|_A$, the fuzziness parameter m and the termination condition $\epsilon > 0$.

Step 2: Give an initial value $U_0 \in M_{\text{fcn}}$.

Step 3: For $t = 1, 2, \ldots, T$

3a. Calculate the c cluster centers $\{v_{i,t}\}$, $i = 1, \ldots, c$

$$v_{i,t} = \frac{\sum_{k=1}^{n} u_{ik,t-1}^m x_k}{\sum_{k=1}^{n} u_{ik,t-1}^m}. \qquad (2.4)$$

3b. Update the membership matrix. Check the occurrence of singularities $(d_{ik,t} = \|x_k - v_{i,t}\|_A = 0)$. Let $\Upsilon = \{1, \ldots, c\}$, $\Upsilon_{k,t} = \{i \in \Upsilon \mid d_{ik,t} = 0\}$, and $\tilde{\Upsilon}_{k,t} = \Upsilon \setminus \Upsilon_{k,t}$. Then calculate

$$
u_{ik,t} =
\begin{cases}
\displaystyle \sum_{j=1}^{c} \left(\frac{d_{ik,t}}{d_{ik,t}} \right)^{-\frac{2}{m-1}} & \text{if } \Upsilon_{k,t} = \emptyset \\[2ex]
\begin{aligned} 0 \quad &\forall i \in \tilde{\Upsilon}_{k,t} \\ a_{ik,t} \quad &\forall i \in \Upsilon_{k,t} \end{aligned} & \text{if } \Upsilon_{k,t} \ne \emptyset
\end{cases}
, \qquad (2.5)
$$

with

$$a_{ik,t} : \sum_{i \in \Upsilon_{k,t}} u_{ik,t-1} = 1, \quad \forall i \in \Upsilon_{k,t} \ne \emptyset. \qquad (2.6)$$

Choose $a_{ik,t} = 1/\#(\Upsilon_{k,t})$, $\forall i \in \Upsilon$; $\#(\cdot)$ denotes the ordinal number.

Step 4: If $E_t = \|U_{t-1} - U_t\| \le \epsilon$ stop, otherwise return to step 3.

This procedure converges to a local minimum or a saddle point of J_m. The FCM algorithm computes the partition matrix U and the clusters' prototypes that will be used in Sects. 4 and 5 in order to derive the fuzzy models from these matrices. Before this is done, one important issue has to be resolved, namely how to determine the number of clusters. This is discussed in the next section.

3. Using Hierarchical Clustering to Preprocess Data

3.1 General view

Cluster validity and determination of the 'optimal' number of clusters is one of the most difficult problems in clustering [Dub88, Bez81, Dun74b, Lop88]. Traditionally, the optimal number of clusters is determined by evaluating a certain global validity measure of the c-partition for a range of c values, and then taking the value of c that optimizes this cluster validity measure in some sense. However, this is a very tedious and computationally expensive process since one needs to cluster the data for a range of c values and this range is not known in advance. Moreover, like all gradient-type methods, different initializations may lead to different local extrema. Thus, we need to try different initializations to assure that the substructure found by FCM is 'stable', in the sense that different initializations do not always lead to different terminal states.

In some approaches, see for example [Gat89], a goal-oriented cluster validity strategy has been proposed. Optimality is there restricted to the notion of optimizing performance measures based on cluster hypervolume and density criteria. It also involves the determination of prototypes for the initialization of the iterative process of clustering by means of the FCM. Other approaches have focused on the improvement of the algorithm by introducing modifications like the combination of different kinds of fuzzy clustering algorithms, or by executing several times the same algorithm though with different parameters [Kri93, Gat89].

Here we propose to determine the initial number of clusters with which to start the FCM algorithm by first using hierarchical clustering for the purpose of preprocessing the I/O sample data. Hierarchical clustering provides multiple (possibly nested) clusters through algorithms of reasonable computational cost. However, its drawback is that it requires the user to use some additional heuristic criteria to determine the most appropriate partition. To solve this problem we propose several measures that allow us to rank the possible partitions.

3.2 Hierarchical Clustering – Similarity Relations

Let $I = \{O_1, O_2, \ldots, O_n\} \equiv \{1, 2, \ldots, n\}$ be a finite set of objects, each object being defined by an r-dimensional vector of attribute values.[1]

Definition 3.1. *A family $H(I)$ of subsets of I is called a hierarchy of parts for (relative to) I when:*

1. *I belongs to $H(I)$.*
2. *The set $\{j\}$ associated to any element j of I belongs to $H(I)$.*
3. *If H and H' belong to $H(I)$ and have nonempty intersection, then H is included in H' or conversely.*

Definition 3.2. *A hierarchy of parts is called a hierarchical clustering when it is indexed, in the sense that for any $H \in H(I)$ there exists an $x_H \in [0, 1]$ such that*

1. *$x_H = 0$ if $H = \{j\}$ for any element j of I.*
2. *$x_{H'} \leq x_H$ if and only if $H' \subseteq H$.*

Work by Dunn, Zadeh, and Bezdek [Dun74a, Zad71, Bez78] has established the equivalence between hierarchical clustering, max-min transitive fuzzy relations, and ultrametric distances [Ben75]. In this context, for any hierarchical clustering $H(I)$ there always exists an ultrametric distance matrix $U = (u_{ij}), i, j = 1, 2, \ldots, n$, that can be normalized in $[0, 1]$.

Zadeh [Zad71] shows that the ultrametric inequality for U is equivalent to the max-min transitivity for $R = (r_{ij}), r_{ij} = 1 - u_{ij}, i, j = 1, 2, \ldots, n$. Thus R is a similarity relation on I or, more generally, an F-indistinguishibility with F being the t-norm *min* [Val85].

For any $b \in [0, 1]$ we will denote $A^b = (a_{ij}^b)$ the matrix that is associated with the b-cut of the similarity relation R, thus representing a relation given by

$$a_{ij}^b = \begin{cases} 1 & \text{if } r_{ij} \geq b \\ 0 & \text{otherwise.} \end{cases} \tag{3.1}$$

It is very easy to show that A^b is a crisp equivalence relation on I for any b. R is a finite matrix and then it is obvious that it contains only a finite number of different elements. Thus for any R (hierarchical clustering) we can always identify a unique finite sequence $\{S_k, k = 1, 2, \ldots, h\}$ such that

(i) $0 = S_1 < S_2 < \ldots < S_h = 1$, and
(ii) for any $k \in \{1, 2, \ldots, h\}$, there exists at least one r_{ij} equal to S_k and conversely.

---

[1] In the case of model identification, the objects O_i will be the I/O data pairs (x_t, y_t) where $t = 1, \ldots, n$.

Let us remark that $A^b = A^{b'}$ for any $b, b' \in (S_{k-1}, S_k]$, and $A^b \neq A^{b'}$ if b and b' belong to two different intervals $(S_{k-1}, S_k]$ and $(S_{\ell-1}, S_\ell]$ respectively, $\ell \neq k$ (see [Vil79]). In other words, the sequence $\{S_k, k = 1, 2, \ldots, h\}$ determines unequivocally the set of all possible different b-cuts of R, that is, the set of all different equivalence relations associated with R.

For any $b \in \{S_k, k = 1, 2, \ldots, h\}$ we denote with $\{C_e^b, e = 1, 2, \ldots, m(b)\}$ the partition induced by the equivalence relation given by A^b. Let b, b' belong to $\{S_k, k = 1, 2, \ldots, h\}$ and assume that $b < b'$. According to the properties of the level sets, the equivalence relation associated with b contains the partition associated with b', that is, the partition generated by $A^{b'}$ is obtained from the one generated by A^b by splitting some of its classes. Moreover, it is easy to show that A^0 generates a single class equal to I, whereas A^1 splits I into single point classes $\{\{j\}, j \in I\}$.

In summary, any hierarchical clustering generates an indexed finite sequence of nested crisp partitions (and conversely) which extends from a single class equal to the whole set of objects, to the set of the single object classes. The problem then is to decide which one of these partitions better represents the structure of the set of objects.[2]

3.3 Unsupervised Learning Procedures to Select 'Good' Partitions

We will now present some procedures used to select suitable partitions from hierarchical clustering, when no previous information about the structure of the I/O data is available [Del94b]. In this case, we will need to introduce some measures to handle selection heuristics based on distance or stability.

3.3.1 Measures of Cluster Dispersion. The usual criteria used to select a partition are those that minimize the distance within each cluster and/or maximize the distance between the clusters – regarding this distance as the separation between their centroids, medians, etc. [Dub88]. We will analyze some selection criteria that are associated with some particular distance functions developed by us.

Let $b \in \{S_k\}$ and let $\{C_e^b, e = 1, \ldots, m(b)\}$ be the clusters induced by the b-cut. For any $i, j \in \{1, \ldots, n\}$ there are two possible alternatives:

1. $\exists e : i, j \in C_e^b$ so that the distance between them is $u_{ij} = 1 - r_{ij}$.
2. $\neg \exists e : i, j \in C_e^b$ so that, considering the properties of the relation R, if $i \in C_e^b$ and $j \in C_{e'}^b$, for $t \in C_e^b$ and $t' \in C_{e'}^b$, then $r_{ij} = r_{it'} = r_{jt}$ and $u_{ij} = u_{jt} = u_{it'}$

Thus u_{ij} assesses the distance between i and j if they belong to the same cluster. If i and j belong to different clusters, then u_{ij} is the distance between these clusters.

---

[2] A step by step description of the hierarchical clustering algorithm may be found in [Dub88].

From the distance between the elements of the clusters and the distance between the clusters, it is possible to define two functions $\mathbf{d^b}$ and $\mathbf{D^b}$ such that the first is a global measure of the distance between the elements of each cluster, while the second one represents a global measure of the distances between the clusters. Each one will be defined over the sequence $\{S_k\}$ and will take values in $[0, 1]$.

We propose two possible definitions. The first one is based on an optimization criterion over average distance values, while the second one is based on the classical max-min and min-max criteria.

Definition 3.3. *Let $b \in \{S_k\}$ and $\{C_e^b, e = 1, \ldots, m(b)\}$ be the induced clusters. For any $e = 1, \ldots, m(b)$, the global average distance within all the possible clusters of level b is*

$$d_1^b = \sum_{e=1}^{m(b)} \frac{d^b(e)}{m(b)} \tag{3.2}$$

with $d^b(e) = \begin{cases} \dfrac{\displaystyle\sum_{i,j \in C_e^b, i \neq j} 2u_{ij}}{|C_e^b| \times (|C_e^b| - 1)} & \text{if } |C_e^b| > 0, \\ 0 & \text{otherwise.} \end{cases} \tag{3.3}$

The global average distance between the clusters for this level is

$$D_1^b = \begin{cases} \dfrac{\displaystyle\sum_{e=1}^{m(b)-1} \sum_{e'>e} 2D^b(e, e')}{m(b) \times (m(b) - 1)} & \text{if } m(b) > 1 \\ 0 & \text{otherwise.} \end{cases} \tag{3.4}$$

with $D^b(e, e') = u_{ij}$, $i \in C_e^b$, $j \in C_{e'}^b$, $\forall e, e' \in \{1, 2, \ldots, m(b)\}$.

Definition 3.4. *Under the previous hypothesis*

$$d_2^b = \max_{e=1,\ldots,m(b)} \left\{ \max_{i,j \in C_e^b} u_{ij} \right\}, \tag{3.5}$$

$$D_2^b = \min_{\substack{e,e'=1,\ldots,m(b) \\ e \neq e'}} D^b(e, e'). \tag{3.6}$$

It is obvious that $\mathbf{d_2^b}$ is the maximum distance between the elements of a cluster, and $\mathbf{D_2^b}$ is the minimum distance between the clusters.

Once we have defined the distance functions, the problem now is to find a certain $S_t \in \{S_k\}$ such that for either $i = 1$ or $i = 2$.

$$D_i^{S_t} = \max_k \left\{ D_i^{S_k} \right\}, \quad d_i^{S_t} = \min_k \left\{ d_i^{S_k} \right\}. \tag{3.7}$$

To solve (3.7) let us point out the following:

1. d_1^b and d_2^b reach minimum over $b = 1$ because for $b = 1$ we have n clusters each containing one element.
2. The elements of the sequence $\{S_k\}$ are sorted in ascending order. In hierarchical clustering, two clusters are joined at a level when in the previous level there was a minimum distance between them. Thus we can assume that for any $S_k > 0$,

$$D_2^{S_k} = d_2^{S_{k-1}} = 1 - S_{k-1}. \tag{3.8}$$

3. There are cases where

$$\max_k D_1^{S_k} \neq D_1^1. \tag{3.9}$$

Under these conditions we can state that

C1: $\min\limits_{k} d_2^{S_k} \neq \max\limits_{k} D_2^{S_k}$ in any case.

C2: $\min\limits_{k} d_1^{S_k} \neq \max\limits_{k} D_1^{S_k}$ except for extreme cases.

Thus the objectives in (3.7) are contradictory for both $i = 1$ and $i = 2$. In this case some suitable multicriteria optimization technique is to be used to obtain a compromise (satisfactory) solution. Here we propose to maximize the weighted combination

$$H_i(S_k) = p D_i^{S_k} - q d_i^{S_k}, \quad p, q \in [0, 1], \text{ for } i = 1 \text{ or } i = 2, \tag{3.10}$$

where p and q are subjective weights reflecting the fact that the clusters need to be separated and that elements within the clusters need to be close to each other. If $p = q$ then the problem is to maximize

$$\max_k H_i(S_k) = D_i^{S_k} - d_i^{S_k}, \ i = 1, 2. \tag{3.11}$$

Proposition 3.1. *Let us consider the objective function $H_2(S_k) = D_2^{S_k} - d_2^{S_k}$. If S_t is the point in the sequence $\{S_k\}$ where the maximum of $H_2(S_k)$ is reached, then the associated partition is maintained within the largest range of similarity within the hierarchical clustering, that is the most stable range.*

3.3.2 Measures Based on Matrix Distances. The largest amount of information about the data points to be clustered together is contained in the fuzzy similarity relation matrix. However, in order to obtain a certain group of clusters we need to make a b-cut over this similarity relation. Thus it seems reasonable to assume that a b-cut would be optimal if the distance between its partition matrix A^b and the fuzzy similarity matrix is minimal. In the following we give an optimal search criterion based on these ideas.

In the space of matrices different distance measures can be defined, but all of these would be equivalent. We select the following one:

$$d(A, B) = \max_{i,j} \mid a_{ij} - b_{ij} \mid, \tag{3.12}$$

where $A = (a_{ij})$ and $B = (b_{ij})$ are matrices of the same dimension.

Because of the properties of the sequence $\{S_k\}$ and its associated sequence of equivalence relations $\{A^k\}$, $k = 1, \ldots, h$, (for simplicity we will use the notation A^k instead of A^{S_k}), our problem transforms into

$$\min_R H_3(S_k) = \min_R d(R, A^k) = \min_R \max_{i,j} | r_{ij} - a_{ij}^b | . \tag{3.13}$$

Observe here that in virtue of the b-cuts construction,

$$|r_{ij} - a_{ij}^b| = \begin{cases} r_{ij} & \text{if } r_{ij} < S_b \\ 1 - r_{ij} & \text{if } r_{ij} \geq S_b \end{cases} , \tag{3.14}$$

for any $b \in \{S_k, k = 1, 2, \ldots, h\}$ and any $i, j \in \{1, 2, \ldots, n\}$.

On the other hand, from [Zad71] we know that

$$r_{ij} = \sup_k \ S_k \ a_{ij}^k. \tag{3.15}$$

Thus,

$$\begin{aligned} H_3(S_b) &= \max(r_{ij} - a_{ij}^b) \\ &= \begin{cases} \max_{i,j} \sup_k S_k a_{ij}^k & \text{if } \sup_k S_k a_{ij}^k < S_b \\ \max_{ij}(1 - \sup_k S_k a_{ij}^k) & \text{otherwise.} \end{cases} \end{aligned} \tag{3.16}$$

Therefore, according to the properties of the sequence $\{S_k\}$ (different values of R arranged in ascending order), it can be verified that

$$\begin{aligned} \max_{i,j} \ \sup_{\{k | \sup_k S_k a_{ij}^k < S_b\}} S_k a_{ij}^k &= S_{b-1}, \\ \min_{i,j} \ \sup_{\{k | \sup_k S_k a_{ij}^k \geq S_b\}} S_k a_{ij}^k &= S_b. \end{aligned} \tag{3.17}$$

Hence,

$$H_3(S_b) = \max(S_{b-1}, 1 - S_b). \tag{3.18}$$

As we have said before, to obtain the optimal partition, according to the criterion considered in this subsection, we must look for the minimum of $H_3(S_b)$. With regard to the relative position of the values in the sequence $\{S_k\}$ the following possibilities arise:

1. $0 < S_{b-1} < S_b < 0.5$. In this case $H_3(S_b) = 1 - S_b > 0.5$.
2. $0 < S_{b-1} < 0.5 \leq S_b$. In this case $H_3(S_b) \leq 0.5$.
3. $0 < 0.5 \leq S_{b-1} < S_b$. In this case $H_3(S_b) = S_{b-1} \geq 0.5$.

Hence, the minimum of $H_3(\cdot)$ will be obtained in the similarity range that contains the value 0.5. In the case where there exists b which fulfills $S_b = 0.5$, the minimal value will be obtained in the whole range $(S_{b-1}, S_{b+1}]$, and there will be two optimal partitions.

An interesting relationship exists between the optimal value of the distance and the amplitude of the similarity range for which it is obtained. Let

$(S_{\text{op}-1}, S_{\text{op}}]$ be the optimal range and $H_3(S_{\text{op}})$ its distance value. Obviously we can write $S_{\text{op}} = S_{\text{op}-1} + z$, and then it follows immediately that:

1. If $H_3(S_{\text{op}}) = S_{\text{op}-1}$ it must be the case that $S_{\text{op}-1} > 1 - S_{\text{op}-1} - z$ and consequently, $H_3(S_{\text{op}}) \geq (1-z)/2$.
2. If $H_3(S_{\text{op}}) = 1 - S_{\text{op}}$ then it must be the case that $1 - S_{\text{op}} > S_{\text{op}} - z$ and thus $H_3(S_{\text{op}}) \geq (1-z)/2$.

Thus the length of the optimal range establishes a lower bound for $H_3(S_{\text{op}})$. We suggest that the larger the range of similarity that contains the value 0.5 is, the more valid this criterion is.

3.3.3 Measures Based on the Fuzzy Sets Associated with a Partition.

In papers dealing with the link between clusters and membership functions [Bac75, Bez92b, Lop88], it is common to interpret the elements of the clusters in terms of membership functions defined on the set $\{1, \ldots, n\}$. Using these ideas and the notion of **the nearest crisp subset to a given fuzzy set** [Zad65], we can develop a new criterion for optimal partition.

Let us consider again the fuzzy similarity relation, its associated sequence $\{S_k\}$, and the partition induced by each value b of this sequence $\{C_e^b, e = 1, \ldots, m(b)\}$. For any $e \in \{1, 2, \ldots, m(b)\}$, we can define a fuzzy subset (membership function) F_e^b associated with C_e^b by the membership function

$$F_e^b(i) = \min_{j \in C_e^b} r_{ij}, \quad \forall i \in \{1, 2, \ldots, n\}. \tag{3.19}$$

This membership function has a number of interesting properties expressed by the following propositions.

Proposition 3.2. *For any* $i, j \in C_{e'}^b, e' \in \{1, 2, \ldots, m(b)\}$,

$$F_e^b(i) = F_e^b(j), \qquad \forall\ e. \tag{3.20}$$

This proposition means that any condition to be imposed on F_e^b, or any operation over it, cannot be verified in a pointwise manner, but only globally on the whole set. Thus we have

$$F^b(e, e') = F_e^b(i) \quad \forall i \in C_{e'}^b \tag{3.21}$$

for each $e, e' \in \{1, 2, \ldots, m(b)\}$, and any level b. Also $F^b(e, e') = F^b(e', e)$, because of the symmetry of the similarity relation R.

Proposition 3.3. *The value* $P_e^b = \{\max_i\ F_e^b(i) - \min_i\ F_e^b(i)\}/2$ *fulfills* $P_e^b = F^b(e, e)/2$.

Thus we have that $P_e^b \geq b/2$. Furthermore, if C_e^b contains only a single point, then $P_e^b = 0.5$.

From the fuzzy set F_e^b, we can define the crisp set $X_e^b = \{i \mid F_e^b(i) \geq P_e^b\}$ that we call the *nearest* to F_e^b in a hierarchical cluster. It follows immediately that $C_e^b \subset X_e^b$ and $C^0, X^0 = \{1, 2, \ldots, n\}$.

On these premises it seems reasonable to consider that the more similar C_e^b is with X_e^b, the more coherent C_e^b is with the I/O data. A measure of the difference between X_e^b and C_e^b can be formulated as

$$H_e(b) = \frac{\sum\limits_{C_{e'}^b \subset X_e^b} \left| C_{e'}^b \right| \left(F^b(e,e) - F^b(e,e') \right)}{\left| X_e^b \right| - \left| C_e^b \right|} \tag{3.22}$$

if $\left| X_e^b \right| \neq \left| C_e^b \right|$ and $H_e(b) = 0$ if $\left| X_e^b \right| = \left| C_e^b \right|$.

Since the objective is the determination of an optimal b-level, it seems necessary to establish a global measure of the differences between $\{C_e^b, e = 1, 2, \ldots, m(b)\}$ and $\{X_e^b, e = 1, 2, \ldots, m(b)\}$ for any b. Taking into account (3.6), this global measure can be defined as

$$H_4(b) = \frac{\sum\limits_{e=1}^{m(b)} H_e(b)}{m(b)}. \tag{3.23}$$

The above global measure must be minimized over $\{S_k\}$. With $H_4(b)$ we measure, for any level in the hierarchy, the cost of the change to the next level, or what is the tendency in the variability of the similarity between levels.

It is easily seen that

$$H_4(S_k) \in [0, 1] \quad \forall k, \text{ and } H_4(0) = 0. \tag{3.24}$$

Proposition 3.4. *If S_t is such that $S_{t-1} < S_t/2$, then $H_4(S_t) = 0$.*

As a consequence of Proposition 3.4 we have that

1. If $S_{h-1} < 0.5$, then $H_4(1) = 0$.
2. $H_4(S_2) = 0$ in any case.

The search for an optimal value for $H_4(\cdot)$ must take into account that it will reach the value 0 for the first two levels of the sequence. In order to choose the optimal value and avoid this difficulty we must consider

Optimal $b = \max \{b \mid H_4(b) = 0\}$. \hfill (3.25)

3.4 Examples

In this section we report the results of using the different cluster validity measures from Sect. 3.3 on the following data:

1. An example of fuzzy classification, Anderson's well known Iris data set.
2. An example of identification of a fermentor fuzzy model.

In order to analyze the performance of our measures we have compared them with different classical cluster validity measures used in fuzzy clustering [Bez81]. In particular,

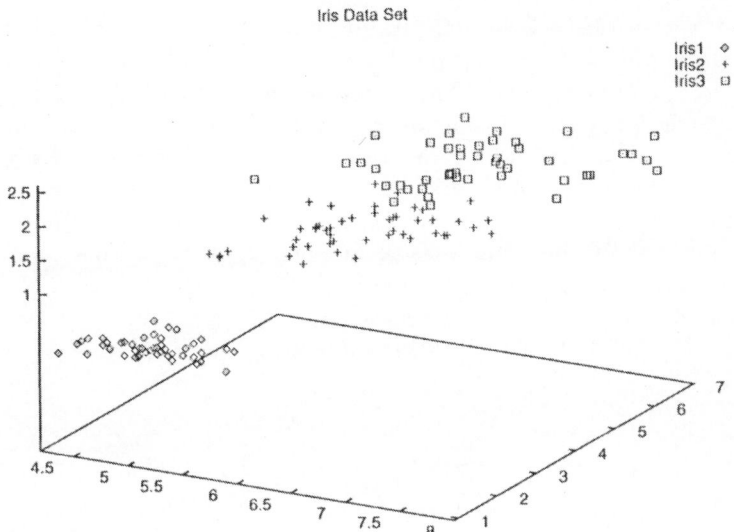

Fig. 3.1. Iris data set

- the separation index D proposed by Dunn which identifies 'compact, separate' (CS) clusters,
- the partition coefficient F introduced by Bezdek to measure the amount of 'overlap' between clusters,
- the fuzzy entropy H introduced by De Luca and Termini,
- the index S proposed by Xie and Beni which measures the overall average compactness and separation of a fuzzy c-partition,
- the index SY (we use here SY instead of the original S in order to differentiate this index from the previous one) used by Sugeno and Yasukawa in [Sug93]. This index measures the variance within each cluster and the variance between the clusters.

Indices D and F are to be maximized, whereas indices H, S, and SY must be minimized.

It should be stressed here that in order to maintain the relation between ultrametric distance, hierarchical clustering, and similarity relation, it is necessary to use a monotonic grouping method like

- the single-linkage that uses the minimum distance between clusters,
- the complete-linkage that uses the maximum distance between clusters,
- Ward's method, also called the minimum variance method.

Several comparative studies [Dub88], and also our own experience, indicate that Ward's method performs better than other hierarchical clustering methods. The method is based on the notion of squared-error and other statistical procedures. The squared-error criterion is also used in the most popular fuzzy clustering algorithm, the fuzzy c-means [Bez81].

3.4.1 Anderson's Iris Data Set. The Iris data set has been used extensively for evaluating the performance of clustering algorithms. It contains 150 four-dimensional feature vectors. The latter belong to three classes representing different Iris subspecies. Each class contains 50 feature vectors. One of the three classes is well separated from the others, which are not easily separable due to the existence of similar feature vectors. The performance of the classification algorithms tested on this data set is usually evaluated by counting the number of clustering errors, i.e., the number of misclassifications. Several studies indicate that unsupervised clustering algorithms fail in the assignment of around 15 feature vectors [Bez92a, Bez92b]. The first, third and fourth features of the Iris data are plotted in Fig. 3.1.

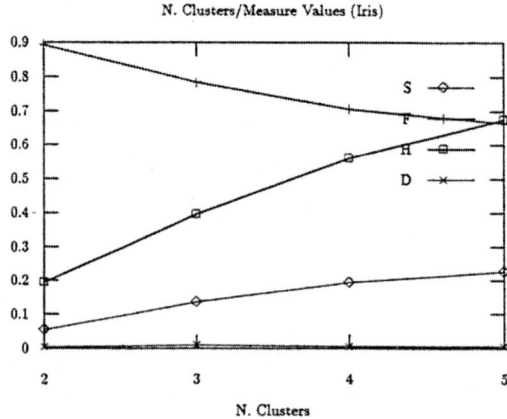

Fig. 3.2. Validity measures used in fuzzy clustering of the Iris data

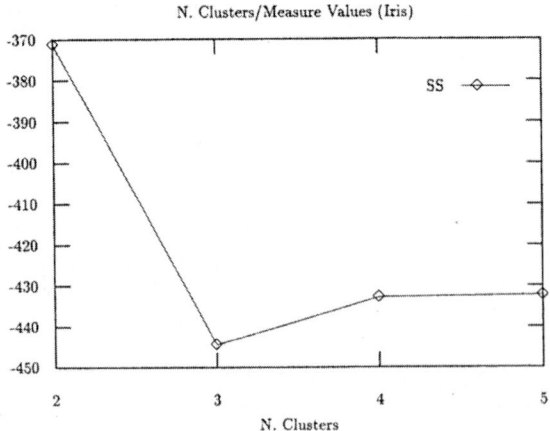

Fig. 3.3. Validity measures used in fuzzy clustering of the Iris data

Depending on the cluster validity measure employed, it is established that the number of clusters in the data is two or three [Bez81, Bez92b]. In Figs. 3.2 and 3.3 we can see the values of the different previously mentioned validity measures. As can be seen, several of the them show that the optimal number of clusters is two and others show that the number of clusters is three. What is also important is that for several of these measures it is not clear what a second option for the number of clusters is. Of course, there are other fuzzy clustering validity measures, but they do not produce better results. For example, the cluster separation measure of Davis and Bouldin failed to uncover the botanically correct number of classes for the Iris data. Furthermore, it exhibited two extra local minima, botanically meaningless, as reported in [Gat89]. Other performance measures are either a monotone function of the number of clusters [Bez81], or show a very slight preference for a certain value of the number of clusters. This is the case of Windham's proportional exponent and the UDF criterion applied to the Iris data.

In our case, by applying different hierarchical clustering methods like Ward's method, complete linkage, and the different cluster validity measures proposed in this chapter, we obtained (in all cases), either two or three clusters. In particular, when using Ward's method with Euclidean distance to assess dissimilarity, the values for the different cluster validity measures are collected in Table 3.1. Here, H_1 and H_2 are the measures of cluster dispersion (2) with distances (d_1, D_1) and (d_2, D_2); H_3 is the the measure based on matrix distance (6); H_4 is the measure based on fuzzy sets (11).

Table 3.1. Optimal partitions on the Iris data set

	No. clusters first option	No. clusters second option
Measure H_1	2	3
Measure H_2	2	3
Measure H_3	2	2
Measure H_4	2	3

Although we indicate two clusters as a first option (a similar result to the one obtained by other authors), in all cases where we can generate a ranking between the partitions (H_1, H_2, and H_4), we obtain three clusters as a second option. Moreover, using the crisp partition obtained in the case of three clusters as initialization value of the FCM algorithm, the number of iterations of the algorithm is reduced. Furthermore, we obtain a missclassification of 14 feature vectors, that correspond to only one class, and obtain 100% success rate for the other two classes. There are some fuzzy clustering algorithms derived from the FCM algorithm, like the FMLE (Fuzzy Maximum Likelihood Estimation), that show better performance in some cases, but always at the expense of more complex algorithms and the need for a predefined initialization (knowledge that indicates a supervised environment). Our approach is more generic and can be used in any situation, with any prior knowledge, and with small computational cost. Moreover, we are not looking for the 'best' or the optimal value of the number of clusters, but rather for a set of 'good' partitions in the largest possible number of cases. We can use these partitions as a first approximation and then obtain a more accurate classification or a fuzzy model.

3.4.2 A Fermentor Fuzzy Model. The fermentor model is presented by Babuška et al. in [Bab94c, Bab95a]. The objective is to identify a fuzzy model for a laboratory fed-batch fermentor pressure dynamics. Figure 3.4 shows a schematic diagram of the process. The pressure in the fermentor vessel is controlled by opening the outlet valve. A first-principles physical model can be derived for this process, based on simplified assumptions. With a constant input flow-rate, the setting of the outlet valve results in a certain transient behavior of the pressure, which can be described by a first-order nonlinear differential equation. However, the identification of the white-box model parameters is difficult and inaccurate. Also, linear black-box models cannot be used due to the highly nonlinear nature of the process. However, in [Bab95a], a TS fuzzy model provides a simple and effective solution to this modeling problem.

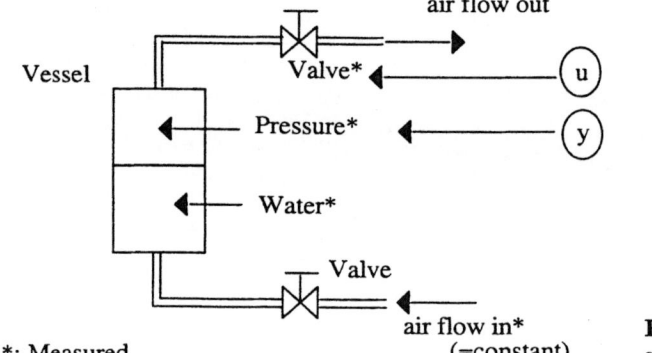

Fig. 3.4. Fermentor experiment

To generate an input-output data for identification, the system input was excited with a sinusoidal signal containing several higher harmonic frequencies. For technical reasons, different input signals were applied around low and high pressure operating points separately. The process can be represented by a nonlinear first-order dynamic model $y(k+1) = f(y(k), u(k))$, where k is discrete time. Figure 3.5 shows the data set used.

This example is well suited for fuzzy clustering, i.e., to detect fuzzy groupings (clusters) in the input-output data that indicate similar input/output

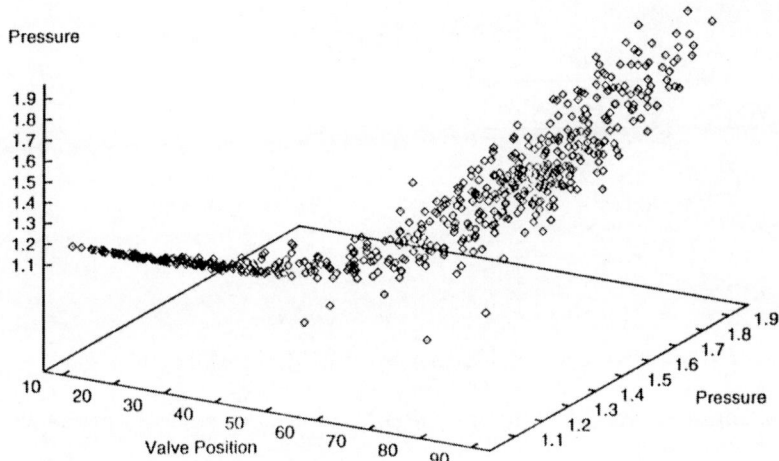

Fig. 3.5. Fermentor data

behavior, and then to associate a fuzzy rule with each one of the obtained fuzzy clusters.

In [Bab95a] the authors consider three fuzzy rules that were obtained by the Gustafson–Kessel clustering algorithm. The membership functions for the input and output variables are obtained by projecting fuzzy clusters on the domains of the inputs and outputs. The linear consequents of the TS fuzzy model are estimated by using the weighted LMS error method [Tak85]. A quadratic-squared error of 3.414×10^{-4} is obtained.

In [Bab95b] the authors proposed a new TS fuzzy model with four clusters, since they show that four fuzzy rules improve the accuracy of the fuzzy model. By applying our cluster validity measures to hierarchical clustering we conclude that the best options are three or four clusters (Table 3.2).

Table 3.2. Optimal partitions fermentor/measure

	No. Clusters first option	No. clusters second option
Measure H_1	3	4
Measure H_2	4	3
Measure H_3	3	3
Measure H_4	4	3

By using the partitions obtained in a similar way as Babuška et al., we reduce the quadratic-squared error to 7.56×10^{-5}. This shows that the clusters obtained from hierarchical clustering can provide a better fuzzy model.

Table 3.3. Number of iterations

	FCM	HIER+FCM
Iris	25	13
3 Cluster	22	9
Fermentor	20	5

3.5 Conclusions on the Use of Hierarchical Clustering

The following enumerates the different use of our cluster validity measures in hierarchical clustering.

1. In the context of hierarchical clustering:
   a) To provide information about the possible levels in the cluster hierarchy that better match the I/O data.
   b) To establish a ranking order between the levels, in order to analyze the crisp partitions and obtain 'optimal partitions'.
   c) To improve statistical classification techniques by means of hierarchical cluster analysis.
2. In the context of fuzzy clustering:
   a) To provide a range for the initial number of clusters to be used for the initialization of a fuzzy clustering algorithm like FCM [Bez81] or KFCM [Bez92a].
   b) For each possible partition, an initialization of the matrix of membership functions can be determined and thus the initial values of the prototypes (centroids) of each cluster can be obtained.
   c) To reduce the range of values of c.

4. Rapid Prototyping of Approximative Fuzzy Models

In the approximative approach the fuzzy model (approximative fuzzy model) is obtained directly (or almost directly) from fuzzy clustering performed on $Z = (X^p \times Y) = (\mathcal{X} \times Y)$. We start here from the assumption that k fuzzy clusters have been determined in the product space Z and these are to be represented by k fuzzy rules.

Taking this into account, we use Mizumoto's simplified reasoning method to compute the overall output of a fuzzy model. Thus the type of fuzzy model we are to deal with is a singleton fuzzy model. With respect to the representation of the antecedent of a fuzzy rule in this type of model, two alternative options are considered. The first option is to consider $\mathcal{X}$ as a global input space and then define an input membership function directly on the product space of the inputs. In this case the fuzzy rules are of the form

$$R_i : \textbf{If } x \textbf{ is } A^i \textbf{ then } y \textbf{ is } b^i, \qquad i = 1, \ldots, k. \tag{4.1}$$

In the second option we will consider a singleton fuzzy model whose antecedent is a conjunction of the membership functions of the inputs, i.e., there is a membership function for each of the inputs, hence the fuzzy rules are of the form

$$R_i : \textbf{If } x_1 \textbf{ is } A_1^i \textbf{ and } \ldots \textbf{ and } x_p \textbf{ is } A_p^i \textbf{ then } y \textbf{ is } b^i. \tag{4.2}$$

In both cases the problem is to determine b^i and the membership functions of the fuzzy sets associated with the antecedents of the fuzzy rules. This is to be done by using the fuzzy clusters C_{XY}^i that we have identified on Z by means of the FCM algorithm.

For any input $\hat{x}$ and using the aforementioned simplified reasoning method, the overall output of the fuzzy model will be computed as:

$$\hat{y} = \frac{\sum\limits_{i=1}^{k} \tau_i \times (b^i)}{\sum\limits_{i=1}^{k} \tau_i} \tag{4.3}$$

where τ_i is to be computed from the membership functions in the antecedent of each fuzzy rule.

In the next subsection we describe how to directly obtain the antecedent membership functions and the consequent singletons from the fuzzy clusters in Z, instead of using the projections of these clusters on the different domains of the inputs and outputs. The latter alternative will be also described in a later subsection. The different methods used to compute the overall output of the fuzzy models in the next subsection are denoted $ESTn$, where EST is short for estimation procedure and n is a cardinal number.

4.1 $\mathcal{X}$ as a Global Input Space

The first and simplest way to obtain the membership functions is to use the centroids and the metric associated with the clusters obtained in Z directly. Thus we consider membership functions generated on the domains $\mathcal{X}$ and Y by the components of the centroids c_{XY}^i, $i = 1, \ldots, k$, in $\mathcal{X}$ and Y. These components we denote as c_X^i and c_Y^i respectively. We must note here that the membership functions are not a projection of the clusters, but are rather induced in $\mathcal{X}$ and Y by the fuzzy clusters. In this way we obtain k fuzzy sets in $\mathcal{X}$ and Y denoted by C_X^i and C_Y^i, $i = 1, 2, \ldots, k$, with membership functions defined as

$$\mu_{C_X^i}(x_t) = \left\{ \frac{\sum_{l=1}^{k} \left\| x_t - c_X^i \right\|^2}{\left\| x_t - c_X^l \right\|^2} \right\}^{-\frac{1}{m-1}}, \tag{4.4}$$

$$\mu_{C_Y^i}(y_t) = \left\{ \frac{\sum_{l=1}^{k} \left\| y_t - c_Y^i \right\|^2}{\left\| y_t - c_Y^l \right\|^2} \right\}^{-\frac{1}{m-1}}. \tag{4.5}$$

Thus k fuzzy rules are obtained:

$$R_i : \textbf{If } x \text{ is } \mu_{C_X^i}(\cdot) \textbf{ then } y \text{ is } \mu_{C_Y^i}(\cdot), \qquad i = 1, \ldots, k. \tag{4.6}$$

According to the agglutinative character of the centroids, we can alternatively generate a set of fuzzy rules where each singleton consequent is induced in Y by the centroids c_Y^i:

$$R_i : \textbf{If } x \text{ is } \mu_{C_X^i}(\cdot) \textbf{ then } y \text{ is } c_Y^i, \qquad i = 1, \ldots, k. \tag{4.7}$$

The overall output of this set of fuzzy rules, for any input x is, obtained according to the expression

$$EST1 = \hat{y} = \frac{\sum_{i=1}^{k} \mu_{C_X^i}(x) \times c_Y^i}{\sum_{i=1}^{k} \mu_{C_X^i}(x)}. \tag{4.8}$$

Usually, the local output from a given fuzzy rule is obtained via the use of the fuzzy relation corresponding to this fuzzy rule. Then the overall output from a set of fuzzy rules can be obtained by aggregating the different fuzzy relations (or the local outputs) without reference to the fuzzy sets (in the antecedent or in the consequent). According to this idea, we develop another method that is similar to the one proposed by Sin and deFigueredo [Sin93]). There the membership functions of the clusters in Z are considered as the most appropriate fuzzy relations to represent the fuzzy model, instead of specifying fuzzy sets on $\mathcal{X}$ and Y. Thus, for any input x the output is computed as

$$EST2 = \hat{y} = \frac{\sum_{i=1}^{k} \mu_{C_{XY}^i}(x, c_Y^i) \times c_Y^i}{\sum_{i=1}^{k} \mu_{C_{XY}^i}(x, c_Y^i)} \tag{4.9}$$

where $\mu_{C_{XY}^i}(x, c_Y^i)$ is the membership degree of the pair (x, c_Y^i) to the i-th cluster in Z.

Both methods are rather simple because they directly 'translate' on $\mathcal{X}$ and Y the information contained in the fuzzy clusters from Z. Observe here that the same output, c_Y^i, is used by both methods. The two methods from above can be improved if the consequents of the fuzzy rules are constructed by using all of the information contained in the clusters from Z. Two possibilities can be immediately seen:

$$y_i^* = \frac{\sum\limits_{t=1}^{n} \mu_{C_X^i}(x_t) \times \mu_{C_{XY}^i}(x_t, y_t) \times y_t}{\sum\limits_{t=1}^{n} \mu_{C_X^i}(x_t) \times \mu_{C_{XY}^i}(x_t, y_t)}, \qquad (4.10)$$

or

$$y_i^* = \frac{\sum\limits_{t=1}^{n} \left[\mu_{C_X^i}(x_t) \times \mu_{C_{XY}^i}(x_t, y_t)\right]^m \times y_t}{\sum\limits_{t=1}^{n} \left[\mu_{C_X^i}(x_t) \times \mu_{C_{XY}^i}(x_t, y_t)\right]^m} \qquad (4.11)$$

where $\mu_{C_{XY}^i}(\cdot)$ and $\mu_{C_X^i}(\cdot)$ are the membership degrees of the data in the fuzzy sets corresponding respectively to the i-th cluster on Z and the cluster induced by it on $\mathcal{X}$.

Let us point out here that the output in (4.10) is obtained by the center-of-gravity defuzzification method performed on each cluster. The output provided by (4.11) can be seen as the projection in the domain Y of the centroids of the clusters obtained by combining the fuzzy clusters in Z with the fuzzy sets in $\mathcal{X}$ induced by the these clusters. In both cases we obtain a collection of k fuzzy rules R_i : **If** x is $\mu_{C_X^i}(\cdot)$ **then** y is y_i^* $i = 1, 2, \ldots, k$. The methods that use this type of a singleton consequent will be denoted as $EST3$ and $EST4$ for the cases of (4.11) or (4.10) respectively.

It should be stressed here that although in all of these methods we have not used the fuzzy sets induced on Y, it is always possible to use these fuzzy sets and obtain

$$\hat{y}' = \frac{\sum\limits_{i=1}^{k} \mu_{C_Y^i}(\hat{y}) \times c_Y^i}{\sum\limits_{i=1}^{k} \mu_{C_Y^i}(\hat{y})}. \qquad (4.12)$$

Although the previous methods use the same distance measure for the fuzzy clustering algorithm, it is necessary to remark that using any other kind of distance measure is possible. This is so because the key results of the clustering algorithm are the position of the centroids and the distribution of the data around them. Once we have this information, we can use it in several ways. For example we can define fuzzy sets with exponential membership functions on $\mathcal{X}$, using the components of the centroids from the clusters in Z and the fuzzy covariances of the sample data in $\mathcal{X}$ [Gat89][Kar94]. In this way we obtain a different method for computing the overall output

$$EST5 = \hat{y} = \frac{\sum\limits_{i=1}^{k} \exp\left(-\frac{1}{2}\left\|x - c_X^i\right\|_{\mathrm{CV}^i}^2\right) \times c_Y^i}{\sum\limits_{i=1}^{k} \exp\left(-\frac{1}{2}\left\|x - c_X^i\right\|_{\mathrm{CV}^i}^2\right)} \qquad (4.13)$$

where CV^i is the covariance matrix associated with the i-th cluster.

The use here of exponential membership functions is motivated by the fact that we can use another kind of fuzzy clustering algorithm, like the possibilistic c-means (PCM), [Kri93] instead of FCM.

Remark: Although we have so far only considered singleton fuzzy models, it is also possible to deal with other types of fuzzy model using the approach described here. For example, if a linear relation between inputs and outputs is considered we can use an affine TS-model and obtain k fuzzy rules of the form

$$R_i : \text{ If } x \text{ is } \mu_{C^i}(\cdot) \text{ then } y \text{ is } f^i(x) \qquad (4.14)$$

where $f^i(x) = a_{1i}x_1 + a_{2i}x_2 + \ldots + a_{pi}x_p + a_{(p+1)i}$, with $i = 1, 2, \ldots, k$. The overall output is then given by:

$$EST6 = \hat{y} = \frac{\sum\limits_{i=1}^{k} \mu_{C^i}(x) \times f^i(x)}{\sum\limits_{i=1}^{k} \mu_{C^i}(x)}, \qquad (4.15)$$

where the $\mu_{C^i}(\cdot)$ is again any exponential membership function.

4.2 Considering $X_1, X_2, \ldots, X_p$

This approach (in the same line of work as in Babuška et al. [Bab95a], and Sugeno et al. [Sug93]) tries to generate fuzzy sets in the domains of discourse of each input/output variable using the fuzzy clusters in Z. To compare this approach with the one presented in the previous section, let us consider the following. Once we obtain the projection of the fuzzy clusters from Z on a given domain we construct the extensional hull of the fuzzy set obtained, and then we approximate it by a trapezoidal fuzzy set. Within this formulation we look for a collection of k fuzzy rules of the form

$$R_i : \text{ If } x_1 \text{ is } A_1^i \text{ and } x_2 \text{ is } A_2^i \text{ and } \ldots \text{and } x_p \text{ is } A_p^i \text{ then } y \text{ is } b^i, \quad (4.16)$$

$i = 1, \ldots, k$, from which the global output can be computed for each input $x = (x_1, x_2, \ldots, x_p)$ as

$$\hat{y} = \frac{\sum\limits_{i=1}^{k} T\left(A_1^i(x_1), A_2^i(x_2), \ldots, A_p^i(x_p)\right) \times c_Y^i}{\sum\limits_{i=1}^{k} T\left(A_1^i(x_1), A_2^i(x_2), \ldots, A_p^i(x_p)\right)}, \qquad (4.17)$$

where T is a t-norm, and c_Y^i is the component in Y of the centroid of the i-th fuzzy cluster in Z. Depending on the type of t-norm two possibilities arise corresponding to T as min and T as a product. That is,

$$EST7 = \frac{\sum_{i=1}^{k} \left[\min_{q=1}^{p} \left(A_q^i(x_q) \right) \right] \times c_Y^i}{\sum_{i=1}^{k} \left[\min_{q=1}^{p} \left(A_q^i(x_q) \right) \right]}, \tag{4.18}$$

$$EST8 = \frac{\sum_{i=1}^{k} \left[\prod_{q=1}^{p} \left(A_q^i(x_q) \right) \right] \times c_Y^i}{\sum_{i=1}^{k} \left[\prod_{q=1}^{p} \left(A_q^i(x_q) \right) \right]}. \tag{4.19}$$

4.3 Summary

This section summarizes the described identification of approximative fuzzy models.

Step 1: *Collect a set of representative data and choose the model structure.* This step is relevant because it will determine the accuracy of the fuzzy model obtained. There are a number of techniques to select the relevant input and output variables. These techniques use, for example, classical optimization approaches [Sug91] or neural networks [Tak91].

Step 2: *Preprocessing the data using hierarchical clustering.* The use of the fuzzy c-means algorithm requires that the initial number of clusters must be defined before clustering is performed. This parameter is important because it is closely related to the number of fuzzy rules. To detect the optimal clusters we use the preprocessing of the I/O data by means of hierarchical clustering and the use of cluster validity measures. Herewith we obtain information about the initial number of clusters and the initial partition matrix both to be used in the consequent application of the clustering algorithm.

Step 3: *Cluster the data and generate the associated fuzzy rules.* The different clustering parameters like the distance metric or m must be defined before clustering. The choice of $m \in [1.5 - 2]$ and Euclidian or diagonal distance have shown good results. Once we have the fuzzy clusters the next step is to generate the fuzzy sets from the antecedents and the singletons from the consequents.

Step 4: *Analysis and comparison of the different EST methods* Although the different methods we have proposed are similar in their structure they correspond to alternative points of view on the fuzzy model to be obtained. Hence, it is important to analyze and compare the results obtained by them in the case of each prototype fuzzy model. Once we have selected a good prototype we can try to tune it using such techniques as genetic algorithms as shown in Sect. 6.

5. Rapid Prototyping of Descriptive Fuzzy Models

In the descriptive approach (descriptive fuzzy models), it is assumed that the linguistic values taken by the input and output variables are predefined via their respective term sets. Also, the membership functions giving the meaning of these linguistic values are predefined. Thus the set of all possible fuzzy rules is just the combination of all linguistic values, i.e., $R = L_1 \times L_2 \times \ldots \times L_p \times T$, where L_i is the fixed set of linguistic values for the i-th input variable and T is the set of linguistic values for the output. Then identification can be understood as selecting the fuzzy rules that better match the input-output data. This match is to be represented by a simple weight attached to each fuzzy rule.

The main problem with the descriptive approaches is the combinatorial explosion in the number of fuzzy rules as a function of the number of inputs and the number of linguistic values. A very large number of possible fuzzy rules makes the identification too complex and thus reduction of the number of fuzzy rules is desired. In this context, our proposal is to work with fuzzy sets defined on the whole input space X, and with fuzzy sets defined in the output space Y. Then we can actually say that we approach the identification problem from a 'quasi-descriptive' point of view.

Thus we assume that collections of k_1 and k_2 fuzzy sets on X and Y respectively are given. The fuzzy model is composed by a collection of $k_1 \times k_2$ fuzzy rules of the form

$$R_{ij} : \textbf{If } x \text{ is } A_i \textbf{ then } y \text{ is } B_j \ [w_{ij}], \qquad (5.1)$$

with $i = 1, \ldots, k_1$, $j = 1, \ldots, k_2$, and $x = (x_1, \ldots, x_p)$. A_i and B_j are linguistic values with corresponding membership functions $A_i(x)$ and $B_j(y)$ in $\mathcal{X}$ and Y respectively. w_{ij} is a weight or certainty value assigned to the rule. In order to compute the output of the fuzzy model we will use a singleton value b_j associated with each $B_j(y)$.

Let us remark that in the descriptive approach fuzzy clustering is mainly used to assign a weight to the fuzzy rules. Thus the membership functions $A_i(x)$ and $B_j(y)$, $i = 1, 2, \ldots, k_1$, $j = 1, 2, \ldots, k_2$, are given and the task is to determine the weight attached to each fuzzy rule by using the fuzzy clustering on Z. We propose here the use of a two-step method:

A. Start with the k clusters in Z and generate a local weight for each fuzzy rule. In this way k matrices W_h, $h = 1, 2, \ldots, k$, are obtained, where each element w_{ij}^h, $i = 1, \ldots, k_1$, $j = 1, \ldots, k_2$, represents the local weight assigned to the corresponding combination of linguistic values:

$W_h\,(i,j) = w_{ij}^h \Rightarrow$

If x is A_i **then** y is B_j $[$weight $w_{ij}^h/$cluster $h]$ $\qquad$ (5.2)

where $[/$cluster $h]$ indicates that the computation of the matrix elements is conditioned on the information provided by the h-th fuzzy cluster. We will later explain how this computation is carried out.

Now methods to estimate the weights are needed. Two basic alternatives are available:

1. *S-T* composition where S is a t-conorm and T is a t-norm:
$$w_{ij} = [S_{t=1}^n(T(A_i(x_t), B_j(y_t)))/\text{cluster } h].\qquad (5.3)$$

2. The fuzzy frequency, defined by Delgado and Gonzalez [Del94a]:
$$w_{ij} = \left[\frac{1}{n}\sum_{t=1}^n(T(A_i(x_t), B_j(y_t)))/\text{cluster } h\right].\qquad (5.4)$$

Recall here that the fuzzy frequency of a fuzzy rule R_h on the set of examples E_n is
$$\psi_{E_n}(R_h) = \frac{1}{n}\sum_{t=1}^n R_h(e_t)\qquad (5.5)$$

with $R_h(e_t)$ being the compatibility degree of the fuzzy rule R_h: **If** x is A_i **then** y is B_j, with the example $e_t = (x_t, y_t) \in (\mathcal{X} \times \mathcal{Y})$ defined as:
$$\begin{aligned} R_h(e_t) &= T(A_i(x_t), B_j(y_t)) \\ A_i(x_t) &= T\left(A_i^1(x_t^1),\dots,A_i^p(x_t^p)\right) \end{aligned}\qquad (5.6)$$
where T is a t-norm.

B. In the second step, the local weights, conditioned on the clusters in Z, are to be aggregated in order to generate a global weight for each fuzzy rule. Two ways of doing the aggregation are possible:

B1. Aggregating the different matrices W_h in one matrix
$$W = \bigoplus_{h=1}^k W_h = F(W_1, W_2, \dots, W_k).\qquad (5.7)$$

In this case we can use different aggregation functions
$$w_{ij} = F_1(w_{ij}^1, w_{ij}^2, \dots, w_{ij}^k) = \max_{h=1}^k w_{ij}^h,\qquad (5.8)$$

$$w_{ij} = F_2(w_{ij}^1, w_{ij}^2, \dots, w_{ij}^k) = \frac{1}{k}\sum_{h=1}^k w_{ij}^h.\qquad (5.9)$$

After the aggregation is performed, one can compute the output of the fuzzy model by using a variant of the simplified reasoning method which takes into account the weights obtained for each fuzzy rule. That is,

$$\hat{y} = \frac{\sum_{i=1}^{k_1}\sum_{j=1}^{k_2} T(A_i(x), w_{ij}) \times v_j}{\sum_{i=1}^{k_1}\sum_{j=1}^{k_2} T(A_i(x), w_{ij})} = \frac{\sum_{i=1}^{k_1}\sum_{j=1}^{k_2} \tau_{ij} \times v_j}{\sum_{i=1}^{k_1}\sum_{j=1}^{k_2} \tau_{ij}}\qquad (5.10)$$

where T is a t-norm, and v_j is the singleton consequent. Note that in the case where the membership functions in the output space have been obtained using fuzzy clustering, these singletons may be the centroids of the fuzzy clusters. This is so since they may coincide with the value obtained by using the center-of-gravity defuzzification method on the membership functions $B_j(y)$.

B2. An alternative procedure is to perform aggregation after having carried out the computation of the output. In this case new value $\hat{y}^h$ is computed for each fuzzy cluster in Z, $h = 1, \ldots, k$, by using the expression (5.10), with w^h_{ij} instead of w_{ij}. Once the k different values $\hat{y}^h$, $h = 1, \ldots, k$, are obtained they are aggregated. For this purpose the fuzzy clusters in Z can be used again, as we will show below.

In the following we will describe the computation performed at steps (A,B1) and (A,B2). Let us recall here that $A_i(x)$ and $B_j(y)$, $i = 1, 2, \ldots, k_1$, $j = 1, 2, \ldots, k_2$, may be available from a direct fuzzy clustering on $\mathcal{X}$ and Y respectively. In this case the weights may be directly estimated and then a learning method can be used, which we will describe too. The computational methods used will be denoted PSn, where PS stands for 'Prototype Solution' and n, $n = 1, 2, ..,$ enumerates each such prototype solution.

5.1 Aggregate and Infer

Different methods that implement the aforementioned steps A and $B1$ will be presented here. Considering the fuzzy rule **If** x is A_i **then** y is B_j, we want to generate a weight $W(i, j) = w_{ij}$ such that it is globally conditioned on the fuzzy clusters in Z.

With the fuzzy frequency assessment of partial weights and using a max-min composition we aggregate as

$$PS1 : w_{ij} = \max w^h_{ij}, \tag{5.11}$$

with

$$w^h_{ij} = \frac{1}{n} \sum_{t=1}^{n} \min\left(\min\left(A_i(x_t), B_j(y_t)\right), \mu_{C_h}(x_t, y_t)\right). \tag{5.12}$$

Because of the properties of the *min* t-norm and the *max* t-conorm, the previous expression can be rewritten as

$$PS1 : w_{ij} = \frac{1}{n} \sum_{t=1}^{n} T\left(T\left(A_i(x_t), B_j(y_t)\right), S^k_{h=1}\left(\mu_{C_h}(x_t, y_t)\right)\right) \tag{5.13}$$

with T being *min* and S being *max*.

Several other aggregation possibilities are also available.

$$PS7 : w_{ij} = \max_{h=1}^{k} \mu_{C_h}(x^*_{A_i}, y^*_{B_j}) \tag{5.14}$$

where $(x_{A_i}^*, y_{B_j}^*)$ are points with $A_i(x) = 1.0$ and $B_j(y) = 1.0$.

$$PS9 \ : \ w_{ij} = \max_{h=1}^{k} T\left(A_i\left(c_X^h\right), B_j\left(c_Y^h\right)\right) \tag{5.15}$$

where c_X^h and c_Y^h are the components on $\mathcal{X}$ and Y of the centroid of the h-th fuzzy cluster.

$$PS2 \ : \ w_{ij} = \max_{h=1}^{k} \mu_{C_h}(x_h^*, y_h^*) \tag{5.16}$$

where (x_h^*, y_h^*) is a new centroid obtained only from those data that have a significant membership degree in the membership functions from the fuzzy rules, that is,

$$(x_h^*, y_h^*) = \frac{\sum_{t=1}^{n_0} \left[\mu_{C_h}\left(x_t^0, y_t^0\right)\right]^m \times \left(x_t^0, y_t^0\right)}{\sum_{j=1}^{n} \left[\mu_{C_h}\left(x_t^0, y_t^0\right)\right]^m} \tag{5.17}$$

where $(x_t^0, y_t^0) \in \{(x_t, y_t) \mid t = 1 \ldots n, \text{ such that } T(A_i(x_t), B_j(y_t)) > \gamma\}$.

5.2 Infer and Aggregate

Here we consider the computation at steps A and $B2$.

A weight is assigned to each possible fuzzy rule and each possible fuzzy cluster in Z by using the fuzzy frequency operator, that is

$$w_{ij}^h = \frac{1}{n} \sum_{t=1}^{n} T\left(T\left(A_i(x_t), B_j(y_t)\right), \mu_{C_h}(x_t, y_t)\right). \tag{5.18}$$

Once the matrices consisting of the above local weights are constructed, a local output is computed and associated with each possible cluster in Z. This can be done in a number of different ways:

1. By considering the degree of firing of the antecedent and the information from the h-th fuzzy cluster

$$\hat{y}^h = \frac{\sum_{i=1}^{k_1} \sum_{j=1}^{k_2} T\left(A_i(x), w_{ij}^h\right) \times v_j}{\sum_{i=1}^{k_1} \sum_{j=1}^{k_2} T\left(A_i(x), w_{ij}^h\right)} \quad h = 1, \ldots, k. \tag{5.19}$$

2. Obtaining a new centroid projected on the output space

$$\hat{y}^h = \frac{\sum_{i=1}^{k_1} \sum_{j=1}^{k_2} \left[\mu_{C_h}\left(x, \hat{y}_{ij}^h\right)\right]^m \times \hat{y}_{ij}^h}{\sum_{i=1}^{k_1^h} \sum_{j=1}^{k_2^h} \left[\mu_{C_h}\left(x, \hat{y}_{ij}^h\right)\right]^m} \tag{5.20}$$

where $\hat{y}_{ij}^h = T\left(A_i(x), w_{ij}^h\right) \times v_j$.

Finally, the aggregation of the different locally computed outputs can be done taking again into account the main source of information we have, i.e., the fuzzy clusters in Z. To achieve this we combine the pair $(x, \hat{y}^h)$ with the 'ideal prototype' represented by the centroids of the fuzzy clusters in $(\mathcal{X} \times Y)$ by using the expression

$$\hat{y} = \frac{\sum_{h=1}^{k} \mu_{C_h}(x, \hat{y}^h) \times y_{C_{XY}^h}}{\sum_{h=1}^{k} \mu_{C_h}(x, \hat{y}^h)}. \tag{5.21}$$

Depending on the expression used to calculate the output for each cluster $\hat{y}^h$, (5.19) or (5.20), the method used will be denoted as $PS3$ or $PS8$ respectively.

5.3 An Alternative Classical Approach

Finally a more classical approach may be used. If $A_i(x)$, $i = 1, 2, \ldots, k_1$, and $B_j(y)$, $j = 1, 2, \ldots, k_2$, are supposed to be obtained from a fuzzy clustering performed on the input and the output spaces respectively, then one may want to assign a weight to each rule (each pair $(A_i(x), B_j(y))$) without considering the information obtained from fuzzy clustering in the product space Z.

The first two alternatives for assigning weights to the fuzzy rules arise depending on the use of the operators:

– max-min composition:

$$PS6 \ : \ w_{ij} = \max_{t=1}^{n} \ \min \left(A_i(x_t), B_j(y_t) \right). \tag{5.22}$$

– fuzzy frequency:

$$PS5 \ : \ w_{ij} = \frac{1}{n} \sum_{t=1}^{n} \min \left(A_i(x_t), B_j(x_t) \right). \tag{5.23}$$

But we can also consider another method. The idea is to assume that the different fuzzy sets in the output space correspond to the different possible classes, and thus the objective is to assess the goodness of the membership functions found in the input space in order to characterize these different fuzzy classes. Thus we arrive at

$$PS4 \ : \ w_{ij} = \frac{\sum\limits_{t=1}^{n} A_i(x_t) \times B_j(y_t)}{\sum\limits_{t=1}^{n} B_j(y_t)}. \tag{5.24}$$

Once the weights are available, (5.10) computes the output for any input.

5.4 Improving the Fuzzy Rules Base

Once we generate the matrix of weights W, we can assign just one or two output membership functions to each input membership function $A_i(x)$. For example, only one membership function from the output space, say $B'(y)$, can be obtained for each $A_i(x)$ as

$$w_{i'} = \max_{j=1}^{s} w_{ij}. \tag{5.25}$$

Now we can assign a confidence factor to this particular choice. Two possible confidence factors are as follows:

$$w_i = FC_i = \frac{1}{2}\left[w_{i'} - \frac{1}{s-1}\sum_{j=1}^{s}(w_{i'} - w_{ij}) \right],$$

$$w_i = FC_i = \frac{w_{i'} - \frac{1}{s-1}\sum_{j=1,j\neq i'}^{s} w_{ij}}{\sum_{j=1}^{s} w_{ij}}. \tag{5.26}$$

We can perform deeper analysis of the fuzzy clusters so determined with the help of the following procedure

1. Generate k fuzzy clusters in Z.
2. For each cluster of Z select the data that fulfill the property $D_h = \{(x_t, y_t)/\mu_{C_h}(x_t, y_t) \geq \delta\}$ $\forall t$, $t = 1, 2, \ldots, n$, $h = 1, 2, \ldots, k$, being $n_h = |D_h|$.
3. For each D_h, perform an independent fuzzy clustering in $\mathcal{X}$ and Y, in order to generate the membership functions $A_i^h(x)$ and $B_j^h(y)$ ($i = 1, 2, \ldots, k_1^h$ $j = 1, 2, \ldots, k_2^h$) with centroids c_{iX}^h and c_{jY}^h.
4. For the different combinations of $A_i^h(x)$ and $B_j^h(x)$ and in relation to the h-th fuzzy cluster in Z, assign the following weight to each rule

$$w_{ij}^h = \frac{1}{n_h}\sum_{t=1}^{n_h} T\left(T\left(A_i^h(x_t), B_j^h(y_t)\right), \mu_{C_h}(x_t, y_t)\right). \tag{5.27}$$

5. For each cluster in Z select the combinations of $A_i^h(x)$ and $B_j^h(y)$ that best describe the fuzzy clusters detected (i.e., the combinations with $w_{ij}^h > \gamma$).
6. Compute a local output for each cluster in $(\mathcal{X} \times Y)$ as

$$\hat{y}^h = \frac{\sum_{i=1}^{k_1^h}\sum_{j=1}^{k_2^h} T\left(A_i^h(x), w_{ij}^h\right) \times c_{jY}^h}{\sum_{i=1}^{k_1^h}\sum_{j=1}^{k_2^h} T\left(A_i^h(x), w_{ij}^h\right)}. \tag{5.28}$$

7. Aggregate the locally computed outputs.

5.5 Summary

The descriptive identification approach is similar to the approximative one (see Sect. 4.3) and thus we will only comment on the differences between these two approaches.

Step 1: *Collect a set of representative data and choose the model structure.*

Step 2: *Preprocess the data using hierarchical clustering.* In the descriptive approach this step can be also applied to the input and the output variables in order to determine the membership functions in these domains if these membership functions are not predefined.

Step 3: *Cluster the data and generate the associated fuzzy rules.* In a similar way to the approximative approach we cluster the data in the product space to detect the groupings that will help to assign a weight to each of the fuzzy rules.

Step 4: *Analyze and compare the different methods for weight computation and aggregation in relation with the prototype fuzzy model obtained.*

6. Examples

In this section we present the results of the different methods (from the approximative and descriptive approaches) applied to the well known problem of the inverted pendulum. The performance index in all the cases considered will be the squared mean error

$$E = \sqrt{\frac{1}{n} \sum_{t=1}^{n} (y_t - \hat{y}_t)^2} \tag{6.1}$$

where n is the number of data, y_t is the actual output, and $\hat{y}_t$ is the fuzzy model output.

6.1 Inverted Pendulum

The inverted pendulum is a well known example in control engineering used to illustrate control and modeling techniques. Figure 6.1 illustrates it.

The state variables are the angle θ and the angular speed ω, and the control variable is the force F. For every (θ_0, ω_0) the objective is to determine the force that must be applied to the center of gravity of the pendulum during a constant time in order to place and/or keep the pendulum in a vertical position.

With the assumption $|\theta| \ll 1$ (radian), the behavior of the pendulum may be described by a nonlinear differential equation. Our model will be

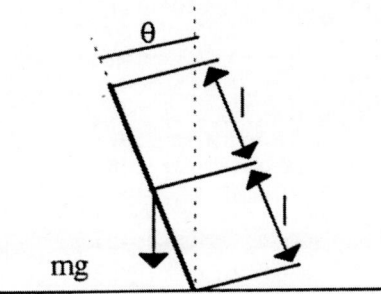

Fig. 6.1. Inverted pendulum

the one used by Yamakawa [Yam89] and consists of a set of seven linguistic fuzzy rules. For the antecedents and the consequents of these fuzzy rules the following term set (set of predefined linguistic values) is used

$$\{NL, NM, NS, ZR, PS, PM, PL\} \qquad (6.2)$$

where N and P stand for Negative and Positive respectively, and where the suffixes L, M, and S mean Large, Medium, and Small respectively. The linguistic fuzzy rules are shown in Table 6.1, where rows correspond to the angle and columns to the angular speed. The entries in the table represent the force to be applied in each case. Let us remark that this is a closed loop system.

Table 6.1. Inverted pendulum rules

$\theta\backslash\omega$	NL	NM	NS	ZR	PS	PM	PL
NL							
NM							
NS			NS		ZR		
ZR		NM		ZR		PM	
PS			ZR		PS		
PM							
PL							

In order to evaluate the performance of our methods we have used the training data set and the test data set proposed by Herrera et al. [Her95]. These authors consider Yamakawa's fuzzy model with trapezoidal membership functions as proposed in [Lia91]. For the semantics of the linguistic values, i.e., their membership functions, see Table 6.2.

From this model, we simulated a pendulum with 5 kg mass, 5 m in length. The force to the pendulum's center of gravity is applied during a constant time of 10 ms. We obtained a set of 348 triples (θ, ω, F) such that

152   M. Delgado et al.

Table 6.2. Trapezoidal membership functions of the seven rules in Yamakawa's model

Angle	Angular speed	Force
[0.17,0.30,0.39,0.52]	[-0.28,-0.07,0.072,0.28]	[1001.26,1752.22,2252.85,3003.80]
[0.00,0.13,0.22,0.35]	[0.00,0.21,0.36,0.57]	[0.00,750.95,1251.58,2002.53]
[0.00,0.13,0.22,0.35]	[-0.57,-0.36,-0.21,0.00]	[-1001.26,-250.31,250.31,1001.26]
[-0.52,-0.39,-0.30,-0.17]	[-0.28,-0.07,0.07,0.28]	[-3003.80,-2252.85,-1752.22,-1001.26]
[-0.35,-0.22,-0.13,0.00]	[-0.57,-0.36,-0.21,0.00]	[-2002.53,-1251.58,-750.95,0.00]
[-0.35,-0.22,-0.13,0.00]	[0.00,0.21,0.36,0.57]	[-1001.26,-250.31,250.31,1001.26]
[-0.17,-0.04,0.04,0.17]	[-0.28,-0.07,0.07,0.28]	[-1001.26,-250.31,250.31,1001.26]

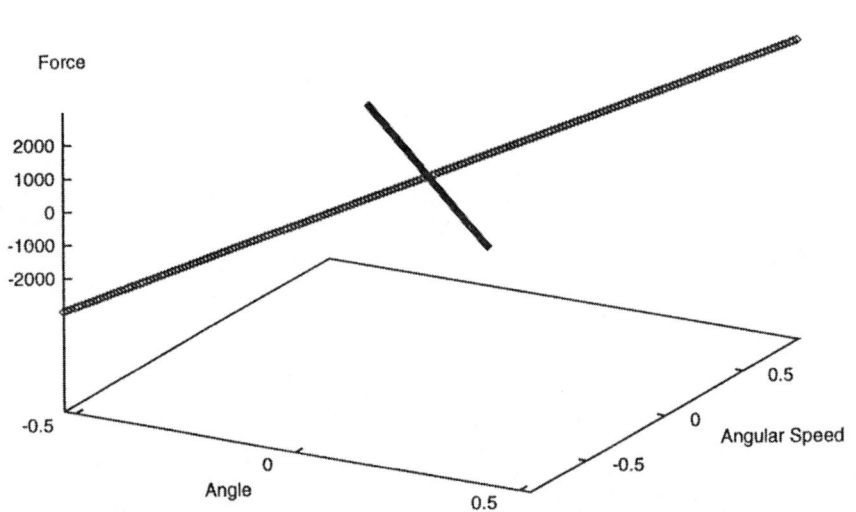

Fig. 6.2. Training data set

$$\theta \in [-0.5240, 0.5240]\,\text{rad}$$
$$\omega \in [-0.8580, 0.8580]\,\text{rad s}^{-1}$$
$$F \in [-2980, 2980]\,\text{N}. \tag{6.3}$$

From the 348 data above, a subset of 280 was used as a training set to reidentify the fuzzy model. The remaining 68 data were used to check the model. A performance error $E = 138.23$ [Her95] was obtained.

We use the same training set both for the descriptive and approximative approaches. In the case of the descriptive approach we start without any known partition of the domains. By using different cluster validity criteria we have established that the more suitable number of fuzzy clusters is equal to seven on $\mathcal{X}$ and five on Y. We also considered the case of nine clusters on $\mathcal{X}$ in order to compare the performance when the number of membership functions increases.

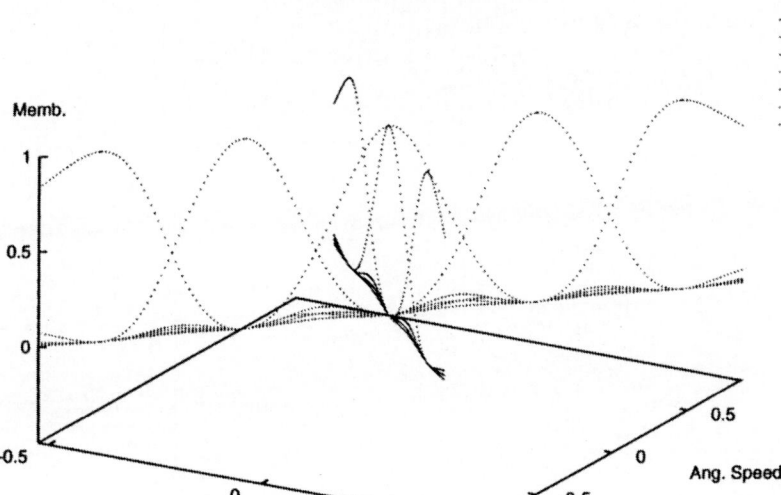

Fig. 6.3. Clusters in the input space

Table 6.3. Trapezoidal membership functions in seven rules of our approach

Angle	Angular speed	Force
[-0.84,-0.49,-0.38,-0.21]	[-1.31,-0.83,-0.71,-0.38]	[-3868.60,-2573.66,-2100.37,-1821.46]
[-0.21,-0.05,0.03,0.23]	[-0.31,-0.00,0.12,0.40]	[-967.29,-288.58,278.30,968.06]
[0.25,0.42,0.60,0.93]	[0.38,0.62,0.95,1.55]	[1776.21,2353.40,3396.59,5493.55]
[-0.36,-0.27,-0.13,0.00]	[-0.64,-0.47,-0.23,-0.00]	[-2252.75,-1689.00,-1071.29,-202.05]
[-0.53,-0.30,-0.07,-0.00]	[-0.18,0.19,0.53,0.90]	[-701.55,-125.32,167.90,676.25]
[-0.23,0.14,0.31,0.63]	[-0.94,-0.52,-0.17,0.22]	[-751.01,-206.97,188.37,764.20]
[-0.00,0.15,0.28,0.41]	[0.00,0.30,0.45,0.61]	[334.22,1136.98,1837.89,2312.51]

Similarly, in the case of the approximative approach we do not assume any known partition to start the fuzzy clustering on $(\mathcal{X} \times \mathcal{Y})$. Using the cluster validity criteria we determine that the most suitable number of elements in the fuzzy partition of the product space (and thus the number of fuzzy rules) is seven or nine.

In Fig. 6.3 we can see the seven fuzzy sets induced on $\mathcal{X}$ by the fuzzy clusters in Z in the approximative approach. Table 6.3 shows the trapezoidal fuzzy sets that correspond to the seven rules with greater weight in the descriptive approach.

After identification we use the same 68 triples from Herrera et al. [Her95] to check the performance of the obtained fuzzy models. The errors obtained in the descriptive approach with either seven or nine membership functions in the antecedent are shown in Fig. 6.4. The errors for the approximative approach with either seven or nine fuzzy rules are shown in Fig. 6.5. The error obtained with the method *EST*6 is not shown since the range for this error is lower that the other ones. The value of the error in this case is $E = 0.34$ which is significantly lower.

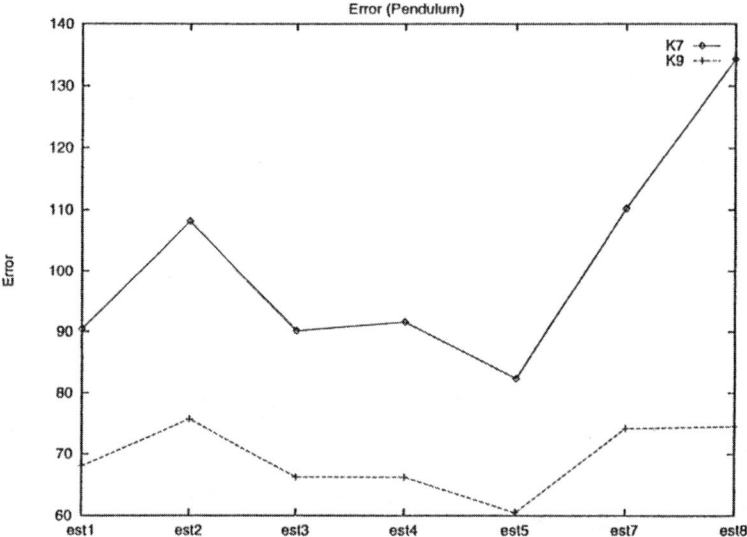

Fig. 6.4. Descriptive method's errors

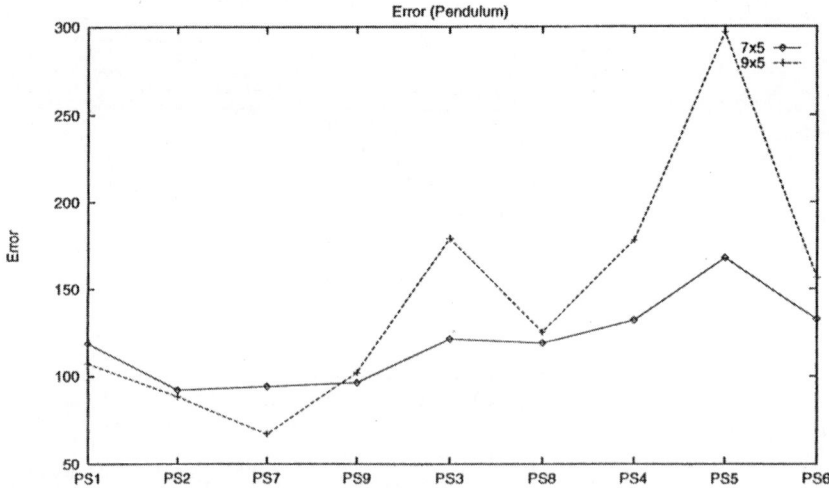

Fig. 6.5. Approximative method's errors

6.2 Tuning the Fuzzy Rules with a Genetic Algorithm

Using the different approaches we have described we obtain a first approximation of the fuzzy model, and thus the next step is to optimize this fuzzy model. There are several possibles techniques that can be used. For example we can use the FLVQ to adjust the membership function parameters as proposed in [Vou94], or we can use optimization techniques like genetic algorithms.

In [Her95] a tuning method using genetic algorithms is presented. In that paper the GA tuning method tunes the membership functions of the fuzzy rules taking into account how the output of the fuzzy model is computed and the defuzzification strategy considered. In our case we have

- t-norm *min*,
- fuzzy implication represented as *min*,
- center-of-gravity defuzzification method.

Table 6.4 shows the results obtained with respect to the performance error when the genetic algorithm is applied to optimize the fuzzy model of the inverted pendulum.

Table 6.4. Tuning with genetic algorithm

	Before tuning	After tuning
Yamakawa rules	138.23	25.48
7 rules	110.30	18.04
9 rules	75.92	3.7

The tuned fuzzy rules in the case of the fuzzy model with seven fuzzy rules obtained by the descriptive approach can be seen in Table 6.5.

Table 6.5. Tuned trapezoidal membership functions for the seven rules model

Angle	Angular speed	Force
[-0.84,-0.49,-0.38,-0.21]	[-1.31,-0.83,-0.71,-0.38]	[-3868.60,-2573.66,-2100.37,-1821.460]
[-0.21,-0.05,0.03,0.23]	[-0.31,-0.00,0.12,0.40]	[-967.29,-288.58,278.30,968.06]
[0.25,0.42,0.60,0.93]	[0.38,0.62,0.95,1.55]	[1776.21,2353.40,3396.59,5493.55]
[-0.36,-0.27,-0.13,0.00]	[-0.64,-0.47,-0.23,-0.00]	[-2252.75,-1689.00,-1071.29,-202.05]
[-0.53,-0.30,-0.07,-0.00]	[-0.18,0.19,0.53,0.90]	[-701.55,-125.32,167.90,676.25]
[-0.23,0.14,0.31,0.63]	[-0.94,-0.52,-0.17,0.22]	[-751.01,-206.97,188.37,764.20]
[-0.00,0.15,0.28,0.41]	[0.00,0.30,0.45,0.61]	[334.22,1136.98,1837.89,2312.51]

7. Practical Considerations and Concluding Remarks

7.1 Use of Prior Knowledge

The rapid prototyping approach described can make efficient use of any prior knowledge about the system under identification. For example, in the analysis phase about what is the best partition, one can use prior qualitative or quantitative knowledge about the real physical system in conjunction with the information obtained from hierarchical clustering. Additionally, in the descriptive approach, one can use qualitative knowledge to characterize the domains of the input and output variables and the composition of the fuzzy rule base.

7.2 Model Complexity

One of the central aspects of the methodology we propose is the use of simple but effective methods for rapid prototyping of a fuzzy model. From this perspective, we have paid special attention to reducing the complexity of our approach in relation to those that appear in the literature. This reduction of complexity has been considered from two points of view:

− We have attempted to decrease the number of fuzzy rules by using the aggregated information generated from fuzzy clustering instead of using just the available data.
− The rapid prototyping approach allows the easy derivation of options with varying degree of complexity (number of clusters).

This reduction of complexity does not imply a significant reduction either in robustness or in performance as can be seen from the examples considered. Furthermore, standard optimization or tuning techniques can be easily applied to the initial fuzzy model in order to obtain a more accurate one.

7.3 Robustness and Generalization Capacity

The results of any clustering approach (including fuzzy clustering) strongly depends on the quality of the available data. Data that is not representative for the system under consideration, polluted by noise, or incomplete will decrease the performance of the fuzzy models obtained.

Fuzzy models derived via fuzzy clustering techniques are able to reflect the global behavior of a system since they take into account overlapping local behaviors characterized by the fuzzy clusters.

7.4 Real-World Applications and Future Developments

The proposed approach has been applied to modeling of control systems [Del95b, Del96], econometric problems, robot navigation [Gas96], and many others.

A. Appendix: Proofs of Propositions

Proposition A.1. *Assume the objective function $H_2(S_k) = D_2^{S_k} - d_2^{S_k}$. If S_t is the point in the sequence $\{S_k\}$ where the maximum of $H_2(S_k)$ is reached, then the associated classification is maintained within the largest range of similarity within the hierarchical clustering, that is, the most stable range.*

Proof: The classification associated with any element of the sequence $\{S_k\}$ is constrained within $(S_{k-1}, S_k]$. Taking this into account we obtain

$$1 - S_k = d_2^{S_k}; 1 - S_{k-1} = D_2^{S_k}, \tag{A.1}$$

and thus

$$\max_k (D_2^{S_k} - d_2^{S_k}) = \max_k (S_k - S_{k-1}). \tag{A.2}$$

$\square$

Proposition A.2. *For any $i, j \in C_{e'}^b$, $e' \in \{1, 2, \ldots, m(b)\}$,*

$$F_e^b(i) = F_e^b(j), \quad \forall e. \tag{A.3}$$

Proof: Let us suppose that

$$F_e^b(i) = r_{it} \; ; \; F_e^b(j) = r_{jt'} \; ; \; t, t' \in C_e^b \tag{A.4}$$

By virtue of the similarity relation properties we have that

$$r_{it} \geq \min(r_{ij}, r_{jt}) \; ; \; r_{jt'} \geq \min(r_{ij}, r_{it'}). \tag{A.5}$$

Then two possible alternatives are available:

1. $e \neq e'$. In this case due to the hierarchical construction we have that

$$r_{ij} > r_{jt} \; ; \; r_{ij} > r_{it'} \tag{A.6}$$

   and thus

$$r_{it} \geq r_{jt}; r_{jt'} \geq r_{it'}. \tag{A.7}$$

2. $e = e'$. In this case (A.6) holds because t and t' are the points where the minimum is reached, and thus (A.7) is true in any case.

If we consider that the first inequality in (A.7) is strict, we could have that $r_{it} > r_{jt'} \geq r_{it'}$. But this is not possible because by definition $F_e^b(i) = r_{it} = \min_{s \in C_e^b} r_{is}$ and thus t is a minimum for i. The same contradiction can be obtained if the second inequality is assumed to be a strict one. This completes the proof. $\square$

Proposition A.3. *The value $P_e^b = \{\max_i F_e^b(i) - \min_i F_e^b(i)\}/2$ fulfills*

$$P_e^b = F^b(e, e)/2. \tag{A.8}$$

Proof: The similarity relation matrix can always be normalized and thus, $\min_i F_e^b(i) = 0$. Then taking into account Proposition A.2 we obtain

$$P_e^b = \{\max_{e'} F^b(e, e')\}/2. \tag{A.9}$$

Moreover, it is evident that $F^b(e, e) = \max_{e'} F^b(e, e')$, because otherwise there will be points in the same cluster of C_e^b whose mutual similarity would be smaller than their similarity with points in other clusters, which is not possible. Thus the proposition holds.   □

Proposition A.4. *If S_t is such that $S_{t-1} < S_t/2$, then $H_4(S_t) = 0$.*

Proof: Let us denote for simplicity $b = S_t$, and let $e \in \{1, 2, \ldots, m(b)\}$. In order to have $X_e^b \neq C_e^b$ there must be $e' \in \{1, 2, \ldots, m(b)\}$, so that, according to the conditions of (A.3) $F^b(e, e') \geq P_e^b \geq b/2$.

In addition $F^b(e, e')$ will belong to the sequence of similarity levels and will be larger than b, so $S_{t-1} \geq F^b(e, e') \geq b/2$. Under these considerations we can state that

$$S_{t-1} < b/2 \text{ implies that } C_e^b = X_e^b, \quad \forall e \in \{1, 2, \ldots, m(b)\}. \tag{A.10}$$

Thus the proposition immediately holds.   □

References

[Ara91] Araki, S.; Nomura; H., Hayashi, I.; Wakami, N., A self-generating method of fuzzy inference rules, *Fuzzy Engineering Toward Human Friendly Systems*, Japan, pp. 1047–1058, 1991.

[Bab94c] Babuška, R.; Verbruggen, H.B., Applied fuzzy modeling, *Proc. IFAC congress AIRTC*, 1994.

[Bab95a] Babuška, R.; Verbruggen, H.B., A new identification method for linguistic fuzzy models, *Proc. FUZZ-IEEE/IFES'95*, Yokohama, Japan, 1995.

[Bab95b] Babuška, R.; Kaymak, U., Application of compatible cluster merging to fuzzy modeling of multivariable systems, *Proc. FUZZ-IEEE/IFES'95*, Yokohama, Japan, 1995.

[Bac75] Backer, E.A., Non-statistical type of uncertainty in fuzzy events, *Colloquia mathematica Societatis Janos Bolyai*, 1975.

[Ben75] Benzecri, J.P., *L'Analyse des donnes I La taxonomie*, Dunod, Paris, 1975.

[Ber93] Berenji, H.R.; Khedkar, P.S.; Clustering in product space for fuzzy inference, *Proc. of the IEEE Int. Conf. on Fuzzy Systems*, pp. 1402–1407, 1993.

[Bez81] Bezdek, J.C., *Pattern recognition with fuzzy objective function algorithms*, Plenum, New York, 1981.

[Bez78] Bezdek, J.C.; Harris, J.D., Fuzzy partitions and relations; An axiomatic basis for clustering, *Fuzzy Sets and Systems*, Vol 1, 111–127, 1978.

[Bez92a] Bezdek, J.C., Tsao, E.C., Pal, N.R., Fuzzy Kohonen clustering networks, *Proc. First IEEE Conf. on Fuzzy Systems*, San Diego, 1035–1043, 1992.

[Bez92b] Bezdek, J.C., Pal, S.K. (eds.), *Fuzzy models for pattern recognition*, IEEE Press, New York, 1992.

[Chi94] Chiu, S.L., A cluster estimation method with extension to fuzzy model identification, *Proc. of the IEEE Int. Conf. on Fuzzy Systems*, pp. 1240–1245, 1994.

[Dav93] Dave, R.N., Robust fuzzy clustering algorithms, *Proc. IEEE Conf. on Fuzzy Systems*, 1281–1286, 1993.

[Del94a] Delgado, M.; González, A., A frequency model in a fuzzy environment, *Int. Journal of Approximate Reasoning*, 11, pp. 159–174, 1994.

[Del94b] Delgado, M.; Vila, M.A.; Gómez-Skarmeta, A.F., An unsupervised learning procedure to obtain a classification from a hierarchical clustering, *Proc. EUFIT'94*, Aachen, Germany, 528–533, 1994.

[Del94c] Delgado, M.; Vila, M.A.; Gomez-Skarmeta, A.F., A validity method of fuzzy clustering processes, *IV Congreso de la Asociación Española sobre Tecnología y Lógica Fuzzy*, Blanes, Spain, 279–284, 1994.

[Del95a] Delgado, M.; Gómez-Skarmeta, A.F., Vila, M.A., Hierarchical clustering to validate fuzzy clustering, *Proc. IEEE Int. Conf. on Fuzzy Systems*, pp. 1807–1812, 1995.

[Del95b] Delgado, M.; Gómez-Skarmeta, A.F.; Martín, F., Generating fuzzy rules using clustering based approach, *Proc. EUFIT'95*, Aachen, Germany, 810–814, 1995.

[Del96] Delgado, M.; Gómez-Skarmeta, A.F.; Martín, F., Using fuzzy clusters to model fuzzy systems in a descriptive approach, *Proc. of the Information Processing and Management of Uncertainty in Knowledge-Based Systems IPMU'96*, , Granada, Spain, 1996.

[Dub88] Dubes, R., Jain, A., *Algorithms that cluster data*, Prentice Hall, Englewood Cliffs, NJ, 1988.

[Dun74a] Dunn, J., A graph theoretic analysis of pattern classification via Tamura's fuzzy relation, *IEEE Trans. SMCC*, 310–313, 1974.

[Dun74b] Dunn, J., A fuzzy relative of the ISODATA process and its use in detecting compact well separated clusters, *J. Cybernet.* 3, 32–57, 1974.

[Gas96] Gasos, J.; Martín, S., Mobile robot localization using fuzzy maps, *Proc. IJCAI 95*, (Ralescu, A., Martín, T. eds.), Lectures Notes in Artificial Intelligence, Vol. 1188, Springer, 1997.

[Gat89] Gath, I.; Geva, A., Unsupervised optimal fuzzy clustering, *IEEE Trans. Pattern Anal. Machine Intell.*, vol. 11, pp. 773–781, 1989.

[Gol89] Goldberg, D.E., Genetics algorithms in search, optimization, and machine learning, Addison-Wesley, 1989.

[Gus79] Gustafson, D. E.; Kessel, W.C., Fuzzy clustering with a fuzzy covariance matrix, *Proc. IEEE Int. Conf. on Fuzzy Systems*, San Diego, CA, pp. 761–766, 1979.

[Gue94] Guély, F.; Siarry, P., A centred formulation of Takagi–Sugeno rules for improved learning efficiency, *Fuzzy Sets and Systems*, vol. 62, pp. 277–285, 1994.

[Her95] Herrera, F.; Lozano, M.; Verdegay, J.L., Tuning fuzzy logic controllers by genetic algorithms, *Int. Journal of Approximate Reasoning*, vol 12, pp. 299–315, 1995.

[Ich91] Ichihashi, H., Iterative fuzzy modeling and a hierarchical network, *IFSA*, Bruselas, pp. 49–51, 1991.

[Kar94] Karayiannis, N.B., MECA: Maximun entropy clustering algorithm, *Proc. IEEE*, pp. 630–1648, 1994.

[Kla93] Klawonn, F.; Kruse, R., Equality relations as a basis for fuzzy control, *Fuzzy Sets and Systems*, vol. 54, pp.147–156. 1993.

[Kri93] Krishnapuram, R.; Keller, J.M., A possibilistic approach to clustering, *IEEE Trans. Fuzzy Systems*, vol. 1, n. 2, pp. 98–110, 1993.

[Kri94] Krishnapuram, R., Generation of membership functions via possibilistic clustering, *IEEE World Congress on Computational Intelligence,* vol. 2, pp. 902–908, Orlando, FL, 1994.

[Lia91] Liaw, C.; Wang, J., Design and implementation of a fuzzy controller for a high performance induction motor drive, *IEEE Trans. on Systems, Man and Cybernetics,* 21, pp. 921–929, 1991.

[Lop88] Lopez de Mantaras, R., Valverde, L., New Results in fuzzy clustering based on the concept of indistinguishability relation, *IEEE Trans. Pattern Anal. Machine Intell.,* vol. 10, 754–757, 1988.

[Mam76] Mamdani, E.H., Advances in the linguistic synthesis of fuzzy controllers, *Int. J. Man-Mach. Std.,* 8, pp. 669–678, 1976.

[Miz89] Mizumoto, M., Method of fuzzy inference suitable for fuzzy control, *J. Soc. Instrument and Control Engrs,* 58, pp. 959–963, 1989.

[Nie93] Nie, J.; Linkens, D. A., Learning control using fuzzified self-organizing radial basis function network, *IEEE Trans. on fuzzy systems,* vol. 1, n. 4, pp.280–287, 1993.

[Nom91] Nomura, H.; Hayashi, I.; Wakami, N., A self-tuning method of fuzzy control by descent method, *IFSA,* Brussels, pp. 155–158, 1991.

[Pal93b] Pal, N. R.; Bezdek, J. C.; Tsao, E. C.-K, Generalized clustering networks and Kohonens self-organizing scheme, *IEEE Trans. on Neural Networks,* vol. 4, n. 4, pp. 549–557, 1993.

[Ped84] Pedrycz, W., An identification algorithm in fuzzy relational systems, *Fuzzy Sets and Systems,* 13, pp. 153–167, 1984.

[Pen88] Peng, X.; Wang, P., On generating linguistic rules for fuzzy models, Lectures Notes in Computer Science, Uncertainty and Intelligent Systems (B. Bouchon, L. Saitta, R.R. Yager eds.), Vol. 313, pp. 185–192, Springer, 1988.

[Sim93b] Simpson, P-K. Jahns G., Fuzzy min-max neural networks for function approximation, *Proc. IEEE Int. Conf. on Fuzzy Systems,* pp. 1967–1972, 1993.

[Sin93] Sin, S.K.; de Figueiredo R.J.P., Fuzzy system design through fuzzy clustering and optimal predefuzzification, *Proc. IEEE Int. Conf. on Fuzzy Systems,* pp. 190–195,1993.

[Sug91] Sugeno, M.; Tanaka, K., Successive identification of a fuzzy model and its applications to prediction of a complex system, *Fuzzy Sets and Systems,* 42, pp. 315–334, 1991.

[Sug93] Sugeno, M.; Yasukawa T., A fuzzy-logic-based approach to qualitative modeling, *IEEE Trans. Fuzzy Systems,* vol 1, n.1, pp. 7–31, 1993.

[Tak85] Takagi, T.; Sugeno, M., Fuzzy identification of systems and its applications to modeling and control, *IEEE Trans. on Systems, Man, and Cybernetics,* vol. SMC-15, n. 1, pp. 116–132, 1985.

[Tak91] Takagi, H.; Hayashi, I., NN-driven fuzzy reasoning, *Int. Journal of Approximate Reasoning,* 5, pp. 191–212, 1991.

[Val85] Valverde, L., On the structure of F-indistinguishability Operators, *Fuzzy Sets and Systems,* vol. 17, 313–328, 1985.

[Vil79] Vila Miranda, A., Notes about the calculus of optimal partitions, *Actas de la XI Reunion Nacional de I.O.,* Sevilla (Spain), 1979 (in Spanish).

[Vou94] Vourimaa, P., Fuzzy self-organizing map, *Fuzzy Sets and Systems,* vol. 66, pp. 223–231, 1994.

[Wan92a] Wang, L.; Mendel, J. M.; Back-propagation fuzzy system as nonlinear dynamic system identifiers, *Proc. IEEE Int. Conf. on Fuzzy Systems,* pp. 1409–1416, 1992.

[Wan92b] Wang, L.; Mendel, J. M.; Generating fuzzy rules by learning from examples, *IEEE Trans. on Systems, Man, and Cybernetics,* vol. 22, n.6, pp. 1414–1427, 1992.

[Wan93] Wang, L., Training of fuzzy logic systems using nearest neighborhood clustering, *Proc. IEEE Int. Conf. on Fuzzy Systems,* pp. 13–17, 1993.

[Yag93a] Yager, R. R.; Filev, D. P., Generation of fuzzy rules by mountain Clustering, *Tech. Report* MII-1318, 1993.

[Yag93b] Yager, R. R.; Filev, D. P., Template based fuzzy systems modeling, *Tech. Report* MII-1310, 1993.

[Yam89] Yamakawa, T., Stabilization of an inverted pendulum by a high-speed fuzzy logic controller hardware system, *Fuzzy Sets and Systems,* 32, pp. 161–180, 1989.

[Yos93] Yoshinari, Y.; Pedrycz, W.; Hirota, K., Construction of fuzzy models through clustering techniques, *Fuzzy Sets and Systems,* 54, pp. 157–165, 1993.

[Zad65] Zadeh, L.A., Fuzzy sets, *Information and Control,* vol. 8, 338–353, 1965.

[Zad71] Zadeh, L.A., Similarity relations and fuzzy ordering, *Information Sciences,* vol 3, 177–200, 1971.

[Zad73] Zadeh, L.A., Outline of a new approach to the analysis of complex systems, *IEEE Trans. on Systems, Man, and Cybernetics,* SMC-1, pp. 28–44, 1973.

Neural Networks

Fuzzy Identification Using Methods of Intelligent Data Analysis

J. Hollatz

Siemens AG, Corporate Technology, D-81730 Munich, Germany

1. Introduction

For complex nonlinear systems like industrial processes, the creation of mathematical models for system identification is a difficult and tedious task. Moreover, many model parameters have to be measured in a painstaking and expensive effort to account for fabrication variations and specific environmental conditions. While complete analytical knowledge is rare in complex technical environments, process measurements provide a powerful source of information about their dynamic behavior.

Adaptive fuzzy systems learn from measured data, and thus are especially useful in applications where plant models are not available or controllers have to adapt to changes in the environment. They have been applied successfully to solve a number of nonlinear control problems in an industrial environment. Several approaches are pursued to use adaptive fuzzy systems in the context of nonlinear control tasks. In data-driven identification, fuzzy systems are trained to model the dynamics of an unknown nonlinear plant. The model provides the basis for controller design or system diagnosis. However, the learning task is very difficult and depends not only on the choice of the model class but also on the training data characteristics: often the data are very noisy and clustered, and only a small number of samples are available. So, in all approaches two issues concerning learning turn out to be crucial. First, the incorporation of prior knowledge greatly simplifies the learning tasks. Often, the plant structure is known a priori, but prior knowledge may also be available for control in the form of fuzzy rules. Second, in many cases prior knowledge implies a problem division into linear and nonlinear subtasks. Elaborate combinations of conventional and fuzzy approaches for model identification have proven to be the basis for successful application of fuzzy identification. Learning is essential for knowledge discovery and system identification, but otherwise prior knowledge in the form of fuzzy rules is useful for learning as well. Fuzzy systems have proven to be an appropriate tool to identify nonlinear models from measured data.

In this chapter fuzzy systems of Takagi–Sugeno type [15] are used for system identification. The enormous advantage of Takagi–Sugeno fuzzy systems is the possibility to incorporate data knowledge very easily via learning techniques. To identify Takagi–Sugeno fuzzy models, three different models are described in this chapter: first, a neural network approach with radial basis functions, second, a probability-driven method with estimation maximization, and third, an identification method via fuzzy clustering.

In their formal structure Takagi–Sugeno fuzzy models [15] are connections of linear relations and component-wise applicable one-dimensional bounded nonlinear functions. Without the nonlinear functions, the whole structure would represent only a simple linear relation. On the other hand the interaction of the variables is realized only in the linear domain. This problem is summarized in the complexity diagram of Fig. 1.1: to realize high-dimensional nonlinear models, classical methods are limited in their computational capabilities. Linear algebra is able to handle problems with high dimensionality and linear relationships, and highly nonlinear problems with only a few variables can be solved by means of calculus. If noise or randomness is introduced into the problem, statistical methods can be used to analyze and model the problem, but only if the problem either is low-dimensional (polynomial regression) or linear (multiple linear regression). Adaptive fuzzy models provide the capability to combine all the features to solve high-dimensional nonlinear and noisy problems.

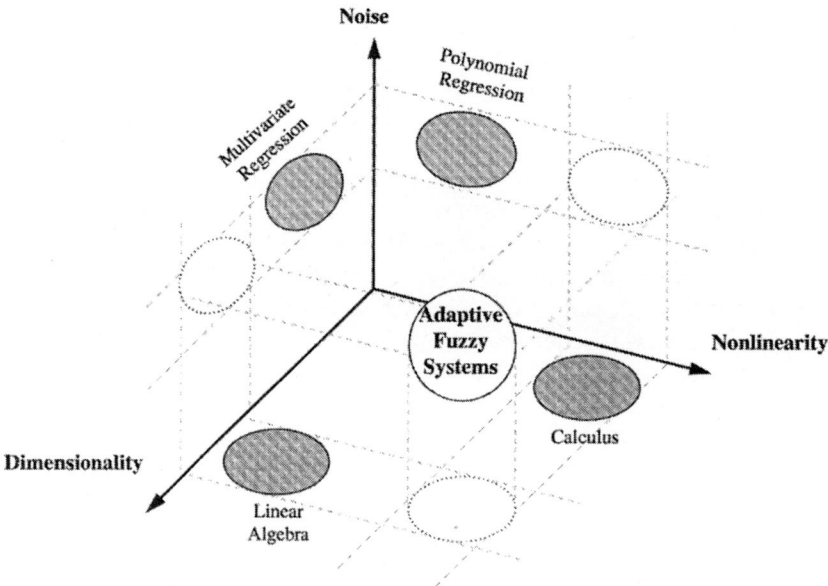

Fig. 1.1. Diagram of problem complexities

2. Neuro-Fuzzy Methods

In the last few years some interesting theoretical results on connectionist systems have been obtained. In particular, it has been shown that different intelligent soft computing methods, like neural networks, fuzzy systems, genetic algorithms, and other machine learning algorithms are able to perform function approximation. Neuro-fuzzy systems are capable of learning, integrating, and extracting rule-based knowledge, by including simultaneously the function approximation property. In this section a fundamental neuro-fuzzy approach is presented, which is discussed in several theoretical investigations and used in a couple of practical applications. In the following, this neural network type and its mathematical foundations are discussed as described in [8].

A radial-basis function network (RBF network) is a two-layer neural network whose output nodes form a linear combination of basis functions. The basis functions in the so-called hidden layer produce a localized response to the input. That is, they produce a significant nonzero response only when the input falls within a small localized region of the input space. For this reason this network is sometimes referred to as the local receptive field network. Although implementations vary, the most common basis is a Gaussian kernel function $b(\cdot)$ of the form

$$b_i(\mathbf{x}) = \exp\left[\frac{(\mathbf{x} - \mathbf{c}_i)^T (\mathbf{x} - \mathbf{c}_i)}{2\sigma_i^2}\right], \quad i = 1, 2, \ldots, M, \tag{2.1}$$

where $b_i(\mathbf{x})$ is the output of the i-th node of the hidden layer, $\mathbf{x}$ is the input pattern, $\mathbf{c}_i$ is the weight vector for the i-th node in the hidden layer, i.e., the center of the Gaussian node for node i; σ_i^2 is the normalization parameter for the i-th node, and M is the number of nodes in the hidden layer. The node outputs are in the range from zero to one so that the closer the input is to the center of the Gaussian, the larger the response of the node. The name *radial basis function* results from the fact that these Gaussian kernels are radially symmetric; that is, each node produces an identical output for inputs that lie a fixed radial distance from the center of the kernel $\mathbf{c}_i$. The output layer node equation is given by

$$y = \sum_i^M \mathbf{w}_i b_i(\mathbf{x}), \tag{2.2}$$

where y is the output of node, $\mathbf{w}_i$ is the weight vector for the node, and $b_i(\mathbf{x})$ is the vector of outputs from the hidden layer. The number of output units can be generalized to a desired value without any restrictions. The output layer node forms a linear combination of the outputs from the hidden layer. Thus, the overall network performs a linear transformation from $\mathbb{R}$ to $\mathbb{R}$ by forming a linear combination of the nonlinear basis function in (2.1).

The RBF network can be used for model identification as well as for classification or general function approximation, just like the often used multilayer

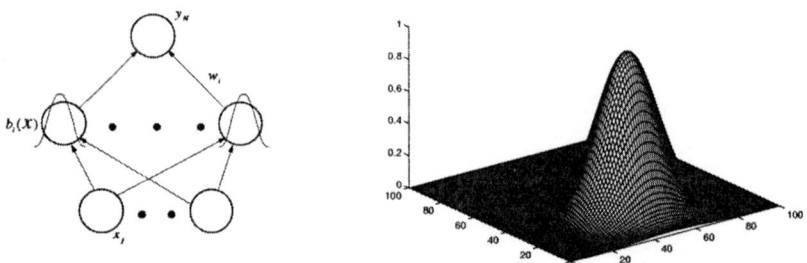

Fig. 2.1. A radial basis function network(left), and a two-dimensional Gaussian basis function (right)

perceptron. The network effectively positions Gaussian kernels at the center of the data, and then weights and thresholds it appropriately to produce the circular decision boundary. In theory, the RBF network, like the multilayer perceptron (MLP), is capable of forming an arbitrarily close approximation of any continuous nonlinear mapping. The primary difference between the two is the nature of their basis functions. The hidden layer nodes in an MLP form sigmoidal basis functions, which are nonzero over an infinitely large region of the input space and form a combination of weighted decision boundaries, while the basis functions in the RBF network cover only small localized regions.

There are a variety of approaches to adapt an RBF network. Most of them start by breaking the problem into two stages: learning in the hidden layer, followed by the learning in the output node. Learning in the hidden layer is typically performed using an unsupervised method, e.g., a clustering algorithm, while learning the output node is supervised. Once an initial solution is found using this approach, a supervised learning algorithm is sometimes applied to both layers simultaneously to fine-tune the parameters of the network. There are numerous clustering algorithms that can be used in the hidden layer. A popular choice is the k-means algorithm. This algorithm is perhaps the most widely known clustering algorithm because of its simplicity and its ability to produce good results. The algorithm is explicitly given in Sect. 4. The normalization parameters σ_i^2 are obtained once the clustering algorithm is complete. They represent a measure of the spread of the data associated with each node. Although they can be determined in a variety of ways, the commonest is to make them equal to the average distance between the cluster centers and the training patterns, that is

$$\sigma_i^2 = \frac{1}{M_i} \sum_{\mathbf{x} \in \Theta_i} (\mathbf{x} - \mathbf{c}_i)^T (\mathbf{x} - \mathbf{c}_i), \tag{2.3}$$

where Θ_i is the set of training patterns grouped with the cluster center $\mathbf{c}_i$, and M_i is the number of patterns in Θ_i. Learning the output node is performed after the parameters of the basis functions have been determined, that is, after learning in the hidden layer is complete. The output node is typically trained using the least mean square (LMS) algorithm. The trained set consists of input/output pairs, but now the input patterns are processed by the hidden layer before being presented to the output node for use in the training algorithm. The LMS algorithm is summarized in Appendix A.

Extensions of the RBF network include variations on the basis functions, the learning algorithm, or both. A common variation on the basis functions is to increase their functionality (in an attempt to decrease the number of hidden nodes) by using the Mahalanobis distance in the Gaussian kernel [6]. The basis function in (2.1) becomes

$$b_i(\mathbf{x}) = \exp[-(\mathbf{x} - \mathbf{c}_i)^T \Sigma_i^{-1}(\mathbf{x} - \mathbf{c}_i)], \quad i = 1, 2, \ldots, M, \qquad (2.4)$$

where Σ_i is the normalization matrix for node i. These basis functions are no longer radially symmetric. There are also numerous variations on learning. For example, once the initial phase of learning is completed as described above, it is sometimes useful to apply supervised learning to both layers simultaneously in an attempt to fine-tune the parameters in the hidden layer. Other variations on learning include techniques that select the centers of the Gaussians as a subset of the training samples. In this method the samples are chosen one at a time in such a way that each new sample maximizes the amount of incremental gain in explaining the variance of the desired output.

Platt [11] postulates a growing network in the hidden layer. Starting with a small number of hidden units a new one is added when a predefined overall error is exceeded. The node's parameters are initialized with the data's values. This algorithm is attractive because it includes a technique for automatically determining the number of hidden nodes.

One of the major advantages of the RBF network is that learning tends to be much faster than the MLP. The main reason for this is that the learning process is broken into two stages, and the algorithm used in each stage can be made relatively efficient. The clustering for a set of data is NP-complete and is the most difficult part of learning. K-means only finds a locally optimal solution, but generally produces good results and is usually very efficient. Once the hidden layer parameters are fixed, learning in the output layer is intrinsically easier. Because the network output is linear in the weights, the learning problem can be made polynomial.

Other similar techniques include Gaussian classifiers and Gaussian mixture methods. They are a consequence of applying the Bayesian decision rule for the case where the probability functions for each class are assumed to be Gaussian. To implement this classifier, one need only form estimates of the mean vector and covariance matrix for each class of data, and substitute these estimates into the decision rule. The decision rule can be simplified to yield a polynomial classifier, which is in general quadratic, but reduces to linear in the case where the covariance matrices for both classes are equal.

More elaborate density estimates can be found using a Gaussian mixture. In this case the density function for each pattern class is approximated as a mixture of Gaussian density functions. Determining the optimal number of Gaussians in the mixture, and the parameters for each of the Gaussians, proves to be a difficult problem. Clustering algorithms like the k-means algorithm can be used to determine the mean vectors of these distributions. Covariance matrices can then be estimated for each cluster once the mean vectors are determined. Such an approach should remind the reader of the radial basis function network. The RBF network can be viewed as a weighted Gaussian mixture where the weights of the Gaussians are determined by the supervised learning algorithm at the output layer. In theory, this approach is capable of forming arbitrarily complex decision boundaries for classification. Now we want to derive a neuro-fuzzy system based on the presented RBF network type.

2.1 Neuro-Fuzzy for Model Identification

We consider a neural network $y = \mathcal{NN}(\mathbf{x})$ which makes an identification about the state of $y \in \mathbb{R}$ given the state of its input $\mathbf{x} \in \mathbb{R}^n$. We assume that an expert provides information about the same mapping in terms of a set of fuzzy rules. The premise of a rule specifies the conditions on $\mathbf{x}$ under which the conclusion can be applied. This region of the input space is formally described by a basis function $b_i(\mathbf{x})$. Instead of allowing only binary values for a basis function (1: premise is valid, 0: premise is not valid), we permit continuous positive values which represent the certainty or weight of a rule given the input.

We assume that the conclusion of the rule can be described in the form of a mathematical expression, such as *conclusion$_i$: the output is equal to $w_i(\mathbf{x})$* where $w_i(\mathbf{x})$ is a function of the input (or a subset of the input) and can be a constant, a polynomial, or even another neural network. This notation corresponds to a fuzzy system of Takagi–Sugeno type as shown in [7].

Since several rules can be active for a given state of the input, we define the output of the network to be a weighted average of the conclusions of the active rules where the weighting factor is proportional to the activity of the basis function given the input

$$y(\mathbf{x}) = \frac{\sum_i w_i(\mathbf{x})\, b_i(\mathbf{x})}{\sum_j b_j(\mathbf{x})}. \tag{2.5}$$

This is a very general concept since we still have complete freedom to specify the form of the basis function $b_i(\mathbf{x})$ and the conclusion $w_i(\mathbf{x})$. If $b_i(\mathbf{x})$ and $w_i(\mathbf{x})$ are described by neural networks themselves, there is a close relationship with the adaptive mixtures of local experts [9]. On the other hand, if we assume that the basis function can be approximated by the multivariate Gaussians presented in 2.4 with a linear adaptive weight factor κ_i,

$$b_i(\mathbf{x}) = \kappa_i \, \exp\left[-\frac{1}{2}\sum_j \frac{(x_j - c_{ij})^2}{\sigma_{ij}^2}\right], \tag{2.6}$$

and if the w_i are constants, we obtain the network of normalized basis functions described by [10] and [14].

In some cases the expert might want to formulate the rules as simple logical expressions. As an example, the rule

If $[x_1$ **is A and** x_4 **is B] or** x_2 **is C then** $y = d \cdot x^2$

is encoded as

$$premise_i : b_i(\mathbf{x}) = \exp\left[-\frac{1}{2}\frac{(x_1 - a)^2 + (x_4 - b)^2}{\sigma^2}\right] + \exp\left[-\frac{1}{2}\frac{(x_2 - c)^2}{\sigma^2}\right]$$

$$conclusion_i : w_i(\mathbf{x}) = d \cdot x^2.$$

A, **B** and **C** denote linguistic values of linguistic variables. The linguistic values are defined by Gaussian membership functions with mean a, respectively b or c. The width of the Gaussian is considered to be equal to σ for all basis functions. This formulation is related to the fuzzy logic approach of [15]. The connection between fuzzy membership functions and Gaussian basis functions is examined by [17]. In the following sections some problems are discussed concerning the integrated knowledge and model complexity.

2.2 Preserving the Rule–Based Knowledge

Equation (2.5) can be implemented as a network of normalized basis functions $\mathcal{NN}^{\text{init}}$ which describes the rule-based knowledge and which can be used for prediction. Actual training data can be used to improve network performance. We consider four different ways to ensure that the expert knowledge is preserved during training.

– *Forget.* We use the data to adapt $\mathcal{NN}^{\text{init}}$ with gradient descent (we typically adapt all parameters in the network). The sooner we stop training, the more of the initial expert knowledge is preserved.
– *Freeze.* We freeze the parameters in the initial network and introduce a new basis function whenever prediction and data show a large deviation. In this way, the network learns an additive correction to the initial network.
– *Correct.* Whereas normal weight decay penalizes the deviation of a parameter from zero, we penalize a parameter if it deviates from its initial value q_j^{init}:

$$E_P = \frac{\alpha_j}{2}\sum_j \left(q_j - q_j^{\text{init}}\right)^2, \tag{2.7}$$

where the q_j is a generic network parameter.

– *Internal teacher.* We formulate a penalty in terms of the mapping rather than in terms of the parameters:

$$E_P = \frac{\alpha}{2} \int \left(\mathcal{NN}^{\text{init}}(\mathbf{x}) - \mathcal{NN}(\mathbf{x}) \right)^2 d\mathbf{x}. \tag{2.8}$$

This has the advantage that we do not have to specify priors on relatively unintuitive network parameters. Instead, the prior directly reflects the certainty that we associate with the mapping of the initialized network which can often be estimated. [12] estimated this certainty from problem-specific knowledge. We can approximate the integral in (2.7) numerically by Monte-Carlo integration, which leads to a training procedure where we adapt the network with a mixture of measured training data and training data artificially generated by $\mathcal{NN}^{\text{init}}(\mathbf{x})$ at randomly chosen inputs. The mixing proportion directly relates to the weight of the penalty α [12].

2.3 Decreasing the Complexity

Although it is relatively easy to generate rules from a trained network, this by itself is not of much use unless the rules are understandable. One approach is to reduce the complexity of the network in a way that minimizes the effect on performance. Complexity can be associated with the number of basis functions (or fuzzy rules) and with the number of inputs (or variables) to the network. Also a network becomes simpler if individual basis functions depend only on a subset of inputs, which means that some of the connections from inputs to basis functions can be pruned. Complexity reduction can be performed in two ways, either after learning has converged to a solution by elimination of the least important components, inputs, or input connections, or during learning by introducing a term that penalizes complexity. In the following sections we pursue the former approach first. The goal is to achieve a network that produces a small number of important rules, each with a small number of assertions. A surprising side effect of these techniques is that, when training data is sparse, such complexity reduction can actually improve performance.

2.4 Pruning Units

The most dramatic pruning can be achieved by removing complete rules or basis functions. One approach is to remove an unimportant unit. So the following pruning strategy for Gaussian basis functions is suggested: Evaluate the relative weight of each basis function at its center $\omega_i = b_i(\mu_i) / \sum_j b_j(\mu_i)$ which is a measure of its importance in the network. Remove the unit with the smallest ω_i.

In some cases, this unit might represent data points far away from the centers of the remaining Gaussians. An alternative way to prune is therefore to tentatively remove a unit and then recalculate the likelihood of the model. Eventually, prune those units whose removal minimally decreases the likelihood. This possibilistic approach is discussed in detail in [16].

2.5 Removing Variables and Connections

In the possibilistic model, it is relatively easy to find out if, according to a trained model, for example, the last variable is independent of the remaining variables. Conditional independence is common if variables are highly correlated. We can rank input j based on this change in the likelihood of the model if input j is removed. The procedure simply consists of tentatively removing an input and examining the subsequent increase of the prediction error either on the training set or on a cross-validation data set (the latter is a subset of the data not used to train the network parameters). We then remove the input whose removal causes the least increase in error.

To remove a connection from a rule, use the following procedure. Successively set the largest σ_{ij} equal to infinity, effectively removing input j from basis function i. Sequentially remove basis functions and conjuncts until the error increases above a threshold. Retrain after a unit or a conjunct is removed.

All constraints, rules, variables, and connection pruning methods can also be implemented by penalty terms during training. We propose the following penalty function:

$$E_P = \frac{\alpha_1}{2} \sum_{ij} \frac{1}{\sigma_{ij}^2} + \frac{\alpha_2}{2} \sum_i |\kappa_i|. \tag{2.9}$$

The first term penalizes small variances and the second term tries to remove basis functions. In our experiments it was very difficult to pick sensible α_1 and α_2.

We found the first introduced procedures more robust. First, instead of penalizing small variances we do not permit variances to decrease beyond a certain threshold. In this way the bandwidth of the function that the network approximates can be limited. Second, the gradient descent algorithm also adapts κ_i. The adaptation adjusts κ_i according to the overall importance of a given rule. By simply removing the basis functions with the smallest κ_i we have a sensible way of gradually removing complexity.

The overall algorithm is given by the following simple steps: first, initialize the parameters α, μ, σ, κ and the weights w. Second, adapt all these parameters until the change of the error is smaller than a certain predefined limit. Third, prune the units, by continuing with the second step. Fourth, remove variables and connections, and continue again with the second step. Repeat these instructions until a minimal error is obtained.

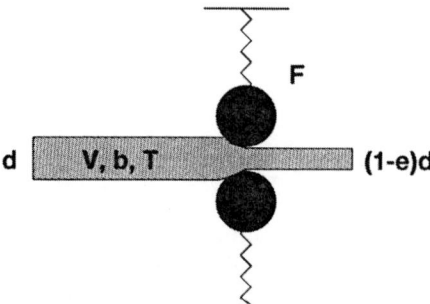

Fig. 2.2. Schematic illustration of a steel rolling mill: rolling force F, relative thickness reduction ϵ, thickness d of the strip entering the mill, width b, temperature T, vanadium concentration V

2.6 Application: Steel Rolling Mill

In the following, a problem is discussed from process automation in steel production. Steel rolling mill presettings are traditionally calculated with the help of mathematical models, which, unfortunately, supply neither a complete nor an exact description of the technical process. This makes continual recalculation of model parameters necessary, to adapt them to actual process events. A further drawback lies in the fact that the refinement potential for this procedure has been all but exhausted. To overcome these shortcomings, it was decided to use neuro-fuzzy systems.

The rolling force is dependent on a number of input values: the relative thickness reduction ϵ, the thickness d of the strip entering the mill, the width b, and temperature T, which influences the plasticity of the material. Additionally, the chemical composition of the strip influences the properties. In the steel production process the vanadium concentration V is important. The task we discuss is to identify the rolling force using the five most important variables V, b, T, d, and ϵ via fuzzy rules.

To decrease the model complexity Takagi–Sugeno rules with constant output are used. For training the RBF network 10 000 data are used. The network is trained by adapting the parameters w_i, c_i, and σ_{ij}. After training the network and finding a minimum of the error function by gradient descent, we obtain the parameters listed in Table 2.1.

There are ten fuzzy rules corresponding to the basis functions in the RBF-network. However, most of the rules are only of slight relevance, as may be seen from the τ_i values. Here we consider only these rules (basis functions) as relevant that have a τ_i value which is at least 20 percent of the highest one. So the number of rules reduces to four: rules 1, 2, 3, and 7. Reducing the number of rules rapidly increases the information level, but it should be noticed that the system precision is decreased as well.

In the following step the inputs and conjunctions of the remaining four rules will be reduced. This will be done by analyzing the σ_{ij}. If σ_{ij} of input x_j is high ($1/\sigma_{ij}$ is small), then a change of the input value x_j results only in a small change of the output of the basis function and also in the overall output of the whole network. In this case, the output value of the basis function is

Table 2.1. The network parameters after training. Each row includes the parameters for basis function Bf_i. Only the rows with a frame are considered with a sufficiently high τ_i.

	Bf_1	Bf_2	Bf_3	Bf_4	Bf_5	Bf_6	Bf_7	Bf_8	Bf_9	Bf_{10}
τ_i	5.4	20.7	4.5	1.2	2.4	1.0	15.1	2.9	2.6	3.6
$1/\sigma_{ij}$										
V	0.1	1.5	0.3	0.4	0.5	0.4	0.0	0.2	0.2	0.0
b	0.4	0.4	0.0	0.0	0.7	0.0	0.3	1.2	0.0	0.1
T	0.8	0.0	0.1	0.7	0.3	0.0	0.9	0.1	0.1	1.2
d	0.0	0.0	1.8	0.2	0.5	0.6	0.0	0.0	0.0	0.6
ϵ	1.1	0.7	0.0	0.4	0.7	0.0	0.9	0.2	0.0	0.0
c_{ij}										
V	0.1	0.1	−0.1	−0.1	−0.1	0.0	−0.1	−0.1	0.0	5.2
b	−0.7	0.9	0.9	−1.4	−0.9	0.4	0.6	−0.5	1.7	−1.4
T	1.1	−1.7	−0.3	−0.2	−0.7	1.2	−1.3	−0.1	0.6	−0.6
d	0.4	0.1	1.4	−0.6	1.2	−0.2	−0.5	−0.7	−0.2	5.0
ϵ	−0.1	2.0	−1.2	0.8	−1.1	0.2	1.2	0.4	−0.2	−2.7
w_i[MN]	−28.9	14.7	−6.8	58.2	−36.3	−87.9	12.6	3.9	9.8	41.2

independent of the considered input and does not have to be integrated in the rule. Here, too, we used 20 percent of the highest σ_{ij} value as the minimum considered as relevant.

After input reduction, the premise of rule 1 includes width b, temperature T and thickness reduction ϵ, the premise of rule 2 includes the values of V, b, and ϵ, the premise of rule 3 includes only d, and the premise of rule 4 includes b, T, and ϵ. The conclusion w_i describes the correction of the rolling force if rule i is valid. To build fuzzy rules, the centers c_{ij} will be transformed into linguistic variables. In this example, we scale the input variables x_j ($j = 1, \ldots, 5$) and the center c_{ij}, to a mean zero and a standard deviation one.

We decided to to use *low*, *medium*, and *high* as linguistic variables: *low* if $c_{ij} < -0.5$, *medium* if $-0.5 < c_{ij} < 0.5$, and *high* if $c_{ij} > 0.5$. This results in the following fuzzy rule base:

1. If b is *low*, T is *high*, and ϵ is *medium* **then** decrease the rolling force by -28.9 MN.
2. If V is *medium*, b is *high*, and ϵ is *high* **then** increase the rolling force by 14.7 MN.
3. If d is *high* **then** decrease the rolling force by -6.8 MN.
4. If b is *high*, T is *low*, and ϵ is *high* **then** increase the rolling force by 12.6 MN.

Note that the fuzzy sets and the membership functions are different for each c_{ij}. For an exact transformation, each linguistic value and membership function needs an index i.

The four rules could be interpreted and related in detail with the expert's experience. The performance of the small rule base is quite consistent with the performance of the classical mathematical model. In this conventional model, the number of parameters is huge and an interpretation is not possible, whereas the fuzzy system is small and interpretable.

3. Density Estimation

3.1 Probabilistic Interpretation of Gaussian Basis Function Networks

An important goal of learning is to discover dependencies between carefully chosen variables z_j based on a number of observed instances, the training data. The perfect prediction of one of those variables from the remaining variables is in many cases impossible. One reason might be that these variables depend on other inaccessible variables such that a model formed on the basis of the available information is necessarily incomplete. Therefore, this section focuses on a framework in which probabilistic predictions are possible to deal with the problem of missing inputs. Furthermore, we assume that it is not clear a priori which of the variables z_j later will be known (will be the queries) and which have to be predicted. Therefore, we would like to guarantee a model which allows us to make prediction about any unknown variable from any combination of known variables. From a statistical point of view, complete knowledge about dependencies contained in the joint probability density can be approximated by a probabilistic mixture model of the form:

$$P(z) = \sum_{i=1}^{N} P(z|s_i) P(s_i). \tag{3.1}$$

According to the model assumptions, a component in the mixture model s_i is selected with a priori probability $P(z|s_i)$. In this paper, we use a mixture of multivariate Gaussians

$$P(z|s_i) = \mathcal{G}(z; c_i, \sigma_i) = \frac{1}{(2\pi)^{\frac{M}{2}} \prod_{j=1}^{M} \sigma_{ij}} \exp\left[-\frac{1}{2} \sum_{j=1}^{M} \frac{(z_j - c_{ij})^2}{\sigma_{ij}^2} \right] \tag{3.2}$$

where c_i is the location of the center of the i-th Gaussian and σ_{ij} corresponds to the width of the i-th Gaussian in the j-th dimension and M is the number of inputs. Since a weighted sum of Gaussians can approximate any continuous function arbitrarily well, we can also expect that the Gaussian mixture can approximate a large class of densities.

Unless known a priori, the parameter in the model, i.e., the components of the center c_{ij}, the widths σ_{ij}, and the a priori component probability $\kappa_i = P(s_i)$ are learned from the training data. A maximum likelihood estimate of these parameters can be obtained effectively by the EM algorithm, which in most cases is used off-line although on-line version also exists. A significant advantage of the EM learning rule is that it can be modified easily to include incomplete and uncertain data.

We can extract simple rules from a trained model by realizing that each Gaussian in a mixture model contributes to the density estimate only in a localized convex region of input space. This is analogous to a boolean expression in disjunctive normal form (where each condition is true for a single region in space) except that a mixture model is probabilistic.

3.2 From Modeling to Prediction

Our mixture model in (3.2) describes or explains the data. In this section we want to show how the model can be used for prediction (prediction in the sense of output computation). Let x be a vector that consists of all the known variables of z and let y be one of the variables we would like to predict. Correspondingly, c_{ix}, c_{iy}, σ_{ix}, and σ_{iy} consist of the components of c_i and σ_i in the dimensions of x and y. Using our mixture, we can easily calculate the conditional density

$$P(y \mid x) = \frac{P(x \mid y)}{P(x)} = \frac{\sum_{i=1}^{N} P(y \mid x, s_i) P(x, \mid s_i) P(s_i)}{\sum_{j=1}^{N} P(x \mid s_j) P(s_j)}. \tag{3.3}$$

We define a basis function as $b_i = P(x|s_i)P(s_i)$ and a normalized function as

$$n_i(x) = P(x \mid s_i) = \frac{P(x \mid s_i) P(s_i)}{\sum_{j=1}^{N} P(x \mid s_j) P(s_j)}. \tag{3.4}$$

The conditional expected value can now be written:

$$\begin{aligned}
\mathcal{E}(y \mid x) &= \frac{\sum_{i=1}^{N} w_i(x) P(x \mid s_i) P(s_i)}{\sum_{j=1}^{N} P(x \mid s_j) P(s_j)} \\
&= \frac{\sum_{i=1}^{N} w_i(x) b_i(x)}{\sum_{j=1}^{N} b_j(x)} = \sum_{i=1}^{N} w_i(x) n_i(x),
\end{aligned} \tag{3.5}$$

where $w_i(x) = \int y P(y|x, s_i) dy$. For our Gaussian model (3.2), $w_i(x) = w_i = c_{ij}$ is a constant and is the center of the i-th Gaussian along dimension y. The marginal distribution of x becomes

$$P(x|s_i) = \int \mathcal{G}(z; c_i, \sigma_i) d\{z_j | z_j \notin x\} = \mathcal{G}(z; c_{ix}, \sigma_{ix}). \tag{3.6}$$

The marginal distribution of a Gaussian is again a Gaussian: it is simply the projection of the Gaussian onto the dimensions of x.

Equation (3.5) can be realized by a neural network of basis functions b_i which are normalized by the sum over the activity of all basis functions. These networks were introduced into the neural network framework by [10], [14], and [9] who used them as regressors and for smooth interpolation between data.

Note that with the mixture model, we can generate the rule base for the prediction of any variable from any set of known variables. The marginal distributions $P(x|s_i)$ can be found by simply projecting the Gaussians onto the set of available dimensions (3.6). The form of (3.5) is equivalent to that of (2.5).

3.3 The EM Algorithm for Missing Inputs

When c_i and Σ_i of the Gaussian representation of a rule, and the a priori probability $P(s_i)$ are all unknown, then the maximum likelihood estimate normally does not result in useful answers, since it is then possible to obtain arbitrarily large likelihoods. If only the largest of the finite local maxima are regarded, however, then it is possible to obtain a useful answer. In [6] an iterative method for finding such solutions is given. For our purpose an adjusted version is needed that takes into account missing values.

In order to accommodate learning with missing values, some assumption has to be made on how missing data can be substituted. Afterwards a new estimation functions has to be found together with a new maximum likelihood function, from which the new EM steps can be derived. The method for filling in missing data during the learning phase, referred to in [1], is to take the expectation of the missing value, given the current model. Using Gaussians as basis functions and substituting Gaussians for $p(\mathbf{x} \mid s_i)$, this can be formulated as

$$E(\tilde{\mathbf{x}} \mid \mathbf{x}) = \frac{\sum_{i=1}^{c} \tilde{\mathbf{c}}_i \mathcal{G}(\mathbf{x}; \mathbf{c}_i, \Sigma_i) P(s_i)}{\sum_{i=1}^{c} \mathcal{G}(\mathbf{x}; \mathbf{c}_i, \Sigma_i) P(s_i)}, \tag{3.7}$$

where the variables indicated with a tilde stand for the part of the vector containing the missing values, the others for known parts. The multivariate Gaussians are simply projected on the known dimensions and evaluated, after which a weighted average can be taken from the centers learned so far.

In the approach developed only the diagonals of the covariance matrix are taken. This implies limitations on the orientation of the Gaussians: either only the values on the diagonal of the covariance matrix are used or only the width of a Gaussian in a certain dimension is specified; rotation of Gaussians is not controlled. Clearly this can result in difficulties when trying to fit a Gaussian on certain clusters of points, for example when a cluster of samples is oriented diagonally in a two-dimensional space. But limitations in speed, memory, and a too-small set of measurements in practice force the use of something less bulky then the covariance matrix. To avoid confusion the variable s will be used to indicate the diagonal elements of Σ.

From now on we call the complete vector containing missing values $\tilde{\mathbf{x}}$ and the unknown true vector $\mathbf{x}$. Then the log-likelihood of the data obtained this way is

$$
\begin{aligned}
\ell &= \sum_{k=1}^{n} \log \int \sum_{i=1}^{c} \mathcal{G}(\mathbf{x}; \mathbf{c}_i, s_i) P(s_i) \mathcal{G}(\tilde{\mathbf{x}}_k; \mathbf{x}, \sigma_k) \, d\mathbf{x} \\
&= \sum_{k=1}^{n} \log \sum_{i=1}^{c} \mathcal{G}(\tilde{\mathbf{x}}_k; \mathbf{c}, S_{ik}) P(s_i).
\end{aligned} \tag{3.8}
$$

Here, $\mathcal{G}(\tilde{\mathbf{x}}_i; \mathbf{x}, \sigma_k)$ stands for the probability $P(\tilde{\mathbf{x}} \mid \mathbf{x})$, where σ_k is a vector indicating each element of pattern $\tilde{\mathbf{x}}$ whether that element is missing ($\sigma_{kj} = \infty$) or not ($\sigma_{kj} = 0$). The new S_{ik} is ∞ if the corresponding element is missing, otherwise its value is s_{ij}. The resulting equation is almost the same in case $\tilde{\mathbf{x}}$ would not contain any missing values (in which the last terms would be $P(\mathbf{x}_k \mid s_i) P(s_i)$), except for the fact that now, for each occurrence of a missing value, the variance of the model is substituted by infinity, which has the effect of simply ignoring that dimension in calculating the Gaussian for that specific model and $\tilde{\mathbf{x}}_k$.

In the mentioned article an EM algorithm was developed that handled both missing and uncertain data, in which case the vector σ could also take values other than 0 or ∞. It is not difficult to derive the required simpler algorithm from it, since one simply uses only the limiting cases. Algorithm 1 is the result of this simplification. As can be seen, the extra steps to handle missing elements produce hardly any extra overhead.

Algorithm 1 (EM adjusted for missing inputs)

Input: *The data set, containing deviations and missing values, $\mathcal{X}$, consisting of the elements $\tilde{\mathbf{x}}_k$, and the initial values $P_1(s_i)$, $\mathbf{c}_{i1}$, and s_{i1}, for each rule i. Furthermore the vectors σ_k must be given, which specify which elements of which vector $\tilde{\mathbf{x}}_k$ are missing.*

Output: *A local maximum likelihood estimate for $P(s_i)$, $\mathbf{c}_i$ and s_i, for each rule i.*

begin

    for $m = 1, \ldots$ **do**

        for *all rules i; all patterns k; all dimensions j* **do**

$$
P_{m+1}(s_i \mid \tilde{\mathbf{x}}_k) = \frac{G(\tilde{\mathbf{x}}_k; \mathbf{c}_{im}, S_{ik}) P_m(s_i)}{\sum_{j=1}^{c} G(\tilde{\mathbf{x}}_k; \mathbf{c}_{jm}, S_{jk}) P_m(s_j)}
$$

$$
P_{m+1}(s_i) = \frac{1}{n} \sum_{k=1}^{n} P_{m+1}(s_i \mid \tilde{\mathbf{x}}_k)
$$

$$
\mathbf{c}_{ijm+1} =
\begin{cases}
\dfrac{\sum_{k=1}^{n} \tilde{\mathbf{x}}_{jk} P_{m+1}(s_i \mid \tilde{\mathbf{x}}_k)}{\sum_{k=1}^{n} P_{m+1}(s_i \mid \tilde{\mathbf{x}}_k)} & : \quad \sigma_{jk} < \infty \\[2mm]
\mathbf{c}_{ijm} & : \quad \sigma_{jk} = \infty
\end{cases}
$$

$$
s_{ijm+1}^2 =
\begin{cases}
\dfrac{\sum_{k=1}^{n} (\tilde{\mathbf{x}}_{jk} - \mathbf{c}_{ijm+1})^2 P_{n+1}(s_i \mid \tilde{\mathbf{x}}_k)}{\sum_{k=1}^{n} P_{m+1}(s_i \mid \tilde{\mathbf{x}}_k)} & : \quad \sigma_{jk} < \infty \\[2mm]
s_{ijm}^2 & : \quad \sigma_{jk} = \infty
\end{cases}
$$

if *either the number of iterations or a satisfying answer is reached*
then stop

end

This algorithm reduces to the standard EM algorithm when every element of σ_k is set to zero for all k. The Gaussian is the usual multivariate normal distribution, simplified for the case of a diagonal covariance matrix:

$$
\mathcal{G}(\mathbf{x}; \mathbf{c}, s) = \frac{1}{(2\pi)^{\frac{p}{2}} \prod_{j=1}^{M} s_j} \exp \left[-\frac{1}{2} \sum_{j=1}^{M} \frac{(\mathbf{x}_j - \mathbf{c}_j)}{s_j^2} \right]. \tag{3.9}
$$

3.4 Example: Problem of the Pulp Production Process

The theory developed in the previous sections will be tested on a pulp production process example. The most important material for making paper is pulp. Before and during the process of making pulp, different measurements are made. They concern controllable attributes, such as boiling time, and uncontrollable attributes, like properties of the mixture. Furthermore, one quality of the resulting pulp is measured. The value can only be measured after the batch has been stopped. The objective is to be able to estimate during the batch, what the value will be after the batch is stopped. The EM algorithm can be used for modeling with missing inputs. The set with complete samples was of size 198; the set containing incomplete samples was of size 325. The training results are displayed in Fig. 3.1. In this case, the expectation value has been used to make the predictions. The best models (lowest test error) of each graph are displayed in Table 3.1. The rule extraction procedure works the same as in the neuro-fuzzy-approach: the premise is defined by the Gaussian multivariate normal distribution (comparable with basis functions of Sect. 2), and the conclusions are given by the parameter $\mathbf{c}$ (comparable with the weights $\mathbf{w}$ of Sect. 2).

Table 3.1. Best results of the EM

	Error	Number of rules
With incomplete samples	21.14	20
Without incomplete samples	16.80	16

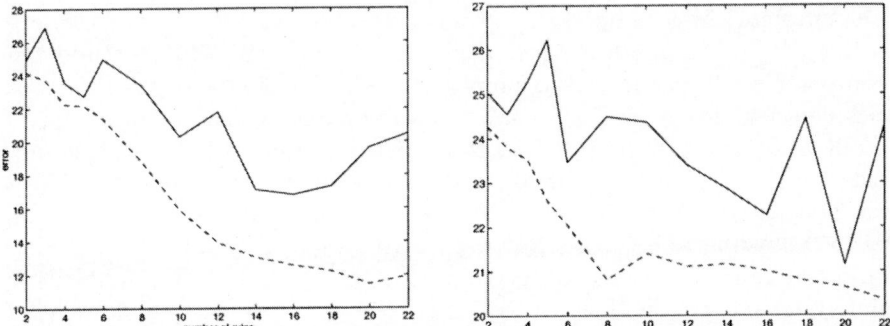

Fig. 3.1. EM: In the left graph, the training results of the models that were trained with complete samples are displayed. In the right graph, the training results of the models that were trained with incomplete samples are displayed. The dashed line denotes the error on the training set, the solid line the generalization error. The predictions have been computed by taking the expectation value.

4. Fuzzy Clustering

Until now we have assumed that the training samples used to design a classifier were labeled to show their category membership. The procedures using labeled samples are said to be supervised. The RBF network is a supervised training paradigm as well as the Gaussian distributions algorithm discussed in Sect. 3. Often methods are formalized as hybrid formalisms between supervised and unsupervised learning methods. Now we shall investigate a typical unsupervised procedure that uses unlabeled samples, that is, a collection of samples is given without their classification or target value.

There are reasons why it is interesting in general to examine the problem of dealing with unlabeled samples and what is possible to learn from them. The first reason is that the collecting and labeling of data sets is very extensive in cost and time. This is in general not the reason here. The second reason is that it is of interest to gain some general insight into the data's structure and nature. This is a very useful side effect in our investigations. Thirdly, a slow change in pattern characteristics in time can be recognized. This is desirable in many practical applications. Finally, the data could be prestructured with an unsupervised clustering method before being included in a supervised learning task. This is the main reason for using the method here. Cluster algorithms are used to build classifiers to identify local regions of the pattern domain. Using hard clustering, patterns belong to only one class. In fuzzy clustering, patterns belong to more than one cluster, but with a certain membership degree.

There are a couple of clustering methods that can be categorized in different groups of algorithms. One possibility to separate them concerns their axiomatic criterion. Using this, three main classes can be assigned: hierarchical, graph theoretic, and iterative or objective functional methods. In hierarchical cluster algorithms samples, are grouped together at certain levels, finally resulting in a corresponding tree, called a dendrogram. This method can be compared with biological taxonomy, where individuals are grouped into species, species into genera, genera into families, and so on. Graph theoretic methods are based on the interpretation of the samples as a node set, and the edge weight between a pair of nodes can be based on a measure of similarity between the pair. Searching for a minimal spanning tree by breaking edges to form subgraphs is an often used clustering strategy. In the last class an objective function is used to formulate an iterative adaptation algorithm. Typically, local extrema of the objective function are defined as optimal clusterings. Starting with an initial random cluster configuration, the position of the clusters is stepwise improved. The clustering methods presented in this investigation are members of the last of these classes of algorithms.

As also in the RBF and EM approaches, fuzzy clustering is useful to divide the input domain into several subregions. In these regions input-output dependencies can be approximated, by, e.g., constant, linear, or polynomial functions. Using the uncertain boundaries of the defined regions, a weighted normalized sum computes the overall output. The idea is the same as in the former section, but the methods produce different results.

Fuzzy clustering has been developed to a well-known method for system identification. A lot of different variants of the algorithm have been developed and successfully implemented. Bezdek [3] has collected and described the basic theory and a number of authors have shown its practicability generating Takagi–Sugeno fuzzy systems, such as [2].

In this section, the basic theory will be described to develop the ability to build application dependent solutions. Afterwards some pointers are given to certain variants of the algorithm. In contrast to other approaches, fuzzy clustering is used here to partition only the input domain and not the whole input-output space. The reason is the separation of the unsupervised and supervised learning steps for a local knowledge representation in the model. Fuzzy clustering is an extension of classical hard c-means clustering (HCM). Hard clustering divide a multi-dimensional space into c subregions with hard boundaries. Hence every element of the data set belongs to exactly one cluster. The cluster chosen is the one with the minimal distance defined by a certain metric. The number of clusters c should be terminated heuristically at the beginning. Visualization methods and nonlinear projections of multidimensional data such as hierarchical clustering or Sammon mapping [6] can help fix the number of clusters. Another possibility to fix the number of clusters automatically is to start with one cluster and increase the number step by step or, the other way round, to decrease the cluster number.

The algorithm for HCM adapts the first randomly initialized cluster centers step by step by minimizing the sum of the distances to the data belonging to the clusters. Membership in a cluster $u_{ik} \in \{0, 1\}$ is determined by the minimal distance to the different cluster centers.

Given a set of data

$$X = \{x_1, \ldots, x_n\} \subset \mathcal{X} \tag{4.1}$$

and a set of c cluster centers

$$V = \{v_1, \ldots, v_c\} \subset \mathcal{X}, \tag{4.2}$$

the hard c-means algorithm determines $u_{ik} \in \{0, 1\}$, the membership of the data x_k in the clusters represented by v_i so that the objective function is minimized. The cost function is given by

$$J(\{u_{ik}\}, \{v_i\}) = \sum_{k=1}^{n} \sum_{i=1}^{c} u_{ik}^m d_{ik}^2. \tag{4.3}$$

The hard-c-means algorithm works as follows:

Step 1: Initialize all u_{ik} and v_i with random values.

Step 2: Calculate the cluster centers as

$$v_i := \frac{\sum\limits_{k=1}^{n} u_{ik}\, x_k}{\sum\limits_{k=1}^{n} u_{ik}}. \tag{4.4}$$

Step 3: Calculate the cluster memberships as

$$u_{ik} := \begin{cases} 1, & d_{ik} = \min\limits_{1 \le j \le c} \{d_{jk}\} \\ 0, & \text{otherwise} \end{cases} \tag{4.5}$$

with the Euclidian distance

$$d_{ik} = \|x_k - v_i\|. \tag{4.6}$$

Step 4: If some u_{ik} changed more than a threshold ε, repeat from step (2).

A simple method to use HCM clustering for determining fuzzy system premises is to set a rule for every cluster and use the inverse of the distance of the data from the cluster centers as membership value.

Fuzzy clustering (FCM) is very similar to HCM. The difference is the range of membership values, which are not only one or zero, but continuous in this interval. The FCM algorithm is the same as the HCM, only with an additional adaptation equation for the membership values u. The FCM clustering is given by the following steps:

Step 1: Initialize all u_{ik} and v_i with random values.

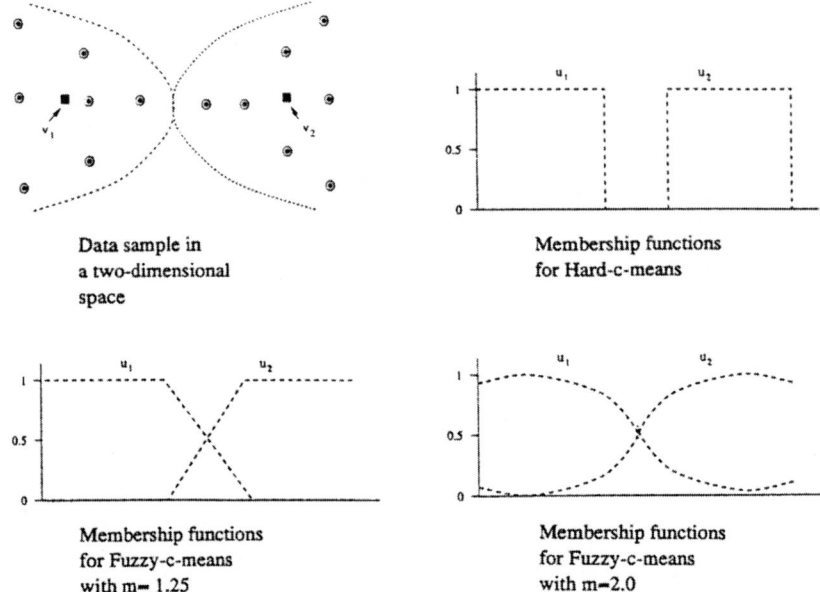

Fig. 4.1. Membership functions using *hard c-means* and *fuzzy c-means* clustering

Step 2: Calculate the cluster centers as

$$v_i := \frac{\sum\limits_{k=1}^{n} u_{ik}^{m}\, x_k}{\sum\limits_{k=1}^{n} u_{ik}^{m}} \tag{4.7}$$

Step 3: Calculate the cluster memberships as

$$u_{ik} := 1 \Big/ \sum_{j=1}^{c} \left(\frac{d_{ik}}{d_{jk}}\right)^{\frac{2}{m-1}} \tag{4.8}$$

with the Euclidian distance

$$d_{ik} = \|x_k - v_i\|. \tag{4.9}$$

Step 4: If some u_{ik} changed more than a threshold ε, repeat from step (2).

The formal derivation of the adaption rule for the membership values is given in Appendix A. As shown in the HCM description, there are a lot of possibilities for computation of the membership values. The derivation is given as an example for deriving an adaptation rule with certain constraints.

While the fuzzy c-means algorithm determines cluster center points, the fuzzy c-lines algorithm uses lines as prototypes [4]. The clusters are described by a point v_i and a direction d_i. The distances D_{ik} between the data points and the clusters are calculated as

$$D_{ik} = \sqrt{\|x_k - v_i\|^2 - ((x_k - v_i)^T d_i)^2},$$ (4.10)

with d_i being the unit eigenvectors of the maximum eigenvalues of the matrix

$$S_i = \sum_{k=1}^{n} u_{ik}^m (x_k - v_i)(x_k - v_i)^T.$$ (4.11)

The prototypes used by the fuzzy c-lines algorithm are infinite lines. The fuzzy c-elliptotypes (FCE) algorithm uses convex combinations of points and lines (or hyperplanes). These prototypes still represent linear structures, but are locally restricted [5]. For one-dimensional elliptotypes the distances D_{ik} are determined as

$$D_{ik} = \sqrt{\|x_k - v_i\|^2 - \alpha ((x_k - v_i)^T d_i)^2},$$ (4.12)

or, more generally, for the elliptotype dimensions p and q, the distances are

$$D_{ik} = \sqrt{\|x_k - v_i\|^2 - \sum_{j=1}^{p} ((x_k - v_i)^T d_{ij})^2 - \alpha \sum_{j=p+1}^{q} ((x_k - v_i)^T d_{ij})^2},$$

with d_{ij} the unit eigenvectors of the j-th largest eigenvalues of the cluster fuzzy scatter matrix.

The factor α determines the sensitivity of the algorithm concerning the direction of the clusters. For $\alpha = 0$, it does not care about the direction and converges to the fuzzy c-means algorithm. For $\alpha = 1$, the cluster location is ignored, i.e., FCE is reduced to the fuzzy c-lines method.

4.1 Process Analysis

Fuzzy clustering is successfully applied in the analysis of paper production processes. To increase productivity and quality, and to minimize energy, raw material, and the environmental influence, data for the process are analyzed. A process model can be generated by fuzzy clustering, which can be used afterwards for process optimization. In the following example, the data originate from the production process of a recycled paper factory. After data acquisition a preprocessing by checking the intervals, filtering, and meaning is done. The data involves measurements, process and machine parameters, and economic and quality parameters of the final product.

After a few steps of preprocessing, the variables are analyzed with fuzzy clustering methods. It can be seen that the examination of the parameter dependencies using a simple correlation analysis is not representative for modeling. Examining the connection between an important measurement and the paper quality with the fuzzy c-lines algorithm results in Fig. 4.2 (a). Here the number of clusters is estimated as two. If a second input variable is added and

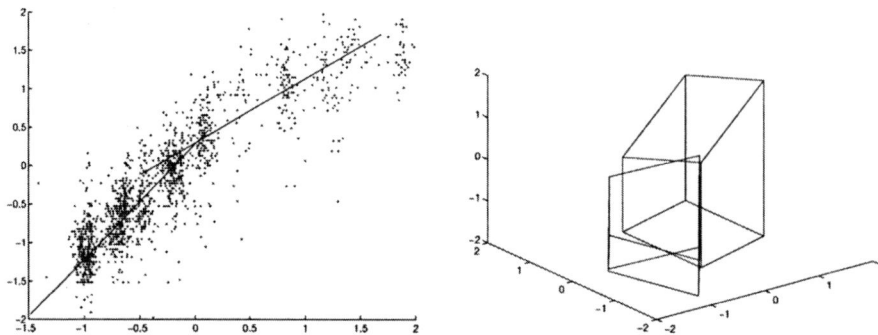

Fig. 4.2. (a) Fuzzy c-lines, (b) fuzzy c-elliptotypes

the space is clustered again, the new result includes the higher-dimensional dependency. The position of the planes shows, that the two-dimensional analysis does not describe the problem sufficiently.

4.2 Extraction of Fuzzy Rules

If the components of a data set represent the input and output variable of a technical system, the fuzzy clusters can be interpreted as fuzzy samples of a system transfer function. A fuzzy cluster could be interpreted as a Mamdani rule. Using the the results of fuzzy c-elliptotype clustering, beside the cluster center, a linear function is available, which can be interpreted as a Takagi–Sugeno conclusion of first order. Practice shows that it is useful to cluster only the input space and to compute afterwards a weighted (concerning the cluster memberships) linear regression for each cluster. This linear regression results in the conclusion of the Takagi–Sugeno rule. The derivation of the membership function of the fuzzy sets in the premise is a little bit more complicated. It is easy to compute for each value a fuzziness, a certain membership to the fuzzy cluster (fuzzy set), but it is more difficult to approximate a whole membership function out of the single membership values. Therefore, standard approximation methods, with combinations of sigmoidal and Gaussian basis functions, are used for membership function approximation.

5. Conclusion

Model identification is very difficult and nontrivial in high-dimensional nonlinear problems. So every available source of knowledge about the real world system should be used for modeling. Fuzzy rule-based systems combined with data-based learning strategies build a practicable solution for a lot of complex problems. The other advantage for integrating fuzzy based and data-driven approaches is based on the possibility of extracting rule-based knowledge after data-based optimization. These two reasons make fuzzy identification using adaptive optimization techniques a practicable solution for modeling. In Sect. 4 we showed how fuzzy systems of Takagi–Sugeno type and learning algorithms with local knowledge representation can be combined for identification. A two-step learning paradigm is postulated, which first identifies certain regions of the domain, and second computes supervised input-output approximations in the subdomains With respect to the different cluster outputs an overall computation is done by weighted normalized summation, corresponding to the defuzzification step in fuzzy systems. The second step is quite simple in computation for low-order approximation functions. For the first step, three different methods were presented to divide the input domain and to extract fuzzy regions. The first is a neuro-fuzzy approach whose practicability is shown in a couple of applications [12]. The advantage of the second probabilistic approach is the reconstruction of missing values in the input data. In the third section fuzzy clustering was explained, the most obvious approach for fuzzy identification. In all methods the technical foundation were given with a theoretical background, so it should be easy to judge each technique, and to develop an application dependent solution.

Appendices

A. From Rules to Networks

We consider Gaussian basis functions with a diagonal covariance matrix Σ_i. Here, σ_{ij}^2 is the j-th diagonal element in Σ_i. The parameter $A_j^i = \mu_{ij}$ defines the position in the j-th dimension of the input space where $Rule_i$ has its largest validity. The parameter $range_{ij} = \sigma_{ij}$ indicates approximately the range in which $Rule_i$ is valid in the j-th input dimension.

Let us consider as an example the case where the input space is 2-dimensional. The $Rule_1$

> **If x_1 is A_1^1 ($range_{11} = R_1^1$) and x_2 is A_2^1 ($range_{12} = R_2^1$) then y is B^1**

corresponds to a Gaussian centered at $\mu_{11} = A_1^1$, $\mu_{12} = A_2^1$ with $\sigma_{11} = R_1^1$ and $\sigma_{12} = R_2^1$ and attached weight $w_1 = B^1$.

The $Rule_2$

> **If x_1 is A_1^2 ($range_{21} = R_1^2$) then y is B^2**

corresponds to a 1-dimensional Gaussian centered at $\mu_{21} = A_1^2$ with $\sigma_{21} = R_1^2$ and weight $w_2 = B^2$. Note, that the basis function is independent of x_2.

Finally, the $Rule_3$

> **If x_1 is A_1^3 ($range_{31} = R_1^3$) or x_2 is A_2^3 ($range_{32} = R_2^3$) then y is B^3**

is first decomposed into the two rules $Rule_{3a}$

If x_1 is A_1^3 ($range_{31} = R_1^3$) **then** y is B^3

and $Rule_{3b}$

If x_2 is A_2^3 ($range_{32} = R_2^3$) **then** y is B^3

and then we can proceed as we did with $Rule_2$. Note that in regions of the input space where $Rule_{3a}$ and $Rule_{3b}$ overlap, the validity of both rules is added and $Rule_3$ has a higher validity there as well.

B. Learning Rule for RBF Networks

In this section the learning algorithm is given for an RBF network derived by a generally used cost function. Let y_k be the output of a unit in the output layer, and t_k the target value of the output; then $(t_k - y_k)$ with e_k and the index k over all training patterns denotes the difference of the outputs and the cost function E is defined by

$$E = \frac{1}{2} \sum_k (t_k - y_k)^2 = \frac{1}{2} \sum_k (e_k)^2, \tag{B.1}$$

The gradient for the weights w_i and the three other adaptive parameters κ, c, and σ of the RBF network are described as follows:

$$\frac{\partial E}{\partial w_i} = -\sum_k e_k \frac{b_i(x_k)}{\sum_j b_j(x_k)} \tag{B.2}$$

$$\frac{\partial E}{\partial q_i} = -\sum_k e_k \frac{w_i - y_k}{\sum_j b_j(x_k)} \frac{\partial}{\partial q_i} b_i(x_k). \tag{B.3}$$

The partial derivatives of the basis function for q_i ($q_i : c_{ij}, \sigma_{ij}, \kappa_i$) are expressed by:

$$\begin{aligned}
\frac{\partial b_i(x_k)}{\partial c_{ij}} &= b_i(x_k) \frac{x_{kj} - c_{ij}}{\sigma_{ij}^2} \\
\frac{\partial b_i(x_k)}{\partial \sigma_{ij}} &= b_i(x_k) \frac{x_{kj} - c_{ij}}{\sigma_{ij}^3} \\
\frac{\partial b_i(x_k)}{\partial \kappa_i} &= \frac{b_i(x_k)}{\kappa_i}.
\end{aligned} \tag{B.4}$$

C. Update Equations for Gaussian Mixtures

We want to maximize ($\mathcal{N}$ denotes a normal distribution):

$$L = \prod_k \left[\sum_i \mathcal{N}(\mathbf{x}^k; \mu_i, \Sigma_i) P(s_i) \right]. \tag{C.1}$$

This leads to the update equations (assuming a diagonal covariance matrix, q_i is a generic parameter):

$$\frac{\partial}{\partial q_i} \log L \quad = \quad \sum_k \frac{1}{p(\mathbf{x}^k)} \frac{\partial}{\partial q_i} p(\mathbf{x}^k)$$

$$\frac{\partial}{\partial \mu_{ij}} p(\mathbf{x}^k) \quad = \quad -\mathcal{N}(\mathbf{x}^k; \mu_i, \Sigma_i) P(s_i) \, (x_{ij}^k - \mu_{ij})/\sigma_{ij}^2$$

$$\frac{\partial}{\partial \sigma_{ij}} p(\mathbf{x}^k) \quad = \quad -\mathcal{N}(\mathbf{x}^k; \mu_i, \Sigma_i) P(s_i) \, [(x_{ij}^k - \mu_{ij})^2/\sigma_{ij}^3 - 1/\sigma_{ij}]$$

$$\frac{\partial}{\partial P(s_i)} p(\mathbf{x}^k) \quad = \quad \mathcal{N}(\mathbf{x}^k; \mu_i, \Sigma_i) P(s_i) \tag{C.2}$$

It is easy to expend these algorithms in the case that we put priors on the parameters.

The solution has to satisfy the following fixed point equations:

$$\mu_{jl} \quad = \quad \frac{\sum_k P(s_i|\mathbf{x}^k) \, x_{jl}^k}{\sum_k P(s_i|\mathbf{x}^k)}$$

$$\sigma_{jl}^2 \quad = \quad \frac{\sum_k P(s_i|\mathbf{x}^k)(x_{jl}^k - \mu_{jl})^2}{\sum_k P(s_i|\mathbf{x}^k)}$$

$$P(s_i) \quad = \quad 1/K \sum_k P(s_i|\mathbf{x}^k)$$

$$P(s_i|\mathbf{x}^k) \quad = \quad \frac{\mathcal{N}(\mathbf{x}^k; \mu_i, \Sigma_i) P(s_i)}{\sum_j \mathcal{N}(\mathbf{x}^k; \mu_j, \Sigma_j) P(s_j)} \tag{C.3}$$

If we put priors on the parameters, in general, the solution cannot be written down in closed form. If we put a quadratic penalty $\alpha/2(\mu_{jl}^{\text{init}} - \mu_{jl})^2$ on a center, however, the closed form solution exists and becomes

$$\mu_{jl} = \frac{1/\sigma_{jl}^2 \sum_k P(s_i|\mathbf{x}^k) x_{jl}^k + \alpha \mu_{jl}^{\text{init}}}{1/\sigma_{jl}^2 \sum_k P(s_i|\mathbf{x}^k) + \alpha}. \tag{C.4}$$

D. Adaptation Algorithm for Fuzzy Clustering

In the following section the derivation of the fuzzy clustering adaptation algorithm is given as it is described in [3].

First fix $\mathbf{v}$ and define $g_m(U) = J_m(U, \mathbf{v})$ (with $J_m(U, \mathbf{v})$ the cost function, which should be minimized) for any U. Solutions for the minimization of $g_m(U)$ is effected with Lagrange multipliers. For each term let

$$g_{mk}(\mathbf{u}_k) = \sum_{i=1}^{c} u_{ik}^m \cdot d_{ik}^2 \tag{D.1}$$

and let its Lagrangian be

$$F_k(\lambda, \mathbf{u}_k) = \sum_{i=1}^{c} u_{ik}^m \cdot d_{ik}^2 - \lambda \left(\sum_{i=1}^{c} u_{ik} - 1 \right). \tag{D.2}$$

$(\lambda, \mathbf{u}_k)$ is stationary for F_k only if $\nabla_{\lambda, \mathbf{u}_k} F_k(\lambda, \mathbf{u}_k) = (0, \theta \in \mathbb{R}^c)$. Setting this gradient equal to zero yields

$$\frac{\partial F_k}{\partial \lambda}(\lambda, \mathbf{u}_k) = \sum_{i=1}^{c} u_{ik} - 1 = 0 \tag{D.3}$$

$$\frac{\partial F_k}{\partial u_{st}}(\lambda, \mathbf{u}_k) = m \cdot u_{st}^{m-1} \cdot d_{st}^2 - \lambda = 0. \tag{D.4}$$

Hence

$$u_{st} = \left(\frac{\lambda}{m \cdot d_{st}^2} \right)^{\frac{1}{m-1}} = \left(\frac{\lambda}{m} \right)^{\frac{1}{m-1}} \cdot \left(\frac{1}{d_{st}^2} \right)^{\frac{1}{m-1}}. \tag{D.5}$$

Using D.3,

$$\begin{aligned}
\sum_{i=1}^{c} u_{it} &= \sum_{i=1}^{c} \left(\frac{\lambda}{m} \right)^{\frac{1}{m-1}} \cdot \left(\frac{1}{d_{it}^2} \right)^{\frac{1}{m-1}} \\
&= \left(\frac{\lambda}{m} \right)^{\frac{1}{m-1}} \sum_{i=1}^{c} \left(\frac{1}{d_{it}^2} \right)^{\frac{1}{m-1}} = 1.
\end{aligned} \tag{D.6}$$

Thus,

$$\left(\frac{\lambda}{m} \right)^{\frac{1}{m-1}} = \sum_{i=1}^{c} \left(\frac{1}{d_{it}^2} \right)^{-\frac{1}{m-1}}. \tag{D.7}$$

Returning to D.5,

$$u_{st} = \sum_{i=1}^{c} \left(\frac{1}{d_{it}^2} \right)^{-\frac{1}{m-1}} \cdot \left(\frac{1}{d_{st}^2} \right)^{\frac{1}{m-1}} = \sum_{i=1}^{c} \left(\frac{d_{st}}{d_{it}} \right)^{-\frac{1}{m-1}}. \tag{D.8}$$

References

1. D. S. Ahmad and V. Tresp. Some solutions to the missing feature problem in vision. In: *Advances in Neural Information Processing Systems 5*,S. J. Hanson, J. D. Cowan and C. L. Jiles, eds.,Morgan Kaufmann, San Mateo, CA, 1993, pp. 393–400.
2. R. Babuška, H. B. Verbruggen. Constructing fuzzy models by product space clustering. This volume.
3. J. C. Bezdek. *Pattern Recognition with Fuzzy Objective Function Algorithms.* Plenum Press, 1981.
4. J. C. Bezdek, C. Coray, R. Gunderson, J. Watson. Detection and characterization of cluster substructure, I. Linear structure: Fuzzy-c-lines. *SIAM Journal on Applied Mathematics*, **40**(2), 339–357, 1981.
5. J. C. Bezdek, C. Coray, R. Gunderson, J. Watson. Detection and characterization of cluster substructure, II. Fuzzy-c-varieties and convex combinations thereof. *SIAM Journal on Applied Mathematics*, **40**(2), 358–372, 1981.
6. R. O, Duda, P.E. Hart. *Pattern Classification and Scene Analysis.* John Wiley and Sons, New York, 1973.
7. J. Hollatz and V. Tresp. A rule-based network architecture. In: *Artificial Neural Networks II*, I. Aleksander, J. Taylor, eds., Elsevier, Amsterdam, 1992, pp. 757–761.
8. D. R. Hush, B. G. Horne, Progress in supervised neural networks? What's new since Lippman ? *Signal Processing*, 1993, pp.8–39.
9. R. A. Jacobs, M. I. Jordan, S. J. Nowlan and G. E. Hinton. Adaptive mixtures of local experts. *Neural Computation,* **3**, 79–87, 1991.
10. J. Moody and C. Darken. Fast learning in networks of locally-tuned processing units. *Neural Computation*, **1**, 281–294, 1989.
11. J. Platt. A resource-allocating network for function approximation. *Neural Computation*, **3**(2), 213–225, 1991.
12. M. Röscheisen, R. Hofmann and V. Tresp. Neural control for rolling mills: incorporating domain theories to overcome data deficiency. In: *Advances in Neural Information Processing Systems 4*, J. E. Moody, S. J. Hanson and R. P. Lippmann, eds., Morgan Kaufmann, San Mateo, CA, 1992, pp. 659–666.
13. T. A. Runkler. *Automatic generation of first order Takagi–Sugeno systems using fuzzy c-elliptotype clustering.* Internal report Siemens Corporate Technology, Munich, 1996.
14. D. F. Specht. Probabilistic neural networks. *Neural Networks,* **3**, 109–117, 1990.
15. T. Takagi and M. Sugeno. Fuzzy identification of systems and its applications to modeling and control. *IEEE Trans. Systems, Man and Cybernetics,* **15**(1), 116–132, 1985.
16. V. Tresp, J. Hollatz, S. Ahmad. Network structuring and training using rule-based knowledge. In: *Advances in Neural Information Processing Systems 5*, S. J. Hanson, J. D. Cowan and C. L. Jiles, eds., Morgan Kaufmann, San Mateo, CA, 1993, pp. 871–878.
17. L.-X. Wang and J. M. Mendel. Fuzzy basis functions, universal approximation, and orthogonal least-squares learning. *IEEE Trans. Neural Networks*, **3**(5), 1992.

Identification of Singleton Fuzzy Models via Fuzzy Hyperrectangular Composite NN

Mu-Chun Su

Tamkang University, Tamsui, Taipei Hsien, 25137, Taiwan, R.O.C.

1. Introduction

In this chapter we present a method for the identification of fuzzy single-ton models based on a class of Fuzzy HyperRectangular Composite Neural Networks (FHRCNNs). The prior knowledge required, besides the available input-output data, is knowledge about the input and output variables and the number of partitions dividing the ranges of the output variables.

In previous work (Su, 1993; Su, 1994) we developed a method for extracting crisp if-then rules directly from input-output data; these crisp rules were then used for pattern recognition. The method was based on applying the supervised decision-directed learning (SDDL) algorithm to train a class of hyperrectangular composite neural networks (HRCNNs). Each generated hyperrectangle which identifies a region of data corresponds to a single crisp if-then rule. Basically, FHRCNNs are a fuzzified version of HRCNNs. Su and Kao (1994; 1995) applied FHRCNNs to function approximation. In a FHRCNN, the synaptic weights of a hidden node define a hyperrectangle which then corresponds to a fuzzy rule. As weighted outputs of hidden nodes propagate to an output node, a defuzzification mechanism is used to provide a crisp output. For the purpose of deriving a fuzzy model of a system, the number of hidden nodes and the synaptic weights have to be identified. The method proceeds as follows. First, we divide the range of an output variable into multiple intervals where each interval represents a particular class of crisp output values. Second, we use the SDDL algorithm to generate hyperrectangles in order to classify the input-output patterns. The most representative hyperrectangles are selected and used in the construction of a fuzzy if-then rule base. Finally, we apply the so-called backpropagation algorithm or the least mean square (LMS) algorithm to tune the FHRCNN as well as to construct the final fuzzy model.

The rest of this chapter is organized as follows. In Section 2, we give a brief review of fuzzy models and then state the motivation for developing the new type of fuzzy model based on FHRCNNs. In Section 3, we introduce the architecture and characteristics of FHRCNNs. Section 4 describes the training procedure for FHRCNNs. In Section 5 we apply our method to a nonlinear system identification and a time series prediction problem to illustrate the performance of the identified fuzzy models. Practical considerations and concluding remarks are presented in Section 6. Finally, Section 7 lists some applications of the proposed identification method in a variety of real-world application domains.

2. Classification of Fuzzy Models

In this section we briefly describe the three basic types of fuzzy models in order to motivate the use of singleton fuzzy models whose identification is the subject of this chapter.

2.1 Linguistic Fuzzy Models

A Mamdani-type linguistic model consists of fuzzy if-then rules of the form

$$R^{(j)}: \text{If } x_1 \text{ is } A_1^j \text{ and } \ldots \text{and } x_n \text{ is } A_n^j \text{ then } y \text{ is } B^j \tag{2.1}$$

where A^j and B^j $(j = 1, 2, \ldots, J)$ are the linguistic values of the crisp input and output variables $\mathbf{x} = (x_1, \ldots, x_n)^T \in \mathcal{U} \subset \mathbb{R}^n$ and $y \in \mathcal{V} \subset \mathbb{R}$ respectively. The linguistic variables in the fuzzy if-then rules are defined by means of fuzzy sets which are characterized by membership functions, $\mu_{A_i^j}$ and μ_{B^j}. Generally, the performance of the fuzzy model depends on the parameters of these membership functions, the type of fuzzification procedure, the fuzzy logic operators, the type of fuzzy inference, and the defuzzification procedure. In the following, we will investigate these performance factors in more detail.

2.1.1 Membership Functions. Basically, any function producing a value between 0 and 1 can be a membership function. The combination of a sigmoidal and Gaussian function provides us with a smooth and differentiable membership function. This allows us to tune the membership function parameters by gradient-based optimization techniques. If the computational aspects are of major concern, then the combination of a trapezoidal and triangular membership function is the one that is widely used in fuzzy control applications. After the types of the membership functions, μ_{A^j} and μ_{B^j}, are selected, the problem is how to use the machinery of fuzzy logic so that the set of fuzzy rules is transformed into a unified mapping from fuzzy sets in $\mathcal{U}$ to fuzzy sets in $\mathcal{V}$.

2.1.2 Interpretation of Fuzzy Rules as Mappings. First, we may interpret the fuzzy rule (2.1) as a fuzzy implication $A_1^j \times \cdots \times A_n^j \to B^j$ defined on the Cartesian product $\mathcal{U} \times \mathcal{V}$. In what follows we consider some commonly used interpretations for a fuzzy rule as a fuzzy implication. For simplicity, we denote $A_1^j \times \cdots \times A_n^j = A_j$ and $B^j = B_j$, and the fuzzy rule (2.1) is therefore denoted as $A_j \to B_j$.

– Product-operation rule of fuzzy implication

$$\mu_{A_j \to B_j}(\mathbf{x}, y) = \mu_{A_j}(\mathbf{x})\mu_{B_j}(y). \tag{2.2}$$

– Min-operation rule of fuzzy implication

$$\mu_{A_j \to B_j}(\mathbf{x}, y) = \min\{\mu_{A_j}(\mathbf{x}), \mu_{B_j}(y)\}. \tag{2.3}$$

– Boolean rule of fuzzy implication

$$\mu_{A_j \to B_j}(\mathbf{x}, y) = \max\{1 - \mu_{A_j}(\mathbf{x}), \mu_{B_j}(y)\}. \tag{2.4}$$

In (2.2–2.4), $\mu_{A_j}(\mathbf{x}) = \mu_{A_1^j \times \cdots \times A_n^j}(\mathbf{x})$ is defined either to be

$$\mu_{A_j}(\mathbf{x}) = \min\{\mu_{A_1^j}(\mathbf{x}), \dots, \mu_{A_n^j}(\mathbf{x})\} \tag{2.5}$$

or the product

$$\mu_{A_j}(\mathbf{x}) = \mu_{A_1^j}(\mathbf{x}) \dots \mu_{A_n^j}(\mathbf{x}). \tag{2.6}$$

2.1.3 Fuzzification. The fuzzification procedure maps a crisp point $\mathbf{x} = (x_1, \dots, x_n)^T \in \mathcal{U}$ into a suitable fuzzy set A' in $\mathcal{U}$. There are two different ways to perform this mapping:

– Singleton fuzzification: A' is a fuzzy singleton with support $\mathbf{x}$, that is, its membership function is

$$\mu_{A'}(\mathbf{x}') = \begin{cases} 1 & \text{if } \mathbf{x}' = \mathbf{x} \\ 0 & \text{otherwise.} \end{cases} \tag{2.7}$$

– Non-singleton fuzzification: As in the singleton case, $\mu_{A'}(\mathbf{x}) = 1$. However, $\mu_{A'}(\mathbf{x}')$ decreases as $\mathbf{x}'$ moves away from $\mathbf{x}$. A possible choice of $\mu_{A'}(\mathbf{x})$ is the Gaussian membership function.

The singleton fuzzification is the one almost always used in practice. It seems that the non-singleton fuzzification may be useful for situations where the inputs are corrupted by noise.

2.1.4 Fuzzy Inference. The next question is: How do we infer a conclusion from the fuzzy rule (2.1) given a fact in the form of a linguistic value $A' \in \mathcal{U}$? In approximate reasoning and fuzzy logic, the most used inference mechanism is generalized modus ponens (GMP). The GMP inference scheme is as follows

Fact: $\mathbf{x}$ is A'
Premise: If $\mathbf{x}$ is A_j then y is B_j
Conclusion: y is B_j'

where A', A_j, B_j, and B_j' are linguistic values with corresponding membership functions. Using the sup-star composition, we thus have that

$$\mu_{B_j'}(y) = \sup_{\mathbf{x} \in \mathcal{U}} \left[\mu_{A_j \to B_j}(\mathbf{x}, y) \star \mu_{A'}(\mathbf{x}) \right] \tag{2.8}$$

where the operator $\star$ could be any operator belonging to the class of the so-called t-norms. For instance, min, product, and bounded-product operators belong to the class of t-norms.

Typically, a fuzzy model consists of a set of, say, J fuzzy rules, each of which is characterized by a fuzzy implication $\mu_{A_j \to B_j}(\mathbf{x}, y)$. We can use (2.8) to infer a conclusion $\mu_{B'_j}(y)$ for each of the J rules. What now remains is to aggregate all conclusions into a single membership function $\mu_{B'}(y)$. That is,

$$\mu_{B'}(y) = \mu_{B'_1}(y) \dot{+} \cdots \dot{+} \mu_{B'_J}(y) \tag{2.9}$$

where the operator $\dot{+}$ could be any operator in the class of the so-called t-conorms. For instance, max, algebraic sum, and bounded-sum operators belong to this class (Zimmermann, 1991).

2.1.5 Defuzzification. Finally, defuzzification is used to perform a mapping from fuzzy sets defined on a universe of discourse $\mathcal{V}$ into a crisp output $y \in \mathcal{V}$. Two of the most used methods are as follows:

– Maximum defuzzification

$$y = \arg \sup_{y \in \mathcal{V}} (\mu_{B'}(y)). \tag{2.10}$$

– Center average defuzzification

$$y = \frac{\sum_{j=1}^{J} \overline{y}^j (\mu_{B'_j}(\overline{y}^j))}{\sum_{j=1}^{J} (\mu_{B'_j}(\overline{y}^j))}. \tag{2.11}$$

In the expression above $\overline{y}^j$ is the center of the fuzzy set B_j, that is, the point in $\mathcal{V}$ at which $\mu_{B'_j}(y)$ achieves its maximum value (Wang, 1994).

Due to the different definitions of fuzzy implication, different types of membership functions, and different fuzzification and defuzzification procedures, the final crisp outputs can be computed in a number of ways. Two such possibilities are given next:

1. {singleton fuzzification} + {product-operation rule with (2.6)} + {product operator for $\star$} + {center average defuzzification} + {$\mu_{B_j}(\overline{y}^j) = 1$}

$$y = \frac{\sum_{j=1}^{J} \overline{y}^j (\prod_{i=1}^{n} \mu_{A_i^j}(x_i))}{\sum_{j=1}^{J} (\prod_{i=1}^{n} \mu_{A_i^j}(x_i))}. \tag{2.12}$$

2. {singleton fuzzification} + {product-operation rule with (2.6)} + {product operator for $\star$} + {max operator for $\dot{+}$} + {maximum defuzzification}

$$y = \arg \sup_{y \in \mathcal{V}} (\max_{1 \leq j \leq J} (\sup_{\mathbf{x} \in \mathcal{U}} (\prod_{i=1}^{n} \mu_{A_i^j}(\mathbf{x}) \mu_{B_j}(y)))). \tag{2.13}$$

Apparently, the crisp outputs from (2.12) and (2.13) will be different and a particular choice between these two options has to be well motivated. Some criteria (e.g., computational effeciency and ease of adaptation) are suggested by Wang (1994).

2.2 Takagi–Sugeno Fuzzy Models

Instead of considering the linguistic fuzzy rules in the form of (2.1), Takagi and Sugeno (1985) extended them to fuzzy rules with consequent parts (then-parts) in the form of linear functions of the variables from the consequent parts (if-parts), for instance

$$R^{(j)} : \text{If } x_1 \text{ is } A_1^j \text{ and } \ldots \text{and } x_n \text{ is } A_n^j \text{ then } y^j = c_0^j + \cdots + c_n^j x_n \quad (2.14)$$

where A_i^j are linguistic values, $\mathbf{x} = (x_1, x_2, \ldots, x_n)^T \subset \mathbb{R}^n$ is a crisp input vector, c_i^j are real-valued parameters, y^j is the crisp system output, and $j = 1, 2, \ldots, J$.

Each such fuzzy rule represents a locally linear model. The advantage of a locally linear model is that the parameters c_i^j of the model can be easily identified from numerical data. A weak point is that the linguistic interpretation of the Takagi–Sugeno fuzzy rules is not so intuitive as in the case of linguistic rules. Namely, it is difficult to incorporate the linguistic description of input-output relationships provided by a human expert with the numerical information from experimental data in order to construct a Takagi–Sugeno (TS) fuzzy model. In addition, the other main disadvantage of the TS fuzzy model comes from the use of trapezoidal membership functions. Due to this type of membership functions, the local linear characteristic may deteriorate the approximation accuracy when the linear model is used to identify a highly nonlinear system. This can be illustrated by using the simple example depicted in Fig. 2.1. On the other hand, if we use Gaussian membership functions and product operators, the output of a TS fuzzy model is the weighted average of the y^j's computed as

$$y(\mathbf{x}) = \frac{\sum_{j=1}^{J} w^j y^j}{\sum_{j=1}^{J} w^j} \quad (2.15)$$

where the weight w^j implies the degree of satisfaction of the antecedent of rule $R^{(j)}$ for the input and is calculated as

$$
\begin{aligned}
w^j &= \prod_{i=1}^{n} \mu_{A_i^j}(x_i) \\
&= \prod_{i=1}^{n} \exp\left[-\frac{1}{2} \left(\frac{x_i - \mu_i^j}{\sigma_i^j} \right)^2 \right] \\
&= \exp\left[-\frac{1}{2} \sum_{i=1}^{n} \left(\frac{x_i - \mu_i^j}{\sigma_i^j} \right)^2 \right]
\end{aligned}
\quad (2.16)
$$

where $\mu_{A_i^j}(x_i)$ represents the grade of membership of x_i in the fuzzy set providing the meaning of the linguistic value A_i^j. Combining (2.15) and (2.16), we get a multivariate nonlinear approximator. Actually, to make a TS fuzzy model identical to a universal approximator the only term we need in the consequence part is the constant term c_0^j. By keeping only c_0^j, the resulting fuzzy model is equivalent to a Radial Basis Function Network (RBFN) which has been proven to be a universal approximator (Mahaskar, 1992; Poggio and Girosi, 1990).

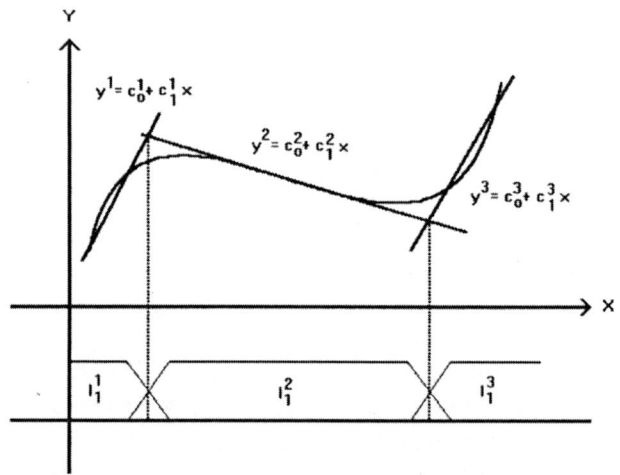

Fig. 2.1. A TS fuzzy model: the local linearization characteristic deteriorates the approximation accuracy in regions where the original curve is nonlinear

2.3 Singleton Fuzzy Models

Our discussion about the linguistic fuzzy models and the TS fuzzy models implies that all we need is a fuzzy rule whose consequent part is a fuzzy singleton. An example of such a fuzzy rule is

$$R^{(j)} : \text{If } x_1 \text{ is } A_1^j \text{ and } \ldots \text{and } x_n \text{ is } A_n^j \text{ then } y \text{ is } c_0^j. \qquad (2.17)$$

A collection of fuzzy rules of the above type is called a singleton fuzzy model. The identification of a singleton fuzzy model is divided into two subproblems: structure identification and parameter identification. Generally speaking, the structure identification deals with two issues. The first one is to specify the inputs to a system where the input variables affect the system's output. Actually, there is no systematic way to solve this particular problem; therefore, a heuristic method is often employed to decide the set of appropriate input variables. The second one is to specify the partitions of the domains of the input and output variables and the number of fuzzy rules. Normally, we take fuzzy grid partitions of these domains, as shown in Fig. 2.2. Accordingly, the number of rules increases exponentially with the

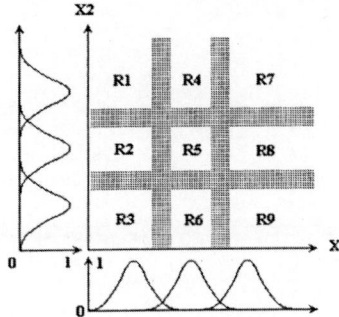

Fig. 2.2. An example of grid fuzzy partitions

number of input and output variables and the number of partitions of the domains of these variables.

Having selected the input variables, the partitions of the input and output domains, and the number of fuzzy rules, the next step is to use input-output data to estimate the unknown fuzzy model parameters by some optimization technique. In a singleton fuzzy model, the parameters are those of the membership functions of the fuzzy sets defining the meaning of the linguistic variables and the coefficients in the consequent parts of the fuzzy rules. Principally, the structure identification and the parameter identification should be performed together so that the performance of the fuzzy model to be constructed is optimal in some sense. But most identification methods treat them separately in order to reduce computational complexity.

One approach to the identification of a singleton fuzzy model is to represent it as a three-layer feedforward neural network and then use the backpropagation algorithm to adjust the parameters of the fuzzy model (Wang, 1994). To overcome the disadvantages of the backpropagation algorithm, (e.g., it may be trapped at a local minimum or have a very low convergence rate) Wang and Mendel (1992) proposed to fix some parameters in the fuzzy model so that it can be represented as a linear combination of the so-called fuzzy basis functions (e.g., Gaussian functions). Then the orthogonal least squares algorithm is utilized to select the significant fuzzy basis functions and the corresponding optimal coefficients. Although these two training algorithms perform successfully, they still require substantial computation for systems of large dimensions.

Let us now focus our attention on the antecedent part. Takagi and Hayashi (1991) stated that fuzzy grid partitions cannot capture the possible correlations between the input variables. In order to overcome this weak point, the input space should be partitioned appropriately in order to reflect the existence of such correlations. Therefore, it is better to use fuzzy rules of the type:

$$R^{(j)} : \textbf{If } \textbf{x} \text{ is } A^j \textbf{ then } y \text{ is } c_0^j \tag{2.18}$$

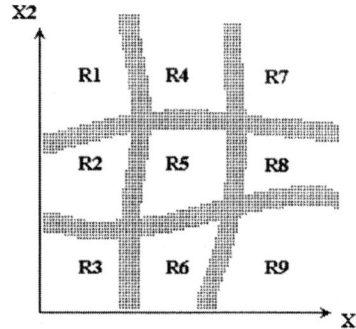

Fig. 2.3. An example of multi dimensional fuzzy partitions

where A^j is a multidimensional fuzzy set defining an arbitrarily shaped fuzzy region in the input space. This type of fuzzy partition offers the possibility of capturing the correlations between the input variables. Figure 2.3 illustrates such a multidimensional fuzzy partition. Accordingly, the complexity of defining this kind of partition increases drastically. Takagi and Hayashi (1991) proposed training feedforward neural networks in order to implement this type of fuzzy partition. The price one has to pay in this case is that the synaptic weights of the trained network have no clear physical meaning. In order to make a satisfactory trade-off between capturing the correlations and the complexity of representation, one option is to use an aggregation of hyperrectangles to fit a multidimensional fuzzy set or a multidimensional fuzzy region. By doing this, we are able to transform the fuzzy rule from (2.18) into the following fuzzy rule:

$$R^{(j)} : \text{If } \mathbf{x} \text{ is } (HR_1^j \cup \ldots \cup HR_k^j) \text{ then } y \text{ is } c_0^j \qquad (2.19)$$

where HR_i^j represents a fuzzy set defining an n-dimensional fuzzy hyperrectangle. A major distinction between the two rules expressed in (2.17) and (2.19) is illustrated in Fig. 2.4. It is apparent that by using the latter type of fuzzy rules we can omit several unnecessary grid partitions and thus greatly decrease the number of fuzzy rules. For the case shown in Fig. 2.4, we need only three large rectangles to fit the ⊃-shaped region instead of using nine smaller rectangles. In this new type of fuzzy model, the goal of structure identification is to decide the number of fuzzy hyperrectangles. The parameter identification is then concerned with the estimation of the parameters defining the corresponding fuzzy hyperrectangles (e.g., sizes and locations) and the coefficients in the consequent parts of the fuzzy rules. To efficiently identify a fuzzy model consisting of the new type of fuzzy rules, we train a two-layer FHRCNN to construct the fuzzy model. The architecture and characteristics of FHRCNNs are explained in the next section.

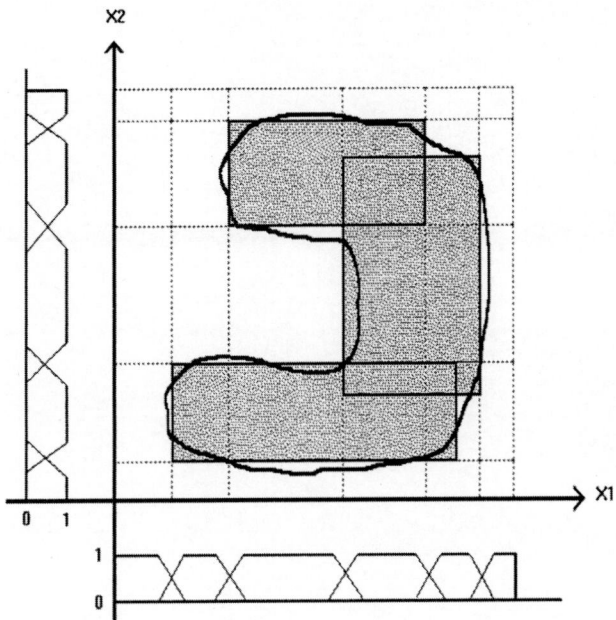

Fig. 2.4. Grid partitions versus multidimensional partitions: dashed lines represent grid partitions, solid lines represent multidimensional partitions, and the ⊃-shaped region represents the region to be approximated.

3. Fuzzy Neural Networks

The class of fuzzy hyperrectangular composite neural networks (FHRCNNs) is the fuzzified version of the class of hyperrectangular composite neural networks (HRCNNs). In (Su, 1993; Su, 1994), we have shown that the values of the synaptic weights of a trained HRCNN can be interpreted as a set of crisp if-then rules. Figure 3.1(a) illustrates a symbolic representation of a two-layer HRCNN. The mathematical description of a two-layer HRCNN is given as follows

$$\text{Out}(\mathbf{x}) = f(\sum_{j=1}^{J} \text{Out}_j(\mathbf{x}) - \eta), \tag{3.1}$$

$$\text{Out}_j(\mathbf{x}) = f(net_j(\mathbf{x})), \tag{3.2}$$

$$net_j(\underline{x}) = \sum_{i=1}^{n} f((M_{ji} - x_i)(x_i - m_{ji})) - n, \tag{3.3}$$

and

$$f(x) = \begin{cases} 1 & \text{if } x \geq 0 \\ 0 & \text{otherwise,} \end{cases} \tag{3.4}$$

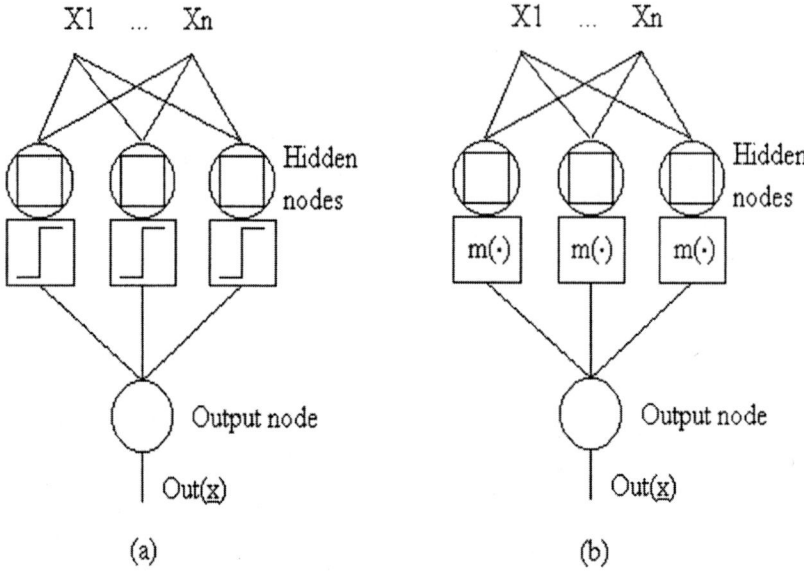

Fig. 3.1. (a) A symbolic representation of a two-layer HRCNN, and (b) a symbolic representation of a two-layer FHRCNN

where

- M_{ji} and $m_{ji} \in R$ are adjustable synaptic weights of the j-th hidden node,
- $\mathbf{x} = (x_1, \ldots, x_n)^T$ is an input pattern and n is the number of input variables,
- η is a small positive real number, and
- $\text{Out}(\mathbf{x}) : R^n \to \{0, 1\}$ is the output function of a two-layer HRCNN with J hidden nodes.

The supervised decision-directed learning (SDDL) algorithm was developed to generate a two-layer HRCNN in a sequential manner by adding hidden nodes as needed (Su, 1993; Su, 1994). As long as there are no identical data over different classes, we can obtain 100% recognition rate for the set of training data. The J crisp rules extracted from the trained network are represented as

If $(\mathbf{x} \in [m_{11}, M_{11}] \times \cdots \times [m_{11}, M_{1n}])$ **then** $\text{Out}(\mathbf{x}) = 1$.

$$\vdots \tag{3.5}$$

If $(\mathbf{x} \in [m_{J1}, M_{J1}] \times \cdots \times [m_{Jn}, M_{Jn}])$ **then** $\text{Out}(\mathbf{x}) = 1$.

In the above, the rule antecedents define a set of n-dimensional hyperrectangles.

A two-layer FHRCNN, shown in Fig. 3.1(b), employs a special member-ship function $m_j(\mathbf{x})$ instead of a hard limiter function $f(x)$ defined in (3.4) as the output function of each hidden node. The membership function $m_j(\mathbf{x})$ measures the degree to which an input pattern is close to the hyperrectangle defined by $[m_{j1,M_{j1}}] \times \cdots \times [m_{jn}, M_{jn}]$. A mathematical representation of a two-layer FHRCNN is of the form

$$\text{Out}(\mathbf{x}) = \sum_{j=1}^{J} w_j m_j(\mathbf{x}) + \theta, \tag{3.6}$$

$$m_j(\mathbf{x}) = \exp\{-s_j^2[vol_j(\mathbf{x}) - vol_j]^2\}, \tag{3.7}$$

$$vol_j = \prod_{i=1}^{n}(M_{ji} - m_{ji}), \tag{3.8}$$

and

$$vol_j(\mathbf{x}) = \prod_{i=1}^{n} \max(M_{ji} - m_{ji}, x_i - m_{ji}, M_{ji} - x_i) \tag{3.9}$$

where w_j is the connection weight from the j-th hidden node to the output node, s_j is the sensitivity factor that regulates the membership value, and θ is a bias term which is adjustable. Apparently, the output function of a two-layer FHRCNN is a linear weighted combination of J local functions. From (3.7)–(3.9), it is easy to find out that the function $m_j(\mathbf{x})$ provides us with more flexibility than the Gaussian function because the former one can be either a step-like function or a Gaussian-like function, as shown in Fig. 3.2(a) and (b). Based on this fact, we can conclude that a two-layer FHRCNN is a universal function approximator whose efficiency is better than that of an RBF network using Gaussian functions.

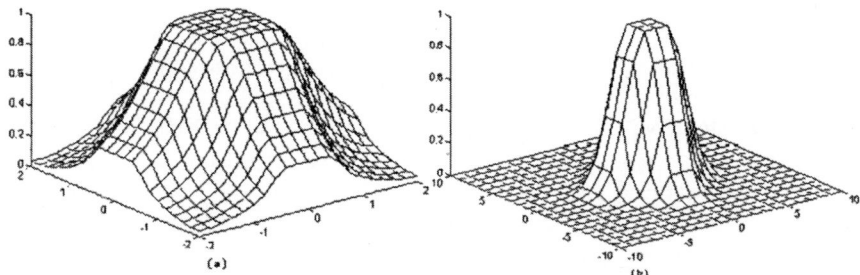

Fig. 3.2. (a) A step-like $m_j(\mathbf{x})$ and (b) a Gaussian-like $m_j(\mathbf{x})$

Basically, a two-layer FHRCNN can be trained either by the backpropa-gation algorithm (Rumelhart et al., 1986; Werbos, 1974) or the real-valued genetic algorithm (Su and Chang, 1995). No matter which method is adopted, if we start from a good initial point then the convergence of the training procedure will be greatly accelerated. In order to give a satisfactory solution to the initialization problem, we have to make full use of both the linguistic information from human experts and the numerical data from experiments. As the architecture of a two-layer FHRCNN provides us with a convenient framework to incorporate linguistic information, we are able to find a good guess for the values of M_{ji}, m_{ji}, and w_j based on given linguistic information. As for the numerical data, the initial weights can be estimated by transforming a function approximation problem into a pattern recognition problem. The training procedure is addressed in the next section.

4. Identification of Singleton Fuzzy Models

The hybrid learning algorithm developed to train a two-layer FHRCNN consists of the following steps. Note that here we focus on multi-input/single-output systems since a large multi-input/multi-output system can be broken into several smaller size multi-input/single-output subsystems.

Step 1. *Partitioning of the output space.* If the problem to be considered is a pattern recognition problem instead of function approximation problem, we skip this step and go directly to Step 2. Here we first have to determine the number of partitions: the greater the number of partitions, the higher the approximation accuracy. After the number of partitions has been set to, say, k, we may uniformly partition the output space into k intervals or use the k-means algorithm to find a more appropriate partition. By doing this, the original input-output pairs are then transformed into quantized input-output pairs.

Step 2. *Transforming a function approximation problem into a pattern recognition problem.* Since the outputs have been labeled as belonging to one of the k intervals, we use the SDDL algorithm to generate k two-layer HRC-NNs to classify the quantized input-output patterns. Assume that h_i^c hidden nodes were generated in the i-th trained HRCNN for $i = 1, 2, \ldots, k$. For each trained HRCNN, corresponding hidden nodes are then ranked in the order of their representativeness. The representativity of a hidden node is based on the number of patterns contained in the corresponding hyperrectangle defined by the values of the synaptic weights of the hidden node. The larger this number is, the more representative a hidden node is.

Step 3. *Initialization and updating of weights.* In order to make a satisfactory compromise between the approximation accuracy and the number of hidden nodes of the FHRCNN to be trained, we select the first n_i $(i = 1, 2, \ldots, k)$ most representative hidden nodes from each of the trained HRCNNs in order to initialize the weights M_{ji} and m_{ji} of the FHRCNN. Therefore the number of hidden nodes of the FHRCNN to be trained is set to $n_1 + n_2 + \cdots + n_k$. The next step is the initialization of the connection weights w_j for $j = 1, 2, \ldots, n_1 + \cdots + n_k$. The idea is that it is reasonable to expect that input patterns whose corresponding outputs were labeled as belonging to the same interval in the first step would indeed produce similar outputs. Therefore we initialize w_j to be the mean value of a corresponding interval. After we have initialized all adjustable weights, we have to tune them so that that the squared error between the system output and the output of the FHRCNN is minimized. Here we have two options. The first one is to use the real-valued GA proposed by Su and Chang (1995) to find a set of approximate weights. The second one is to use gradient-based methods to train the FHRCNN. We also have two different approaches to using gradient-based methods. The first one is to fix the values of M_{ji} and m_{ji} and update only the connection weights w_j using a recursive least mean square (LMS) algorithm. However, it is better to update all weights, M_{ji}, m_{ji}, and w_j simultaneously because then the modeling capability of the FHRCNN will be significantly improved. Therefore the more efficient method is to use the backpropagation algorithm for adaptation.

Step 4. *Interpreting the trained FHRCNN.* After sufficient training, the squared error can be minimized to arbitrary accuracy. Now, the values of the weights are interpreted into fuzzy if-then rules. Since the function $m_j(\mathbf{x})$ measures the degree to which an input pattern $\mathbf{x}$ is close to the n-dimensional hyperrectangle defined by $[m_{j1}, M_{j1}] \times \cdots \times [m_{jn}, M_{jn}]$, we may define a fuzzy set HR_j characterized by $m_j(\mathbf{x})$ in the input space. Here, $m_j(\mathbf{x})$ represents the grade of membership of $\mathbf{x} \in \mathbb{R}^n$ to the fuzzy set HR_j. Then the presence of an input pattern to the j-th hidden node is equivalent to firing the following fuzzy rule:

$$R^{(j)} : \textbf{If } (\mathbf{x} \text{ is } HR_j) \textbf{ then } \text{Out}(\mathbf{x}) \text{ is } w_j. \tag{4.1}$$

The total number of extracted fuzzy rules is $n_1 + \cdots + n_k$ since this is the number of hidden nodes in the trained FHRCNN. In order to numerically combine these fuzzy rules and to compute a crisp output we use the following defuzzification procedure:

$$\text{Out}(\mathbf{x}) = \sum_{j=1}^{J} w_j m_j(\mathbf{x}) + \theta \tag{4.2}$$

The defuzzification procedure defined by (4.2) is computationally equivalent to the most popular "center average defuzzification," as shown in (2.11), except for the bias term θ. The reason for including θ in the architecture of FHRCNNs is that we can increase the modeling capability of FHRCNNs.

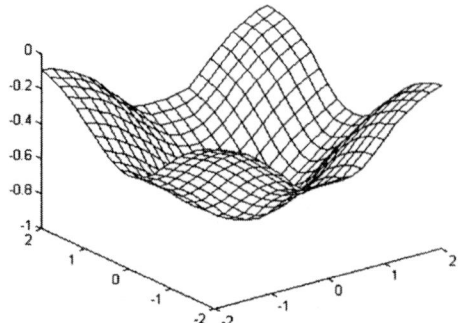

Fig. 5.1. The training patterns presented to the FHRCNN for learning

5. Simulation Results

5.1 Example 1: System Identification

The plant to be identified is governed by the difference equation:

$$y(k+1) = \left(\frac{1}{1+\exp\{-[2y(k)-1]\}} - \frac{1}{1+\exp\{-[2y(k)+1]\}}\right) + \left(\frac{1}{1+\exp\{-[2u(k)-1]\}} - \frac{1}{1+\exp\{-[2u(k)+1]\}}\right) \quad (5.1)$$

where $u(k)$ and $y(k)$ are the input and output of the plant, respectively. The training patterns are in the two-dimensional input space ($y(k)$ and $u(k)$) and the one-dimensional output space $y(k+1)$, as shown in Fig. 5.1. The input parts of the training data are chosen so that they scatter evenly over a square region $[-2, 2] \times [-2, 2]$. The total number of training data is 441.

First, we partitioned the range of the output variable into three intervals, $[-0.92, -0.64]$ (with mean $= -0.78$), $[-0.64, -0.36]$ (with mean $= -0.5$), and $[-0.36, -0.08]$ (with mean $= -0.22$). In the following step, three HRC-NNs were trained to recognize the resulting three-classes pattern recognition problem. After training, three HRCNNs with 5, 22, and 20 hidden nodes, corresponding to these three different classes, were generated. A FHRCNN with $(1 + 4 + 0)$ hidden nodes was trained to identify the plant. Figure 5.2 illustrates the performance of the FHRCNN initialized by the proposed scheme. The LMS algorithm employed had the learning rate 0.01 and continued for 500 training steps. Figure 5.3 depicts the three-dimensional representatives learned by the FHRCNN. The five extracted fuzzy if-then rules are represented as follows:

R^1 : **If** $((y(k), u(k))^T$ is $HR_1 : [-0.03, 0.03] \times [-0.12, 0.02])$
then $y(k+1)$ is -0.41
R^2 : **If** $((y(k), u(k))^T$ is $HR_2 : [-0.05, 0.05] \times [-2.59, -2.41])$

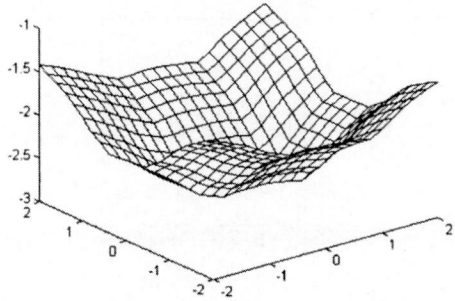

Fig. 5.2. The performance of the FHRCNN initialized by the proposed scheme

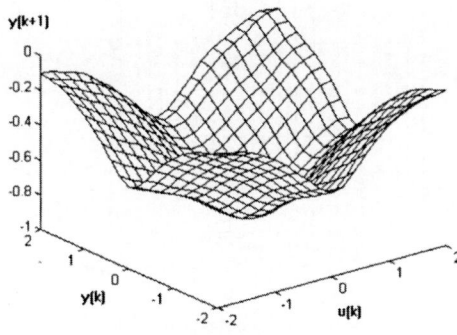

Fig. 5.3. Three-dimensional representatives learned by the FHRCNN

then $y(k + 1)$ is $- 0.16$

R^3 : **If** $((y(k), u(k))^T$ is $HR_3 : [-2.59, -2.41] \times [0.00, 0.10])$

then $y(k + 1)$ is $- 0.16$ $\qquad$ (5.2)

R^4 : **If** $((y(k), u(k))^T$ is $HR_4 : [2.31, 2.49] \times [0.06, 0.14])$

then $y(k + 1)$ is $- 0.14$

R^5 : **If** $((y(k), u(k))^T$ is $HR_5 : [-0.08, 0.01] \times [2.08, 2.22])$

then $y(k + 1)$ is $- 0.12$

5.2 Example 2: Time Series Prediction

Generally, it is very difficult to predict chaotic time series by traditional regression techniques. This problem provides a good example to illustrate the performance of the identified fuzzy model implemented as a two-layer FHRCNN. In general, the time series prediction problem can be formulated as:

$$x(k) = f(x(k - 1), x(k - 2), \ldots, x(k - p)) \qquad (5.3)$$

where $x(k-1), \cdots, x(k-p)$ are the past values and $x(k)$ is the future value to be predicted. The chaotic time series to be predicted is generated by a simple deterministic equation (Yu and Chen, 1993; Lapedes and Farber, 1987)

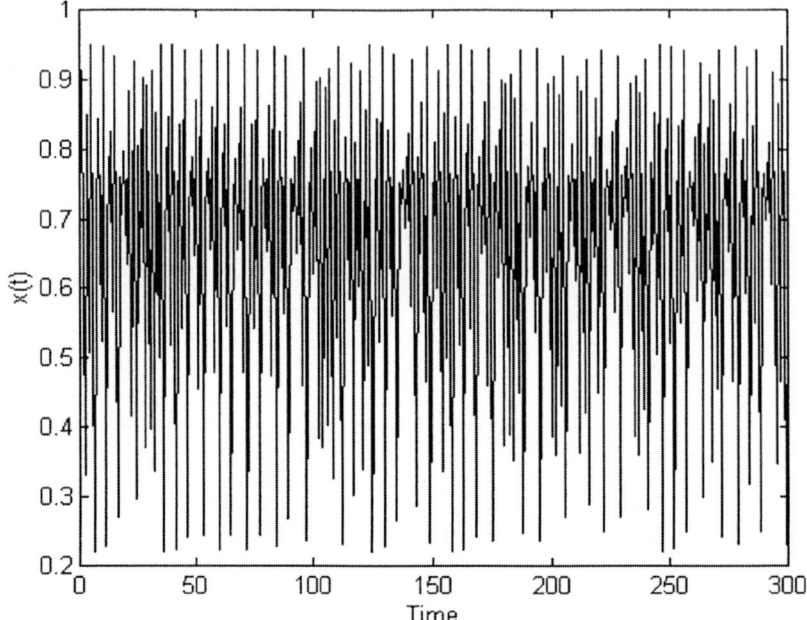

Fig. 5.4. Patterns used to train the FHRCNN with three hidden nodes

$$x(k) = 3.6x(k-1)[1 - x(k-1)] + 0.05 \qquad (5.4)$$

Figure 5.4 illustrates an example of the chaotic time series. Of course, the value of p in (5.3) has a direct effect on the prediction performance. The first 200 observations are used as training data for the FHRCNN to be trained in order to capture the unknown complex relation between the past and future values. The next 100 observations are used as test data to validate the prediction capability of the trained FHRCNN. In our experiment, one-step prediction was conducted. First, we let p be equal to one, that is, the number of nodes in corresponding input and output layer are both one. According to the learning scheme, we first divide the output value into three intervals: [0.221, 0.464] (with mean = 0.3425), [0.464, 0.707] (with mean = 0.585), and [0.707, 0.95] (with mean = 0.8285). Finally, we generated a two-layer FHRCNN with three hidden nodes. After 50 iterations, the mean squared error was down to 0.000249. Figure 5.5 illustrates the prediction performance of the FHRCNN. Apparently, the convergence speed is rather fast and the prediction performance is very encouraging. The three extracted fuzzy rules are

$$R^1 : \textbf{If } (x(k-1) \text{ is } HR_1 \text{ defined by } [0.81, 0.94]) \textbf{ then } x(k) \text{ is } 0.11$$
$$R^2 : \textbf{If } (x(k-1) \text{ is } HR_2 \text{ defined by } [0.67, 0.78]) \textbf{ then } x(k) \text{ is } 0.65 \quad (5.5)$$

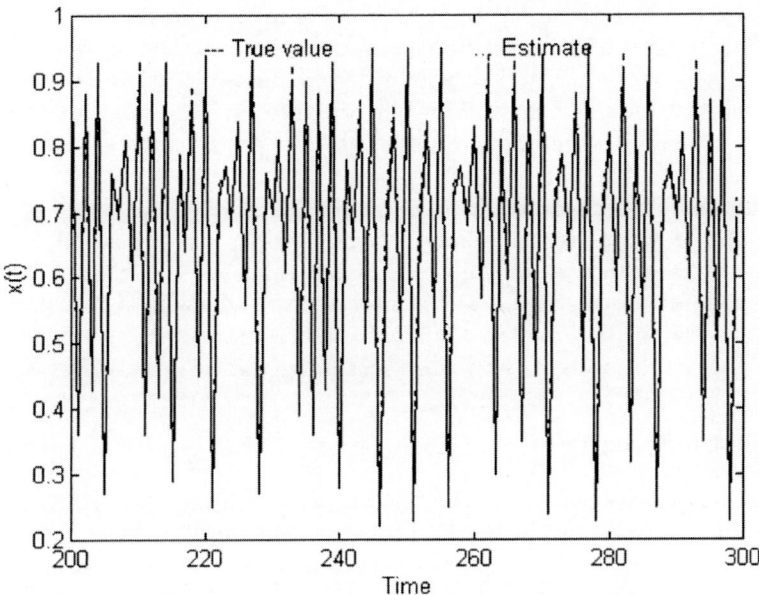

Fig. 5.5. The prediction performance of the FHRCNN: the dash line represents the true value and the dotted line represents the estimated value.

R^3 : **If** $(x(k-1)$ is HR_3 defined by $[-0.005, 0.5])$ **then** $x(k)$ is 1.19

Then, we let p be five, that is, we use five past values to predict the future value. The mean squared error was down to 0.000486 after 500 iterations by training another FHRCNN with five hidden nodes. Theoretically, the larger the value of p is, the better the prediction performance is. The results seems contradictory; however, they are not surprising because the time series was generated by the linear model from (5.4).

6. Practical Considerations and Concluding Remarks

We have described an approach to the identification of singleton fuzzy models from numerical data and linguistic information. This section summarizes the advantages and drawbacks of the approach, focusing mainly on its practical aspects.

6.1 Use of Prior Knowledge

Prior knowledge about the system is needed to determine the input and output variables of the model. Once we have decided the number of divisions of the range of an output variable, the SDDL algorithm can be applied to automatically determine the number of hyperrectangles from the input-output data set. The number of hyperrectangles then determines the number of fuzzy rules in the fuzzy model. Practically, in order to make a satisfactory compromise between the requirements on its performance and the number of fuzzy rules, we often select several most representative hyperrectangles to construct the fuzzy model.

As compared to backpropagation networks, the FHRCNNs have two advantages: (i) the values of the synaptic weights of a trained FHRCNN are easily utilized to generate interpretable fuzzy rules on the basis of preselected meaningful coordinates, (ii) the architecture of a FHRCNN provides us with a convenient framework to incorporate human expert knowledge in order to initialize the weights and thus accelerate the training phase.

6.2 Model Complexity

As compared to the method based on grid fuzzy partitions, fewer fuzzy rules are generated by applying our method for constructing a fuzzy model based on a two-layer FHRCNN. Takagi and Yasukawa (1993) introduced the fuzzy c-means (FCM) method to generate fuzzy rules directly from numerical data. They first perform fuzzy clustering of the output space and then induce corresponding fuzzy clusters in the input space. The advantage of this approach is that unnecessary grid fuzzy partitions can be avoided, which is also achieved by the method presented here. However, the computational complexity for the clustering of the output data, deciding the appropriate number of clusters, and deriving the shape of the membership functions may be very high. In our approach, instead of clustering the output data we directly divide the domain of an output variable into multiple intervals. In this way, we do not waste computational resources on the clustering of the output data. Using the input data and the output intervals, the SDDL algorithm automatically generates hyperrectangles corresponding to fuzzy rules. We can then trim the fuzzy rule base by omitting less representative hyperrectangles and thus construct a less complex fuzzy model. Abe and Lan (1995) also used hyperrectangles to represent fuzzy rules. In this method, fuzzy rules are recursively defined by activation and inhibition hyperrectangles. However, the way they define membership functions is not as straightforward as in the method presented here. It is also very likely that our method will generate less complex models than the method proposed by Abe and Lan.

6.3 Robustness of the Identification Method

Practical experience shows that the method presented here still works well for data sets containing hundreds of sixty-dimensional data points. However, if the existence region of the data is too large and/or the dimensionality of inputs is too high we encounter difficulties in training the FHRCNNs. To overcome this problem, we usually scale-translate the input data into the unit hypercube $[0, 1]^n$ for the n-dimensional case and use the $\sum$ operator to replace the $\prod$ operator appearing in (3.8) and (3.9).

6.4 Real-World Applications

The applications of the presented singleton fuzzy model identification method range widely from biomedical applications (Su, 1996a; Su et al., 1996b), to speech segmentation (Hsieh et al., 1996) and recognition (Hsieh et al., 1994), to transient stability prediction for power systems (Tsay, 1996). The biomedical applications are small-size nonlinear systems and the rest of the applications are medium-complexity systems. In all cases, the results are very encouraging.

References

Abe, S. and M.S. Lan (1995), Fuzzy rules extraction directly from numerical data for function approximation, *IEEE Trans. on Systems, Man, and Cybernetics*, **25**(2.1), pp. 119–129.

Babuska, R. and H.B. Verbruggen (1995), A new identification method for linguistic fuzzy models, *FUZZ-IEEE*, Japan, pp. 905–912.

Hsieh, C.H., M.C. Su, and C.T. Tseng (1994), A mandarin digits recognition system based on a novel class of hyperrectangular composite neural networks, *IASTED Int. Conf. on Modeling, Simulation and Identification*, Japan, pp. 241–244.

Hsieh, C.T., M.C. Su, and C.H. Hsu (1996), Continuous speech segmentation based on a self-learning neuro-fuzzy system, *IEICE Trans. Fundamentals of Electronics, Communications and Computer Sciences*, E79-A (8), 1180–1187, 1996.

Lapedes, A. and R. Farber (1987), *Nonlinear signal processing using neural networks: prediction and system modeling*, Tech. Rep. LA/UR 87/2662, Los Alamos National Lab., Los Alamos, NM.

Mahaskar, H. N. (1992), Approximation by superposition of sigmoidal and radial functions. *Advances in Applied Mathematics*, **13**(3), pp. 350–373.

Poggio, T. and F. Girosi (1990), Networks for approximation and learning, *Proc. of the IEEE*, **78**(9), pp. 1481–1497.

Rumelhart, D.E. and J.L. McCleland, eds. (1986), *Parallel distributed processing*, MIT Press, Cambridge, MA.

Su, M.C. (1993), *A neural network approach to knowledge acquisition*, Ph.D. Dissertation, University of Maryland, August.

Su, M.C. (1994), Use of neural networks as medical diagnosis expert systems, *Computers in Biology and Medicine*, **24**(6), pp. 419–429.

Su, M.C. (1996), A neuro-fuzzy approach to medical diagnosis expert systems, *Biomedical Engineering-Applications, Basis & Communications*, **8**(2), pp. 138–144.

Su, M.C. and C.J. Kao (1994), A neuro-fuzzy approach to system identification, *Int. Symposium on Artificial Neural Networks*, Taiwan, pp. 495–500.

Su, M.C. and C.J. Kao (1995), Time series prediction based on a novel neuro-fuzzy system, *4th Golden West Int. Conf. on Intelligent Systems*, San Francisco, CA, pp. 229–233.

Su, M.C. and H.T. Chang (1995), A real-vauled GA-based approach to extracting fuzzy rules for system identification, *Int. Joint Conf. on Fuzzy Theory and Applications*, pp. 41–46.

Sugeno, M. and T. Yasukawa (1993), A fuzzy-logic-based approach to qualitative modeling, *IEEE Trans. Fuzzy Systems*, **1**(2.1), pp. 7–31.

Takagi, H. and I. Hayashi (1991), NN-driven fuzzy reasoning, *Int. J. Approximate Reasoning*, **5**(3), May, pp. 192–212.

Takagi, T. and M. Sugeno (1985), Fuzzy identification of systems and its application to modeling and control, *IEEE Trans. Systems, Man, and Cybernetics*, **15**, pp. 116–132.

Tsay, S.S. (1996), *The application of synchronized phasor measurements and neuro-fuzzy networks to transient stability prediction.* M.S. thesis, Taiwan University, R.O.C.

Wang, L.X. (1994), *Adaptive fuzzy systems and control: design and stability analysis,* NJ, Prentice Hall.

Wang, L.X. and J.M. Mendel (1992), Fuzzy basis functions, universal approximation, and orthogonal least square learning, *IEEE Trans. Neural Networks*, **3**(5), pp. 807–814.

Werbos, P. (1974), *New tools for predictions and analysis in the behavioral science,* Ph.D. thesis, Harvard University.

Yu, E.S. and C.Y.R. Chen (1993), Traffic prediction using neural networks, *Proc. IEEE Globecom*, pp. 991–995.

Zimmermann, H.-J. (1991), *Fuzzy set theory and its applications* (2nd ed.). Kluwer, Boston.

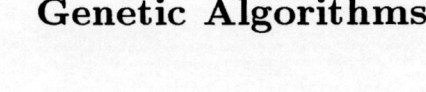

Genetic Algorithms

Identification of Linguistic Fuzzy Models by Means of Genetic Algorithms*

Oscar Cordón and Francisco Herrera

Dept. of Computer Science and Artificial Intelligence, E.T.S.I. Informatica, University of Granada, E-18071 Granada, Spain

1. Introduction

In this chapter, we deal with the identification of linguistic fuzzy models (or Mamdani fuzzy models) for multiple-input/single-output (MISO) systems. We consider a variant of the classical linguistic fuzzy model in which there does not exist a pre-determined relationship between the linguistic values of the input and output variables and the membership functions used to define the meaning (semantics) of these linguistic values. We call this type of linguistic fuzzy model *an approximative linguistic fuzzy model* [4, 10].

Once an approximative linguistic fuzzy model has been chosen to represent a MISO system, the next step is to determine its structure and estimate its parameters. This is done in three steps. First we obtain a initial model structure and parameters. That is, we generate an initial set of fuzzy rules (or initial fuzzy rule base) and initial membership functions for the antecedent and consequent parts of the fuzzy rules. Second, the initial fuzzy rule base is simplified by removing redundant fuzzy rules and, thus, the final structure of the approximative linguistic fuzzy model is determined. Third, we determine the final membership functions in order to maximize the accuracy of the approximative linguistic fuzzy model. Therefore, the proposed identification technique deals with both *structure* and *parameter identification*, and is based on learning from available input-output data.

The learning techniques used for the purpose of the identification of fuzzy models normally deal with the problem of designing and optimizing a fuzzy rule base and/or the parameters of membership functions using on- and/or off-line input-output data. These techniques include inductive learning [13, 38], descent methods [31], neural networks [26, 28], clustering techniques [39], genetic algorithms [6, Sect. 3.13], etc. On the other hand, a large class of methods known under the name of *evolutionary computation* (EC) use computational models of evolutionary processes as key elements in the design and implementation of identification and optimization algorithms. There is a variety of evolutionary computational models referred to as *evolutionary algorithms* (EAs). There are four well-defined EAs which serve as the basis for much of the activity in the field of EC: *genetic algorithms* (GAs), *evolution*

---

* This paper has been partially supported by CICYT PB96-0778.

strategies (ESs), *evolutionary programming* (EP), and *genetic programming* (GP). In this chapter, we make use of the first two types of EAs.

The best-known EAs are the GAs, i.e., search algorithms that use operations found in natural genetics to guide the search in complex search spaces. GAs have been theoretically and empirically proven to have robust and computationally efficient search capabilities. They also have been demonstrated to be a powerful tool for automating the construction of the fuzzy rule bases, since learning and self-organization may be considered in a lot of cases as optimization and/or efficient search problems. The GA-based approaches used in the context of automating and optimizing the construction of fuzzy rule bases are known under the name of *genetic fuzzy systems* (GFSs) [4]. A short description of these approaches is included in Sect. 2. Using the more general term *evolutionary* instead of *genetic* they are also called *evolutionary fuzzy systems*.

The identification method presented in this chapter uses GAs, and can be described as three-stage inductive learning process. These three stages are the following:

1. GA-based generation of approximative fuzzy rules, based either on prior fuzzy partions of the domains of the input and output variables, or no prior fuzzy partitions at all. In the first case, ESs are used for a local tuning of the fuzzy rules. At this stage the initial fuzzy model structure and parameters are obtained.
2. GA-based simplification of the initial fuzzy rule base, thereby avoiding possible overlearning, and removing redundant fuzzy rules. At this stage the final model structure is obtained, i.e., all the fuzzy rules constituting the approximative linguistic fuzzy model.
3. GA-based tuning for adjusting the membership functions in the fuzzy rules using fitness criteria. At this last stage, the final model parameters are estimated.

In addition, the *genetic fuzzy identification method* (GFIM) presented in this chapter, permits the incorporation of prior, qualitative knowledge, and is capable of blending this knowledge with the available input-output data.

The remaining part of this chapter is structured as follows. Section 2 serves as a brief introduction to EAs and GFSs. Section 3 considers the fuzzy model identification problem. Section 4 is devoted to the genetic fuzzy identification method. Section 5 contains an example illustrating the application of GFIM. Some practical aspects and concluding remarks are presented in Section 6.

2. Evolutionary Algorithms and Genetic Fuzzy Systems

In the following we briefly review the GAs and the ESs, both of which shall be used in this contribution.

2.1 Genetic Algorithms

GAs are both theoretically and empirically proven to provide the means for efficient search in complex spaces [17].

Any GA starts with a population of randomly generated solutions, chromosomes, and advances towards better solutions by applying genetic operators such as crossover (recombination) and point mutation. These algorithms maintain a population of solutions for a given problem. The population undergoes 'evolution' in a form that resembles natural selection. In each generation, relatively good solutions reproduce to give 'birth' to offspring that replace relatively bad solutions which in turn eventually 'die'. Fitness criteria play the role of the environment to distinguish between good and bad solutions. The process of going from the current population to the next population constitutes one generation in the execution of a GA.

Although there are many possible variants of the basic GA, its fundamental underlying mechanism operates on a population of chromosomes (individuals) representing possible solutions to a problem, and consists of three basic operations:

1. Evaluation of chromosome's fitness.
2. Formation of a chromosome pool (intermediate population).
3. Recombination and mutation.

The structure of a GA is the following:

Procedure Genetic Algorithm
begin (1)
    $t = 0$;
    initialize $P(t)$;
    evaluate $P(t)$;
    While (**Not** *termination-condition*) **do**
    begin (2)
        $t = t + 1$;
        select $P(t)$ *from* $P(t - 1)$;
        recombine $P(t)$;
        evaluate $P(t)$;
    end (2)
end (1)

A fitness function must be constructed for each particular problem to be solved. Given a particular chromosome, the fitness function returns a single numerical fitness score which is proportional to the utility, or the adaptation ability, of this same chromosome.

There are a number of ways to perform selection. One might view the population as a mapping onto a roulette wheel, where each chromosome is represented by a space that is proportional to its fitness. By repeatedly spinning the roulette wheel, chromosomes are chosen using *stochastic sampling with replacement* to form the intermediate population. The selection procedure proposed in [3], and called *stochastic universal sampling*, is one of the most efficient. Here the number of offspring of any population is bound by the floor and ceiling of the expected number of offspring. After selection has been carried out, the construction of the intermediate population is complete and recombination and mutation can occur.

The crossover operator combines the features of two parent chromosomes to form two similar offspring. It is applied at a random position with a probability of performance, the so-called crossover probability, P_c. The mutation operator arbitrarily alters one or more components of a selected chromosome so as to increase the structural variability of the population. Each position of each chromosome vector in the population undergoes a random change according to a probability defined by a mutation rate, the so-called mutation probability, P_m.

The next figures illustrate the basic operations: reproduction, crossover, and mutation.

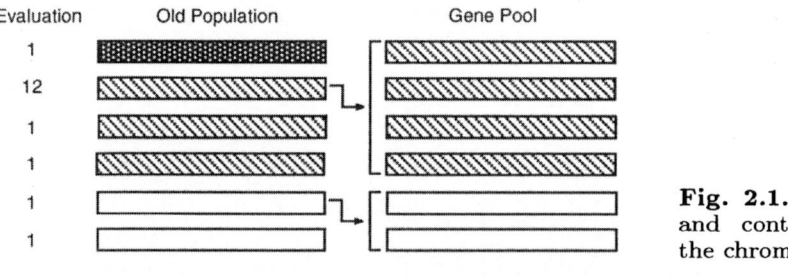

Fig. 2.1. Evaluation and contribution to the chromosome pool

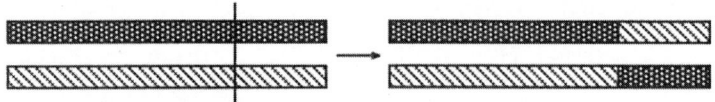

Fig. 2.2. Recombination (one-point crossover)

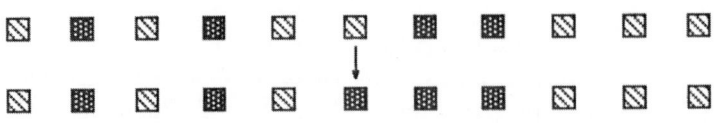

Fig. 2.3. Mutation

The basic principles of the GAs were first laid down rigorously by Holland [25], and are well described in many texts such as [17, 30].

Binary coded strings as solutions for the representation problem have been extensively used. But GAs are not solely dependent on the use of bit strings. Nonbinary representations which are more suitable for a variety of application problems have emerged. One of the most important nonbinary representations is the real coding scheme which seems particularly natural when dealing with optimization problems with variables in continuous search spaces. In this context, a chromosome is a vector of floating point numbers whose size is kept the same as the length of the vector. GAs based on real numbers representation are called *real-coded GAs* (RCGAs). RCGAs have been mainly used for numerical optimization in continuous domains. Using real coding, the representation of the chromosomes is very close to the natural formulation of many problems, e.g., there are no differences between the *genotype* (coding) and the *phenotype* (search space). The use of real coding also makes easier the design of other operators incorporating problem specific knowledge. RCGAs provide greater precision especially in the case of large domains where binary coding would require prohibitively long representation [24, 30].

It is generally accepted that a GA must take into account the five following components in solving any given problem:

1. A genetic representation of the problem solutions.
2. A way to create an initial population of solutions.
3. An evaluation function which computes the fitness of each chromosome.
4. Genetic operators that alter the genetic composition of offspring during reproduction.
5. Values for the parameters that the GA uses (population size, probabilities for applying genetic operators, etc.).

Numerous GA applications have been presented over the last years. Some of these can be classified as numerical function optimization, combinatorial optimization, image processing, fuzzy control and classification, engineering processes, biology, artificial life, machine learning, etc. There is an exceptionally large number of applications of GAs for the design of learning systems [18] and learning fuzzy systems [4, 20]. The interested reader can find free GA software in [16].

2.2 Evolution Strategies

ESs were developed with a strong focus on building systems capable of solving difficult real-valued parameter optimization problems. The natural representation is a vector or real-valued chromosomes which are manipulated primarily by mutation operators designed to perturb their real-valued parameters in a purposeful way.

ESs were initially developed by Rechenberg and Schwefel in 1964 as experimental optimization techniques. The first ES algorithm, the so-called *(1+1)-ES*, was based on working with only two individuals per generation, one parent and one descendent (offspring). Other more complex versions, based on considering higher number of parents ($\mu > 1$) and descendents ($\lambda > 1$), have appeared in last few years. These constitute the so-called $(\mu + \lambda)$-*ES* and (μ, λ)-*ES* algorithms. Also several new generalized ESs have been successfully developed [2, 33].

Without lack of generality, we will use in this chapter *(1+1)-ES*, the simplest ES algorithm. In the following we briefly describe this particular ES algorithm [2, 33]

(1+1)-ES is based on representing the possible optimization problem solution as a real-coded string. This parent string is evolved by applying a mutation operator over each one of its components. The mutation strength is determined by a parameter σ, the standard deviation of a normally distributed random variable. This parameter is associated with the parent and is also evolved in each step of the optimization process. If the evolution has been successful, i.e., the offspring obtained by mutation is better adapted than its parent, then the descendent substitutes the parent in the next generation. The individual adaptation is measured by using a fitness function. The process is iterated until a finishing condition is satisfied.

The main component of the ES algorithm is the mutation operator, **mut**. It is composed of two components: **mu**$_\sigma$, which updates the value of the parameter σ, and **mu**$_x$, which evolves the real coded string. The first component is based on Rechenberg's 1/5-success rule, which evolves the standard deviation according to the current value of the relative frequency p of successful mutations in the following way

$$\sigma' = \mathbf{mu}_\sigma(\sigma) = \begin{cases} \dfrac{\sigma}{\sqrt[n]{c}}, & \text{if } p > \frac{1}{5} \\[2mm] \sigma \cdot \sqrt[n]{c}, & \text{if } p < \frac{1}{5} \\[2mm] \sigma, & \text{if } p = \frac{1}{5}. \end{cases} \qquad (2.1)$$

The second component mutates each element of the real-coded string by adding normally distributed variations with standard deviation σ' to it

$$x' = \mathbf{mu}_x(x) = (x_1 + z_1, \ldots, x_n + z_n) \qquad (2.2)$$

where $z_i \sim N_i(0, \sigma'^2)$.

The final algorithm structure is as follows:

Procedure Evolution Strategy (1+1)
begin (1)
    $t = 0$;
    initialize $P(t) \leftarrow (x, \sigma)$;
    evaluate $f(x)$;

```
While (Not termination-condition) do
begin (2)
      t = t + 1;
      (x', σ') ← mut(x, σ);
      evaluate f(x');
      If Better  (f(x'), f(x))
      then P(t + 1) ← (x', σ')
      else P(t + 1) ← P(t).
   end (2)
end (1)
```

ES software is provided in [33].

2.3 Genetic Fuzzy Systems

In many cases, the identification of a fuzzy model may be considered as an optimization or a search process. GAs have the ability to explore and exploit a given complex search space using an available performance measure and are known to be capable of finding near-optimal solutions in such a search space. The prior knowledge one may wish to use, in addition to input-output data, may be in the form of known linguistic variables, fuzzy membership function parameters, fuzzy rules, number of rules, etc. The generic code structure and independent performance features of GAs make them suitable candidates for incorporating this type of prior knowledge.

These properties of GAs make them suitable candidates for the design and optimization of fuzzy rule bases. In particular, the design, learning, and tuning of fuzzy rule bases have produced quite promising results. Figure 2.4 illustrates this idea.

GAs are applied to modify/learn the model parameters, i.e., the shapes of the membership functions stored in the fuzzy data base, and/or the model structure, i.e., the fuzzy rules composing the fuzzy rule base. It is possible to distinguish three different groups of *genetic fuzzy model design approaches* according to the type of fuzzy identification performed in the learning process. They are briefly described in the following subsections. For a detailed description see [4] and for an extensive bibliography see [6, Sect. 3.13]. Different approaches may be found in [20].

2.4 Genetic Estimation of the Fuzzy Model Parameters (DB)

A fuzzy model has a number of parameters, such as the shapes of the membership functions, the scaling factors, the number of the linguistic values in the term sets associated with the linguistic variables from the fuzzy rules. All these fuzzy model parameters constitute the fuzzy data base (DB) of the fuzzy model and the fuzzy model quality is highly dependent on all of them

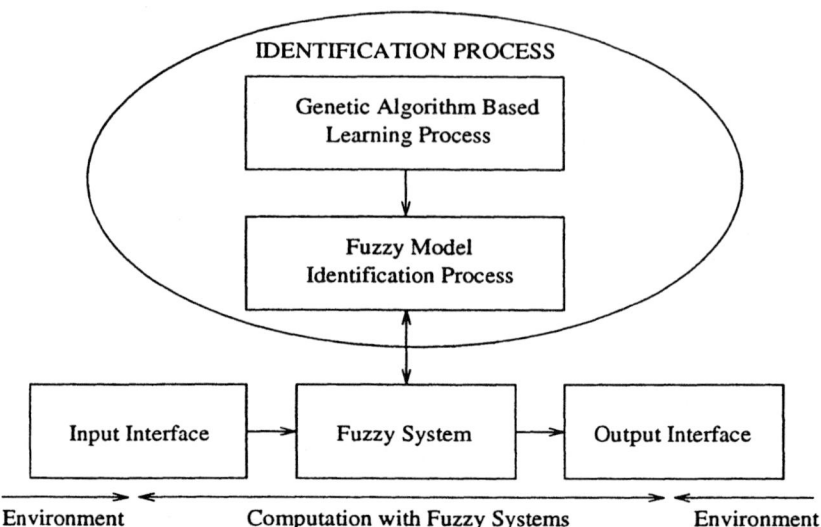

Fig. 2.4. Genetic fuzzy systems

[14, 19, 27, 37]. Therefore, the proper definition of the membership functions is an important task in fuzzy model identification. The parameter estimation method using GAs tunes the membership functions by adjusting their parameters according to a given fitness function.

Several methods have been proposed in order to construct the DB using GAs. All of them are based on the existence of a given set of fuzzy rules (or rule base (RB)) defining the fuzzy model structure and an initial definition of the model parameters. Each chromosome involved in the evolution process represents a different DB definition, i.e., each chromosome will contain a coding of the membership functions. A chromosome's degree of adaptation is measured using a fitness function. This fitness function is based on the quality of the fuzzy model, represented by the given RB, and the model parameters encoded in the chromosome.

2.5 Genetic Derivation of the Fuzzy Model Structure (RB)

All the methods belonging to this family assume that the model parameters are known in advance, i.e., they suppose the existence of a DB. Different GA-based methods for the derivation of the fuzzy model structure exist, depending on the representation chosen for RB: a set of fuzzy rules, a decision table, or a relational matrix. Most of these methods consider an RB represented in the form of a *decision table* (also called *look-up table*). As is well known, an RB consisting of fuzzy rules with n input variables and a single output variable may be represented by using an n-dimensional decision table, where each dimension corresponds to one input variable. Table 2.1 shows an example of a decision table for the control of an inverted pendulum. Every dimension has associated an array containing the linguistic values of the particular input variable. A cell in the decision table contains the linguistic value which the output variable takes for the combination of the linguistic values of the input variables corresponding to this cell. Therefore, each cell represents a fuzzy rule that may belong to the final model structure.

Table 2.1. Decision table for the control of inverted pendulum

Angle		NL	NM	NS	ZR	PS	PM	PL
	NL							
	NM							
Change	NS			NS		ZR		
of	ZR		NM		ZR		PM	
angle	PS			ZR		PS		
	PM							
	PL							

The above structure is encoded in the individuals (chromosomes) forming the GA population. If there is no value representing an empty cell in the decision table, then it is not possible to derive a fuzzy model structure with an optimal number of rules, because for this purpose all the possible fuzzy rules have to be considered.

2.6 Genetic Learning of the Fuzzy Model Structure and Parameters (RB and DB)

There is a multiplicity of approaches in genetic learning all aimed at the identification of the fuzzy model structure and parameters.

Amongst these some use variable chromosomal length, others use fixed chromosomal length encoding a fixed number of fuzzy rules together with the membership functions defining the linguistic values of the input and output variables, others use chromosomes each encoding a single fuzzy rule and its corresponding membership function parameters, etc.

Many of these approaches define the fitness function simply as an error measure, whereas others include a variety of objectives to be optimized in order to obtain more robust fuzzy models.

3. The Fuzzy Model Identification Problem

As we have mentioned earlier, in this chapter we focus on the linguistic type fuzzy model for MISO systems, where the structure of the fuzzy model consists of a collection of Mamdani-type fuzzy rules (with the logical connective ALSO between the fuzzy rules). Thus each fuzzy rule is of the form

R_i: **If** x_1 is A_{i1} **and** ...**and** x_n is A_{in} **then** y is B

where $x_1, \ldots, x_n$ are input variables, y is the output variable, and $A_{i1}, \ldots, A_{in}$, B are the linguistic values of the input and output variables in the i-th fuzzy rule. The input and output variables take their values in the universes of discourse $U_1, \ldots, U_n$, and V respectively. The meaning (semantics) of the linguistic values is characterized by the membership functions $\mu_{A_{ij}}(x_j)$ and $\mu_{B_i}(y)$ defined on the universes of discourse U and V respectively. In this chapter we consider triangular membership functions. A computationally efficient way to characterize this type of membership function is by using a parametric representation achieved by means of the 3-tuple (a_{ij}, b_{ij}, c_{ij}), (a_i, b_i, c_i), $j = 1, \ldots, n$.

The number of linguistic values for each input and output variable, the scaling factors, and the shapes of the membership functions constitute the fuzzy model parameters or DB. In [1] the following can be found about the linguistic fuzzy model:

> This representation is suitable for incorporating a priori knowledge by formulating the typical input-output situations in terms of rules. Since there is no structure assumed, virtually any system can be represented by the linguistic model. For this flexibility one has to pay by exponentially increasing model complexity, i.e., many rules may be needed to approximate a system to a given degree of accuracy, especially with many input variables. Also, the identification of the linguistic model from numerical data is not straightforward because one has to estimate both the membership functions and the relation between them (the rules). It is not trivial to estimate the membership functions from the data, since without any prior information one does not know where the 'important points' lie. Once the membership functions are found, rules can be identified quite easily.

The above difficulties can be said to motivate our particular identification strategy: once initial fuzzy model parameters have been derived and then the corresponding initial fuzzy model structure has been identified, the quality of the so obtained initial fuzzy model may be improved by deriving final fuzzy model parameters using the already identified initial fuzzy model structure.

In applying this identification strategy we consider an input-output data set without noise. This data set E_p is composed of p numerical input-output tuples $e_\ell \in E_p$, called *examples*, each example having the form

$$e_\ell = (ex_1^\ell, \ldots, ex_n^\ell, ey^\ell) \quad , \quad \ell = 1, \ldots, p. \tag{3.1}$$

In a conventional linguistic fuzzy model the set of linguistic values taken by the input and output variables is defined in advance. Furthermore, the meaning (semantics) of each linguistic value A_{ij} is determined by the membership function $\mu_{A_{ij}}(x_j)$ and one and the same linguistic value may appear in a number of fuzzy rules. However, in every fuzzy rule in which this linguistic value appears it has the same semantics, i.e., the same membership function. We call this type of linguistic fuzzy model a *descriptive* linguistic fuzzy model since a given membership function describes the semantics of an a priori defined linguistic value and, furthermore, one and the same linguistic value has the same semantics in all fuzzy rules in the RB in which it is encountered.

In this chapter we consider a linguistic fuzzy model in which the input and output variables variables do not take a priori defined linguistic values. This type of linguistic fuzzy model is called an *approximative* linguistic fuzzy model [4, 10]. When considering this type of fuzzy model we say that the fuzzy rules have *free semantics*. The difference between descriptive and approximative linguistic fuzzy models is illustrated is in Figure 3.1.

a) Descriptive Mamdani-type fuzzy model

NB NM NS ZR PS PM PB   NB NM NS ZR PS PM PB

X ⟨diagram⟩ Xr   Y ⟨diagram⟩ Yr

R1: If X is NB then Y is NB   R5: If X is PS then Y is PS
R2: If X is NM then Y is NM   R6: If X is PM then Y is PM
R3: If X is NS then Y is NS   R7: If X is PB then Y is PB
R4: If X is ZR then Y is ZR

b) Approximative Mamdani-type fuzzy model

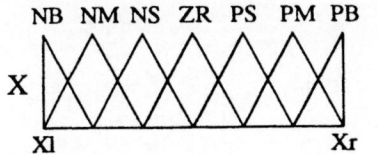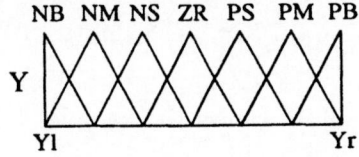

R1: If X is △ then Y is ◣
R2: If X is △ then Y is △
R3: If X is ◢ then Y is △
R4: If X is △ then Y is △

Fig. 3.1. Descriptive versus approximative linguistic fuzzy models

For the purpose of identification we consider fuzzy rules with free semantics, i.e., no a priori defined linguistic values are associated with the input and output variables from the fuzzy rules. However, we further consider two types of free semantics:

1. *Unconstrained free semantics.* The identification method proceeds by learning the fuzzy rules and the initial shapes of the membership functions associated with these rules. This is done without any prior fuzzy partitioning being available. That is, no restrictions are placed on the membership functions' locations and shapes.
2. *Constrained free semantics.* The identification method uses an initial fuzzy partitioning of the domains of the input and output variables par-

tition and the initial membership function parameters are locally (on a rule by rule basis) adjusted during the identification process.

In the case of constrained free semantics, each universe of discourse, U, is partitioned into a finite number of overlapping regions, each region labeled by a linguistic value. For instance, if X is a variable taking its values in U and denoting temperature, then one may define A_1 as "*low temperature*," $A_i (1 < i < r)$ as "*medium temperature*," and A_r as "*high temperature*," etc.

These referential linguistic values are characterized by their membership functions $\mu_{A_i}(u) : U \to [0,1], i = 1, \ldots, r$. To ensure good performance of the fuzzy model it is essential that all the referential membership functions are normal and convex, and they should satisfy the following completeness condition:

$$\forall u \in U \; \exists j, \; 1 \le j \le r, \text{ such that } \mu_{A_j}(u) \ge \delta$$

where δ is a fixed threshold, called the *completeness degree* of a universe of discourse. Figure 3.2 shows an example of a fuzzy partitioning with $\delta = 0.5$.

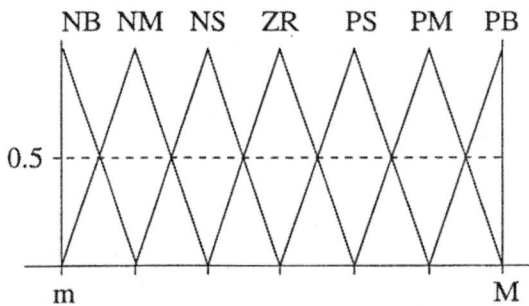

Fig. 3.2. Graphical representation of a fuzzy partitioning

Based on this type of fuzzy partitioning, an interval of performance is associated with each one of the three parameters (a_t, b_t, c_t) defining the membership functions $\mu_{A_t}(\cdot)$:

$$[a_t^\ell, a_t^r] = [a_t - \frac{b_t - a_t}{2}, a_t + \frac{b_t - a_t}{2}] \tag{3.2}$$

$$[b_t^\ell, b_t^r] = [b_t - \frac{b_t - a_t}{2}, b_t + \frac{c_t - b_t}{2}] \tag{3.3}$$

$$[c_t^\ell, c_t^r] = [c_t - \frac{c_t - b_t}{2}, c_t + \frac{c_t - b_t}{2}] \tag{3.4}$$

These intervals of performance are then used for locally adjusting the membership functions parameters during identification. Figure 3.3 shows the intervals of performance associated with each one of the parameters.

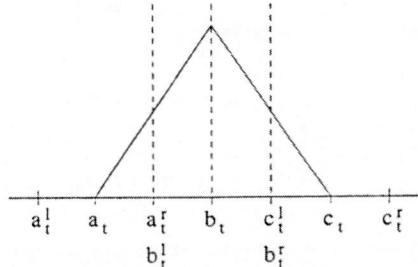

Fig. 3.3. Membership function and intervals of performance

Therefore, the fuzzy rules identified will have their semantics within the performance intervals established by the initial fuzzy partitioning. As already mentioned, the identification procedure is concerned with

- the *fuzzy model structure identification* which allows us to obtain the RB, and
- the *fuzzy model parameters estimation* which allows us to obtain the DB.

The phases of genetic simplification and tuning involved in the identification of the approximative linguistic fuzzy model use the following fuzzy logic operations and an inference procedure [37]. Consider an input vector $\mathbf{x} = (x_1, \ldots, x_n)$ and a fuzzy RB constituted by m linguistic fuzzy rules R_i, $i = 1, \ldots, m$,

If X_{11} is A_{11} and ... and X_{1n} is A_{1n} then Y is B_1

...

If X_{m1} is A_{m1} and ... and X_{mn} is A_{mn} then Y is B_m

1 The logical *and* connective is used to connect the antecedents in each individual rule. This connective is interpreted by the *min* operator

$$\mu_{A_i}(\mathbf{x}) = min(\mu_{A_{i1}}(x_1), \mu_{A_{i2}}(x_2), \ldots, \mu_{A_{in}}(x_n)). \tag{3.5}$$

2 The *if-then* fuzzy implication is defined as a *conjunction* of the membership functions from the antecedent and consequent part of an if-then fuzzy rule. It is interpreted by the *min* operator

$$\mu_{B_i'}(\mathbf{x}, y) = min(\mu_{A_i}(\mathbf{x}), \mu_{B_i}(y). \tag{3.6}$$

3 The logical connective *also* aggregating the fuzzy rules is given as the *weighted sum* of the defuzzified values of the output $\mu_{B_i'}(y)$ of each individual fuzzy rule. The final defuzzified output, y, generated by the fuzzy model, is computed as

$$y = \frac{\sum_{i=1}^{m} \mu_{A_i}(\mathbf{x}) \cdot CG(\mu_{B_i'}(y))}{\sum_{i=1}^{m} \mu_{A_i}(\mathbf{x})} \tag{3.7}$$

where $CG(\mu_{B_i'}(y))$ is the center of gravity of the fuzzy set B_i' [11].

4. The Genetic Fuzzy Identification Method

The GFIM is based on work presented in [5, 21, 22, 23]. It is composed of the following three parts:

I   A genetic fuzzy rule construction method based on EAs, and a covering method for the input-output example set. This part results in an initial fuzzy model structure and parameters, i.e., a set of approximative fuzzy rules with their associated fuzzy sets covering the training input-output data set in an adequate manner.

II   A genetic simplification of fuzzy rules, based on a binary coded GA and a fuzzy model performance measure. This part has the purpose of avoiding overlearning and removing redundant fuzzy rules.

III A genetic tuning of fuzzy model parameters, based on a RCGA and a fuzzy model performance measure. This part results in a final fuzzy model by tuning the membership functions for each fuzzy rule, i.e., it will estimate the final fuzzy model parameters taking into account the fuzzy model structure identified in the previous part.

Therefore, the first two parts form the genetic structure identification part of GFIM and provide the fuzzy model structure or RB together with initial fuzzy model parameters. The third part of GFIM adjusts the DB parameters giving us the final fuzzy model parameters.

The GFIM may be used in a number of ways according to the information available about the system under identification:

1. *Available example set:* The GFIM is applied for identifying a fuzzy model with unconstrained free semantics. No restrictions are imposed on the membership functions of inputs and outputs.

2. *Available example set and a fuzzy partition:* When initial fuzzy partitions (initial fuzzy model parameters) are available for all inputs and outputs, the GFIM is applied for constructing an RB using the initial fuzzy partitions and then locally tuning the membership functions associated with the identified fuzzy rules by means of an ES. Hence, the initial fuzzy partitions are used to guide the genetic search, restricting the shapes and the domains in which each one of the membership functions involved can be locally adjusted.

3. *Available partial RB and an example set:* In this case a fuzzy partitioning may or may not be available. Furthermore, an incomplete linguistic fuzzy model has been derived by a human expert. In this case the GFIM augments the partial RB with fuzzy rules learned from the example set, and then genetic simplification is applied to obtain the final fuzzy model structure. Finally, genetic tuning is applied for adjusting the membership functions of the final RB, thus obtaining the final fuzzy model parameters.

4. *Available RB:* In this case a complete linguistic fuzzy model is made available i.e., the fuzzy model structure and parameters are derived from a

human expert. Then, a genetic tuning process is used to obtain a more accurate fuzzy model by adjusting the already available fuzzy model parameters.

In the next sections we describe the three parts of the GFIM. First, we will present some requirements with respect to the RB.

4.1 Requirements

To identify a set of fuzzy rules R_i, describing the structure of a linguistic fuzzy model, one has to 'cover' all possible input-output pairs, $e_\ell \in E_p$, so that the so-called *completeness property* [14, 27] is achieved. The formulation of this requirement requires a constant $\tau \in [0, 1]$, the nonempty union of the membership functions $\mu_{A_i}(\cdot)$, $\mu_{B_i}(\cdot)$, and is formulated as

$$C_R(e_\ell) = \bigcup_{i=1}^{T} R_i(e_\ell) \geq \tau \quad \ell = 1, \ldots, p, \tag{4.1}$$

$$R_i(e_\ell) = *(\mu_{A_i}(ex^\ell), \mu_{B_i}(ey^\ell)),$$

$$\mu_{A_i}(ex^\ell) = *(\mu_{A_{i1}}(ex_1^\ell), \ldots, \mu_{A_{in}}(ex_n^\ell)),$$

where $*$ is a t-norm, and $R_i(e_\ell)$ is the *compatibility degree* between the rule R_i and the example e_ℓ.

Given a set of fuzzy rules R_i, the *covering value* of an example e_ℓ is defined as

$$CV_R(e_\ell) = \sum_{i=1}^{T} R_i(e_\ell), \tag{4.2}$$

and we require that

$$CV_R(e_\ell) \geq \epsilon \quad \ell = 1, \ldots, p. \tag{4.3}$$

The set of fuzzy rules must satisfy both of the conditions presented above, i.e., it has to have the completeness property and an adequate covering value.

4.2 Genetic Construction of Fuzzy Rules

The different parts of the genetic fuzzy rule construction method have been presented in [5, 22, 23]. In this chapter we describe their integration.

The genetic fuzzy rule construction consists of a *construction method* together with a *covering method*, both working on a given set of examples.

– The construction method is realized by means of a GA encoding of a single fuzzy rule in each chromosome. The GA finds the best fuzzy rule in every run over the set of examples according to the GA fitness function. When

constrained free semantics is considered, an ES is used for locally tuning the best fuzzy rules obtained during the iterations involved in the genetic search.
- The covering method is realized as an iterative process. It allows the construction of a set of fuzzy rules such that they cover the set of examples. In each iteration, the construction method chooses the best chromosome (fuzzy rule), considers the relative covering value this fuzzy rule has with respect to the example set, and removes the examples with a covering value greater than ϵ.

The above methods were separately presented in [22] and [5]. Here we introduce a new fitness criterion for the case of free semantics. This is the so-called *niche interaction rate* which defines the degree of overlapping between a newly constructed fuzzy rule and the previously constructed fuzzy rules. This fitness criterion was introduced in the case of constrained semantics in [5]. We also introduce some modifications to the fitness criteria presented in [22]. The first subsection presents these fitness criteria, the next two sections describe identification with constrained and unconstrained free semantics, and the fourth subsection describes the covering methods.

4.2.1 Fitness Criteria. The fitness functions employed in GFIM are designed according to different fitness criteria.

High frequency value. The frequency of a fuzzy model rule, R_i, on the set of examples, E_p, is defined as

$$\Psi_{E_p}(R_i) = \frac{1}{p}\sum_{\ell=1}^{p} R_i(e_\ell) \tag{4.4}$$

where $R_i(e_\ell)$ is the *compatibility degree* between R_i and e_ℓ.

High average covering degree on positive examples. The set of positive examples for R_i with compatibility degree greater than or equal to ω is defined as

$$E_\omega^+(R_i) = \{e_\ell \in E_p / R_i(e_\ell) \geq \omega\} \tag{4.5}$$

where $n_\omega^+(R_i)$ is equal to $|E_\omega^+(R_i)|$. The *average covering degree* on $E_\omega^+(R_i)$ can be defined as

$$G_\omega(R_i) = \sum_{e_\ell \in E_\omega^+(R_i)} R_i(e_\ell)/n_\omega^+(R_i). \tag{4.6}$$

Small negative example set. The set of the negative examples for R_i is defined as

$$E^-(R_i) = \{e_\ell \in E_p / R_i(e_\ell) = 0 \quad \text{and} \quad A_i(ex^\ell) > 0\}. \tag{4.7}$$

An example is considered negative for a fuzzy rule when it better matches some other fuzzy rule with the same antecedent (if-part), but a different consequent (then-part). The negative examples are always considered on the complete training set of examples.

With $n_{R_i}^- = |E^-(R_i)|$ being the number of negative examples, the *penalty function on the negative example set* is

$$g_n(R_i^-) = \begin{cases} 1 & \text{if } n_{R_i}^- \le k \cdot n_\omega^+(R_i) \\ \dfrac{1}{n_{R_i}^- - k n_\omega^+(R_i) + \exp(1)} & \text{otherwise,} \end{cases} \tag{4.8}$$

where we permit, up to a percentage of the number of positive examples, $k \cdot n_\omega^+(R_i)$, a number of negative examples per fuzzy rule without any penalty. This percentage is determined by the parameter $k \in [0, 1]$.

Small membership function width. The variable width (RW) of a fuzzy rule R_i is defined as

$$RW_i = \frac{1}{h} \sum_{j=1}^{h} \frac{WVR_{ij}}{DW_j} \tag{4.9}$$

where

$$WVR_{ij} = c_{ij}^3 - c_{ij}^1 \tag{4.10}$$

and $(c_{ij}^1, c_{ij}^2, c_{ij}^3)$, $j = 1, \ldots, h = n+1$, are the parameters associated with the membership function. DW_j is the *domain interval width* per input/output variable.

We define the *membership width rate, MWR*, of the rule R_i as a function of RW_i

$$MWR(R_i) = g(RW_i) \tag{4.11}$$

where the function g represents requirements with respect to the size of the width. We require a small width by considering the function

$$g(x) = \frac{e^{1-x} - 1}{e - 1}. \tag{4.12}$$

Highly symmetrical membership functions. The rate of symmetry is defined in order to achieve the symmetry of the membership functions and to prevent the bad covering of extreme points. It is defined as

$$RS(R_i) = \frac{1}{d^i} \tag{4.13}$$

where

$$d^i = \max_{j=1,\ldots,n+1} \{d_i^j\}, \tag{4.14}$$

with

$$d_i^j = \max\left\{\frac{d_{i1}^j}{d_{i2}^j}, \frac{d_{i2}^j}{d_{i1}^j}\right\} \tag{4.15}$$

and

$$d_{i1}^j = c_{ij}^2 - c_{ij}^1, d_{i2}^j = c_{ij}^3 - c_{ij}^2. \tag{4.16}$$

Clearly, $RS \leq 1$, and if all the membership functions of the fuzzy rule are symmetric, then $RS = 1$.

Low niche interaction rate. This criterion is introduced in [5]. Let $N_i = (N_i x, N_i y)$ be the centers of the fuzzy rules (niches) determined until now ($i = 1, \ldots, d$, where d is the number of runs of the construction method developed until now). Let C be a chromosome from the current population. Then the *niche interaction rate* penalizes the fitness score associated with C in the following way:

$$LNIR(C) = 1 - NIR(C), \tag{4.17}$$

$$NIR(C) = \max_i\{h_i\}, \tag{4.18}$$

$$h_i = *(A(N_i x), B(N_i y)), i = 1, \ldots, d, \tag{4.19}$$

$$A(N_i x) = *(A_1(N_i x_1), \ldots, A_n(N_i y)), \tag{4.20}$$

$$C \sim R_i: \text{ If } x_1 \text{ is } A_1 \text{ and } \ldots \text{and } x_n \text{ is } A_n \text{ then } y \text{ is } B. \tag{4.21}$$

Hence $LNIR(C)$ is defined on $[0, 1]$. It gives the maximum value (no penalization) when the fuzzy rule encoded in C does not interact with any of the fuzzy rules constructed previously. The minimum value (maximum penalization) is obtained when the fuzzy rule encoded in C is identical to a fuzzy rule constructed previously.

Therefore, the combination of the niche scheme and the covering method will allow us to verify the two following fundamental aspects of the GFIM:

- GFIM will ensure that fuzzy rules are identified for each available example. The completeness property is verified in this manner.
- GFIM will maintain an adequate rule distribution in each one of the niches existing in the space of examples. It is known that the good performance of a fuzzy model is due to its interpolatation ability. A particular input usually fires more than one fuzzy rule and the interaction between the simultaneously fired fuzzy rules is what allows the fuzzy model to determine the best output for this particular input. Hence, an adequate interaction rate between neighboring rules will improve the fuzzy model quality. Fuzzy rules too close to each other may cause an undesirable *overlearning* due to the fact that their excessive interaction makes the inferred output move away from the optimal. Remote rules make the fuzzy model lose its interpolation capability.

4.2.2 Identification with Unconstrained Free Semantics. The construction method for fuzzy rules is developed by means of an RCGA, where each chromosome represents a fuzzy rule. The RCGA obtains the best fuzzy rule according to a set of fitness criteria, which are included in the fitness function of the RCGA. We describe the RCGA components [22] below.

Representation. In the RCGA population, a candidate chromosome C_r, $r = 1, \ldots, M$, represents a fuzzy rule

If x_1 is A_{r1} and ...and x_n is A_{rn} then y is B_r

where the reals (a_{rj}, b_{rj}, c_{rj}), $(a_r, (b_r, c_r)$ are the parameter vectors of the membership functions of A_{rj}, $j = 1, \ldots, n$, and B_r respectively. C_r codes these vectors as

$$(a_{r1}, b_{r1}, c_{r1}, \ldots, a_{rn}, b_{rn}, c_{rn}, a_r, b_r, c_r).$$

As was justified in [21], we propose approaching the identification problem with real coded chromosomes together with special genetic operators developed for them. Then a fuzzy rule will be a chromosome vector coded as a vector of floating point numbers [22].

Finally, we represent a population of M chromosomes (fuzzy rules) by C, which is set up as follows:

$$C = (C_1, \ldots, C_M). \qquad (4.22)$$

Now, the fundamental underlying mechanisms of a GA, formation of an initial chromosome pool, fitness function, and genetic operators are introduced.

Initial chromosome pool. We denote the closed intervals of reals constituting the domains of the input and output variables, X_j and Y, as U_j and V, respectively. Furthermore, we consider an extension of these intervals in order to define the membership functions covering their extreme values. The intervals will increase their width by 10% for each extreme value. This permits us to cover the extreme values of the domains in an adequate form. In this way, the extended intervals are $U_j = [u_j^1, u_j^2]$ and $V = [v^1, v^2]$.

The initial chromosome pool is created partially from $E_t \subseteq E_p$ (t chromosomes). The remaining ($M - t$ chromosomes) are initialized randomly, as follows:

- Let $t = \min\{|E_p|, M/2\}$. We choose at random t examples from E_p and for each example we determine the chromosome (fuzzy rule) belonging to the initial chromosome pool as follows. Consider the example $e^\ell \in E_t$ and its component $ex_j^\ell \in [u_j^1, u_j^2]$, $\Delta ex_j^\ell = \min\{ex_j^\ell - u_j^1, u_j^2 - ex_j^\ell\}$. Let $\delta(ex_j^\ell)$ be a random value in the range $[0, \Delta ex_j^\ell]$. Then we construct the membership function by means of the triple

$$(ex_j^\ell - \delta(ex_j^\ell), ex_j^\ell, ex_j^\ell + \delta(ex_j^\ell)).$$

The procedure is the same for the remaining components of e_ℓ.

- The remaining $M - t$ chromosomes of the initial population are chosen at random, each chromosome in its respective interval,

$$C_r = (c_{r1}, \ldots, c_{r\ell}), \tag{4.23}$$

$\ell = 3 \cdot (n + 1)$, with requirements $c_{3s+1} \leq c_{3s+2} \leq c_{3s+3}$, $s = 0, \ldots, n$.

Evaluation of chromosome fitness. As we commented earlier, we define the fitness function either according to the five criteria employed in [22] or the six presented in this work.

An *evaluation function* for the fuzzy rule R_i, and therefore a fitness function for the associated chromosome C_i is defined as

$$Z_1(R_i) = \Psi_{E_p}(R_i) \cdot G_w(R_i) \cdot g_n(R_i^-) \cdot MWR(R_i) \cdot RS(R_i), \tag{4.24}$$

$$Z_2(R_i) = \Psi_{E_p}(R_i) \cdot G_w(R_i) \cdot g_n(R_i^-) \cdot MWR(R_i) \cdot RS(R_i) \cdot LNIR(R_i). \tag{4.25}$$

The objective in both of the above cases is the maximization of the fitness function.

Genetic operators. During the GA reproduction phase we use two classical genetic operators, mutation and crossover. The ones selected are the non-uniform mutation proposed by Z. Michalewicz [30] and the max-min arithmetical crossover used in [21]. A short description of them is given below.

- *Non-uniform mutation*

  If $C_v^t = (c_1, \ldots, c_k, \ldots, c_H)$ is a chromosome and the element c_k was selected for this mutation (the domain of c_k is $[c_{k\ell}, c_{kr}]$), the result is a vector $C_v^{t+1} = (c_1, \ldots, c_k', \ldots, c_H)$, with $k \in 1, \ldots, H$, and

  $$c_k' = \begin{cases} c_k + \Delta(t, c_{kr} - c_k) & \text{if } a = 0 \\ c_k - \Delta(t, c_k - c_{k\ell}) & \text{if } a = 1 \end{cases} \tag{4.26}$$

  where a is a random number that may have a value of zero or one, and the function $\Delta(t, y)$ returns a value in the interval $[0, y]$. This value is such that the probability of $\Delta(t, y)$ being close to 0 increases as t increases:

  $$\Delta(t, y) = y(1 - r^{(1 - \frac{t}{T})^b}). \tag{4.27}$$

  In the above expression, r is a random number in the interval $[0, 1]$, T is the maximum number of generations, and b is a parameter chosen by the user, which determines the degree of dependency between the strength of the mutation and the number of iterations developed so far. This property of the probability of $\Delta(t, y)$ forces the non-uniform mutation operator to perform uniform search in the initial chromosome space when t is small, and local search for larger t.

– *Max-min-arithmetical crossover*
If $C_v^t = (c_1, \ldots, c_k, \ldots, c_H)$ and $C_w^t = (c_1', \ldots, c_k', \ldots, c_H')$ are to be crossed, we generate the following four offspring:

$$
\begin{aligned}
C_1^{t+1} &= aC_w^t + (1-a)C_v^t, \\
C_2^{t+1} &= aC_v^t + (1-a)C_w^t, \\
C_3^{t+1} &\text{ with } c_{3k}^{t+1} = \min\{c_k, c_k'\}, \\
C_4^{t+1} &\text{ with } c_{4k}^{t+1} = \max\{c_k, c_k'\}.
\end{aligned}
\tag{4.28}
$$

This operator can use a parameter a which is either a constant, or a variable whose value depends on the age of the population. The resulting descendents are the two best of the four offspring.

With regard to the *selection procedure*, it is a stochastic universal sampling [3], in which the number of offspring of a parent chromosome is limited by the floor and ceiling of the expected number of offspring, together with an elitist selection.

4.2.3 Identification with Constrained Free Semantics. This construction method for fuzzy rules is developed by means of a special GA, where a chromosome encodes a fuzzy rule, and an ES that locally tunes the fuzzy rules. In the following, we describe this method. For more detail see [5].

Representation. Here, a chromosome C is composed of two different parts, C_1 and C_2, each one corresponding to each one of the fuzzy model components. The first part of the chromosome encodes the linguistic fuzzy rule (belonging to the RB), and the second one the membership functions parameters for the input and output variables involved in the fuzzy rule (belonging to the DB).

In order to represent the first part there is a need to number the linguistic values belonging to each one of the term sets for the input and output variables. A variable x_i taking linguistic values in a term set $T(x_i) = \{L_1(x_i), \ldots, L_{n_i}(x_i)\}$ has associated with it the set $T'(x_i) = \{1, \ldots, n_i\}$. On the other hand, the second part adopts the same representation as in the case of unconstrained free semantics. Hence, the fuzzy rule

If x_1 **is** $L_{i_1}(x_1)$ **and** $\ldots$ **and** x_n **is** $L_{i_n}(x_n)$ **then** y **is** $L_{i_{n+1}}(y)$

is encoded into a chromosome C of the following form:

$$
\begin{aligned}
C_1 &= (i_1, \ldots, i_n, i_{n+1}), \\
C_2 &= (a_{L_{i_1}(x_1)}, b_{L_{i_1}(x_1)}, c_{L_{i_1}(x_1)}, \ldots, a_{L_{i_n}(x_n)}, b_{L_{i_n}(x_n)}, c_{L_{i_n}(x_n)}, \\
&\qquad a_{L_{i_{n+1}}(y)}, b_{L_{i_{n+1}}(y)}, c_{L_{i_{n+1}}(y)}) \\
C &= C_1 C_2.
\end{aligned}
\tag{4.29}
$$

Initial chromosome pool. Part of the initial chromosome pool is obtained making use of the examples contained in the training set, E_p, and the remainder of it is generated at random. However, a third possibility can be considered. With M being the GA population size and $t = \min\{|E_p|, \frac{M}{3}\}$, let t examples be selected at random from E_p. Then, the initial population is generated in three steps as follows:

1. Using fuzzy partitions, generate t chromosomes by taking the fuzzy rule which covers best each one of the t randomly selected examples. Initialize C_1 and C_2 by coding the fuzzy rule linguistic values and their semantics.
2. Generate another t chromosomes by initializing C_1 in the same way as in the previous step, and compute the values of C_2 at random, letting each chromosome vary in its respective interval.
3. Generate the remaining $M - 2 \cdot t$ chromosomes by computing at random the values of C_1, and making use of these for randomly generating the C_2 part, letting again each chromosome vary in its respective interval.

Evaluation of chromosome fitness. Due to the fact that permitted chromosome variations are restricted to being performed in intervals smaller than those considered in the case of unconstrained free semantics, several previously employed fitness criteria lose their meaning and others become necessary. Hence, the membership function symmetry and width rates are not used anymore, but the niche interaction rate is required for obtaining an adequate fuzzy rule interaction. The fitness function is finally defined in the following way

$$Z_3(R_i) = \Psi_{E_p}(R_i) \cdot G_w(R_i) \cdot g_n(R_i^-) \cdot LNIR(R_i). \tag{4.30}$$

Genetic operators. Due to the special nature of the chromosomes involved in the case of constrained free semantics, the design of special genetic operators is required. For a more detailed description of these see [5].

With respect to mutation, two different operators are used, each one of them acting on a different chromosome part. Since C_2 corresponds to the representation employed in the case of unconstrained free semantics, the same mutation operator designed for this case is used for C_2. Thus, Michalewicz's non-uniform mutation operator is employed.

The mutation operator selected for C_1 is similar to the one proposed by Thrift in [36]. When mutation for the C_1 part of the chromosome is going to be performed, a local modification is developed by changing the current linguistic value to one of its neighboring linguistic values (the decision is made at random). When the linguistic value to be changed is the first or last one in the term set, the only possible change is to substitute it with its right or left neighbor respectively. Obviously, a mutation in C_1 provokes a change in C_2. When an input/output variable changes its linguistic value value from one term to another, the semantics (membership function) associated with it is automatically updated in the second part of the chromosome by the default values in the corresponding fuzzy partitioning.

With regard to recombination, two different crossover operators are employed. If the fuzzy rule encoded by both parent chromosomes is the same, then the max-min arithmetical crossover operator is applied to C_2 and obviously the parent C_1 values are maintained in the offspring. On the other hand, when the parent chromosomes encode different rules, it makes no sense to apply this operator. Instead, a standard crossover operator is applied on both parts of the parent chromosomes. This operator performs as follows. A crossover point cp is randomly generated in C_1 and the two parent chromosomes are crossed at the cp-th and $n + 1 + 3 \cdot cp$ crossover points in each chromosome part respectively. The crossover is thus performed in both chromosome parts, C_1 and C_2, thereby producing two meaningful offspring.

Let us consider an example in order to clarify the standard crossover operator. Since

$$C_t = (c_1, \ldots, c_{cp}, c_{cp+1}, \ldots, c_{n+1}, a_{c_1}, b_{c_1}, c_{c_1}, \ldots, \tag{4.31}$$
$$a_{c_{cp}}, b_{c_{cp}}, c_{c_{cp}}, a_{c_{cp+1}}, b_{c_{cp+1}}, c_{c_{cp+1}}, \ldots, a_{c_{n+1}}, b_{c_{n+1}}, c_{c_{n+1}})$$

and

$$C'_t = (c'_1, \ldots, c'_{cp}, c'_{cp+1}, \ldots, c'_{n+1}, a_{c'_1}, b_{c'_1}, c_{c'_1}, \ldots, \tag{4.32}$$
$$a_{c'_{cp}}, b_{c'_{cp}}, c_{c'_{cp}}, a_{c'_{cp+1}}, b_{c'_{cp+1}}, c_{c'_{cp+1}}, \ldots, a_{c'_{n+1}}, b_{c'_{n+1}}, c_{c'_{n+1}})$$

are the chromosomes to be crossed at the point cp, the two resulting offspring are

$$C_{t+1} = (c_1, \ldots, c_{cp}, c'_{cp+1}, \ldots, c'_{n+1}, a_{c_1}, b_{c_1}, c_{c_1}, \ldots, \tag{4.33}$$
$$a_{c_{cp}}, b_{c_{cp}}, c_{c_{cp}}, a_{c'_{cp+1}}, b_{c'_{cp+1}}, c_{c'_{cp+1}}, \ldots, a_{c'_{n+1}}, b_{c'_{n+1}}, c_{c'_{n+1}}),$$
$$C'_{t+1} = (c'_1, \ldots, c'_{cp}, c_{cp+1}, \ldots, c_{n+1}, a_{c'_1, b_{c'_1}}, c_{c'_1}, \ldots, \tag{4.34}$$
$$a_{c'_{cp}}, b_{c'_{cp}}, c_{c'_{cp}}, a_{c_{cp}}, b_{c_{cp}}, c_{c_{cp}}, \ldots, a_{c_{n+1}}, b_{c_{n+1}}, c_{c_{n+1}}).$$

The last genetic operator is based on *(1+1)-ES*. This optimization technique has been selected and integrated into the genetic recombination process in order to perform a local tuning of the best chromosomes (fuzzy rules) obtained in each run. Each time a GA generation is performed, the ES will be applied over a percentage α of the best chromosomes from the current genetic population and will adjust their C_2 parts.

The ES employed was briefly presented in Sect. 2.2. In our case, the step size σ can not be a single value because each one of the membership functions encoded in the second part of the chromosome is defined over different universes of discourse and thus requires mutations of a different order. Following the modus operandi presented in [5], each parent component x_i varying in the interval of performance $[x_i^\ell, x_i^r]$ will have its own associated step size $\sigma_i = \sigma \cdot s_i$ with $s_i = (x_i^r - x_i^\ell)/4$. When the mutated value $x'_i = x_i + z_i$ does not belong to the interval of performance, it is assumed equal to the interval extent, x_i^ℓ or x_i^r, closer to $x_i + z_i$.

The *selection procedure* is again based on Baker's stochastic universal sampling and elitist selection.

4.2.4 Covering method. The covering method is presented in detail in [22]. It is developed as an iterative process that allows one to obtain a set of fuzzy rules covering the example set. In each iteration, it considers the relative covering value the best fuzzy rule (chromosome) obtained from the genetic construction method has for the given training set, and removes from it the examples for which the covering value is greater than ϵ.

Let R^e be the set of fuzzy rules provided by a human expert (expert fuzzy rules). For any one of the cases of unconstrained free semantics and constrained free semantics, the covering method proceeds as follows:

1. Initialization stage:
   - Get k, ω and ϵ from the user.
   - Merge the fuzzy rules in R^e with the fuzzy rules, R^g, which are obtained via identification from input-output data.
   - Assign $CV[\ell] \leftarrow CV_{R^e}(e_\ell)$, $\ell = 1, \ldots, p$.
   - If $CV[\ell] \geq \epsilon$, remove e_ℓ from E_p, $\ell = 1, \ldots, p$.
2. Apply the specific construction method for the given set of examples E_p,
3. Select the best chromosome C_r encoding the fuzzy rule R_r.
4. Merge R_r with R^g.
5. For every $e_\ell \in E_p$ do
   $$CV[\ell] \leftarrow CV[\ell] + R_r(e_\ell),$$
   If $CV[\ell] \geq \epsilon$ then remove it from E_p.
6. If $E_p = \emptyset$ then Stop, else return to Step 2.

Since there may be similar fuzzy rules in R^g, or a fuzzy rule from R^g may be similar to a fuzzy rule in R^e, it is necessary to simplify the RB obtained.

4.3 Genetic Simplification

Due to the iterative nature of genetic identification, it may result in redundant fuzzy rules and/or overlearning. This latter occurs when some examples are covered to degree higher than the desired one and in this case the performance of the identified RB is negatively affected.

The genetic simplification was proposed in [23]. It is based on a binary coded GA, in which the selection of chromosomes is performed by using the stochastic universal sampling procedure together with an elitist selection scheme. Also, recombination is put into effect by using the classical binary multipoint crossover (performed on two points) and uniform mutation operators.

The coding scheme generates fixed-length chromosomes. Let us now consider the fuzzy rules in the RB, constructed at the previous step and numbered from 1 to m. An m-bit string $C = (c_1, \ldots, c_m)$ represents a subset of candidate fuzzy rules that is to be obtained as the output, B^s, of genetic simplification, and such that

If $c_i = 1$ then $R_i \in B^s$ else $R_i \notin B^s$.

The initial population of chromosomes is obtained by introducing a chromosome representing the complete rule set R^g, that is, $c_i = 1$ for all i. The remaining chromosomes are selected at random.

The fitness function, $E(\cdot)$, is based on an application-specific measure usually employed in the design of GFSs: the mean square error (SE) over a training data set, E_{TDS}, given as

$$E(C_j) = \frac{1}{2|E_{TDS}|} \sum_{e_\ell \in E_{TDS}} (ey^\ell - S(ex^\ell))^2 \tag{4.35}$$

where $S(ex^\ell)$ is the output value obtained from the identified fuzzy model using the model structure (RB) coded in C_j, $R(C_j)$, when the input values are ex^ℓ, and ey^ℓ is the known desired output value.

During genetic simplification, there is a need to keep the completeness property considered previously: the model must always be able to infer a proper output for every system input. We will ensure this condition by forcing every example contained in the training set to be covered by the encoded RB in a degree greater than or equal to τ,

$$C_{R(C_j)}(e_\ell) = \bigcup_{j=1..T} R_j(e_\ell) \geq \tau, \quad \forall e_\ell \in E_{TDS} \text{ and } R_j \in R(C_j) \tag{4.36}$$

where τ is the minimal degree of completeness of the training set allowed for the genetic simplification. Usually, τ is less than or equal to ω, the compatibility degree used in the identification. Therefore, we define a *training set completeness degree* of $R(C_j)$ on the set of examples E_{TDS} as

$$TSCD(R(C_j), E_{TDS}) = \bigcap_{e_\ell \in E_{TDS}} C_{R(C_j)}(e_\ell). \tag{4.37}$$

The fitness function, penalizing the lack of the completeness property, is

$$F(C_j) = \begin{cases} E(C_j) & \text{if } TSCD(R(C_j), E_{TDS}) \geq \tau \\ \frac{1}{2} \sum_{e_\ell \in E_{TDS}} (ey^\ell)^2 & \text{otherwise.} \end{cases} \tag{4.38}$$

4.4 Genetic Tuning

The genetic tuning process is presented in depth in [21]. It is based on the existence of an initial fuzzy model, that is, an initial estimation of the model parameters (DB), and an initially identified fuzzy model structure defined by an RB and composed of m fuzzy rules.

Each chromosome forming the genetic population encodes the whole RB, R^s, with different fuzzy model parameters associated with it.

The GA designed for the tuning uses real numbers coding, stochastic universal sampling as selection procedure, and Michaelewicz's non-uniform mutation operator. The max-min arithmetical crossover operator is employed again.

As we commented before, the membership functions have a triangular form. Thus, each one of them has an associated parametric representation based on a triple of real values. Each one of the fuzzy rules will be encoded in parts of the chromosome C_{ri}, $i = 1, \ldots, m$, in the following way:

$$C_{ri} = (a_{i1}, b_{i1}, c_{i1}, \ldots, a_{in}, b_{in}, c_{in}, a_i, b_i, c_i). \tag{4.39}$$

Therefore, the complete RB with an associated DB is represented by a complete chromosome C_r:

$$C_r = C_{r1} \, C_{r2} \, \ldots \, C_{rm}. \tag{4.40}$$

Each chromosome in the population represents a complete fuzzy model i.e., both RB and DB. In particular, all of them encode the identified RB, R^s, i.e., the model structure identified in the previous two stages, and the only difference between the different chromosomes are the different membership functions, i.e., the different DBs.

The initial fuzzy model is encoded directly into a chromosome, denoted as C_1. The remaining ones are generated by associating an interval of performance, $[c_h^\ell, c_h^r]$ with every c_h in C_1, $h = 1 \ldots (n+1) \cdot m \cdot 3$. Each interval of performance will be the interval of adjustment for the corresponding variable, $c_h \in [c_h^\ell, c_h^r]$.

If $(t \bmod 3) = 1$ then c_t is the left value of the support of a membership function. The latter is defined by the three parameters (c_t, c_{t+1}, c_{t+2}) and the intervals of performance are

$$c_t \in [c_t^\ell, c_t^r] = [c_t - \frac{c_{t+1} - c_t}{2}, c_t + \frac{c_{t+1} - c_t}{2}], \tag{4.41}$$

$$c_{t+1} \in [c_{t+1}^\ell, c_{t+1}^r] = [c_{t+1} - \frac{c_{t+1} - c_t}{2}, c_{t+1} + \frac{c_{t+2} - c_{t+1}}{2}], \tag{4.42}$$

$$c_{t+2} \in [c_{t+2}^\ell, c_{t+2}^r] = [c_{t+2} - \frac{c_{t+2} - c_{t+1}}{2}, c_{t+2} + \frac{c_{t+3} - c_{t+2}}{2}]. \tag{4.43}$$

Figure 4.1 shows these intervals of performance.

Therefore, we create a population of chromosomes containing C_1 as its first individual and where the remaining ones are initiated randomly, with each gene being in its respective interval of performance.

The fitness function of a chromosome is defined by using the training input-output data set, E_{TDS}, and a specific error measure, the mean square error. In this way, the adaptation value associated with a chromosome is obtained by computing the error between the outputs given by the fuzzy

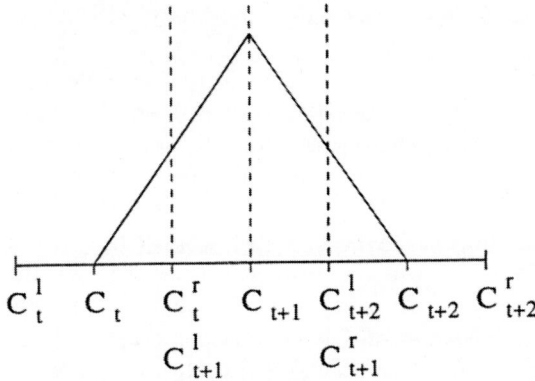

Fig. 4.1. Membership function and intervals of performance for genetic tuning

model contained in the chromosome and the outputs contained in the training input-output data set. The fitness function is given as

$$E(C) = \frac{1}{2|E_{TDS}|} \sum_{e_\ell \in E_{TDS}} (ey^\ell - S(ex^\ell))^2. \tag{4.44}$$

4.5 Summary of the Identification Procedure

This section summarizes the presented GFIM, showing some guidelines for its use depending on the information available about the system under identification.

I. **Available knowledge.** Once the input-output data set is available, the existence of any other kind of prior knowledge for the purpose of identification should be studied: the possibility of defining the model parameters (in the sense of initial fuzzy partitions), the availability of a partial fuzzy model (in the sense of an incomplete RB with or without an initial definition of the DB), or a complete one (in the sense of a complete RB and an initial definition of the DB).

II. **Using the GFIM.** Once the the existing knowledge is made available, the particular identification method to be applied has to be chosen.

1. *No prior knowledge:* In this case, we only have the input-output data set for performing identification. Therefore, the unconstrained free semantics type of identification is first applied for obtaining an initial definition of the fuzzy model structure and parameters. Genetic simplification is then applied for identifying the final fuzzy model structure. Finally, genetic tuning obtains the final fuzzy model parameters.

2. *Initial fuzzy model parameters:* When initial fuzzy partitions, a very preliminary definition of the model parameters, are provided, the overall learning process is applied for identifying a fuzzy model taking the previous semantics as a base. The only change with respect to the

previous case is that the genetic construction method used will be the constrained free semantics one.

3. *Partial fuzzy model:* When an incomplete linguistic fuzzy model has been derived from a human expert, the GFIM permits the incorporation of this partial RB by merging it with the one obtained from genetic identification (constrained or unconstrained free semantics). Genetic simplification and tuning are then applied to obtain the final model structure and parameters.

4. *Complete fuzzy model:* When a complete linguistic fuzzy model has been derived from an expert, i.e., the fuzzy model structure and the fuzzy model parameters are both available, genetic tuning may be used to obtain a more accurate fuzzy model by adjusting the available fuzzy model parameters.

III. **Model validation.** The GFIM proposed takes into account some fuzzy rule base properties in order to guarantee good quality linguistic fuzzy models. Thus, the analysis of how well the fuzzy rules cover the data space should be done with care since a bad coverage decreases the quality of the linguistic fuzzy model. Anyway, there is always the need to validate the identified fuzzy model through numerical simulation and comparisons with system data in order to improve its validity.

To conclude this section we would like to note here that any other type of membership functions may be incorporated in the GFIM with minor modifications. Furthermore, in [7] a different GFIM for the case of unconstrained free semantics is presented where the construction method for fuzzy rules uses an inductive algorithm and an ES. The structure of this GFIM may be used for identifying another type of linguistic fuzzy model, namely a descriptive linguistic fuzzy model [8, 10].

5. Example

In order to analyze the accuracy of the GFIM, we will show now how two n-dimensional functions can be used to identify three-dimensional surfaces. Three different ways of identifying fuzzy models will be compared:

1. Two GFIMs for the case of unconstrained free semantics using the fitness functions Z_1 and Z_2 respectively, and
2. The GFIM for the case of constrained free semantics using the fitness function Z_3.

The n-dimensional functions and the universes of discourse considered are shown below. The *spherical model*, F_1, is a unimodal function, while the *generalized Rastrigin function*, F_2, is a strongly multimodal one. These are illustrated in Figs. 5.1 and 5.3.

$$F_1(x_1, x_2) = x_1^2 + x_2^2, \tag{5.1}$$
$$x_1, x_2 \in [-5, 5], \quad F_1(x_1, x_2) \in [0, 50].$$

$$F_2(x_1, x_2) = x_1^2 + x_2^2 - \cos(18x_1) - \cos(18x_2), \tag{5.2}$$
$$x_1, x_2 \in [-1, 1], \quad F_2(x_1, x_2) \in [2, 3.5231].$$

For each function, an input-output training data set, uniformly distributed in the three-dimensional space, has been obtained experimentally. In this way, two sets with 1681 values have been generated by taking 41 values for each of the two state variables considered to be uniformly distributed in their respective intervals.

Two other data sets have been generated for use as test sets. These are to be used for evaluating the performance of the learning method, thus avoiding any possible bias related to the data in the training set. The size of these data sets is ten percent of the size of the corresponding training set. The data are obtained by generating the state variable values randomly in the concrete universes of discourse for each of them, and then computing the associated output value. Hence, two test sets formed by 168 data are used to measure the accuracy of the fuzzy models identified by computing the mean square error for these fuzzy models.

The initial fuzzy model parameters used in the GFIM for the case of constrained free semantics are defined in terms of three initial fuzzy partitions (two corresponding to the input variables and one associated with the output). The fuzzy partitions have *seven linguistic values* and the semantics of these is defined via the use of triangular membership functions (as shown in Figure 3.1). Adequate scaling factors are used to translate the generic universe of discourse into the one that is associated with each system variable.

The following parameters, corresponding to the first two stages of the identification, are combined for determining the number of runs for the three different GFIMs: $\epsilon = 1.5$, $\omega = 0.05$, $k = 0.1$, and $\tau = \{0.25, 0.5\}$. This leads to a total of two runs per function and GFIM. The remaining parameters used in the three GFIMs to be compared are: the t-norm $*$ used in the fuzzy rule construction method is the *min* operator; GAs run over 50 generations and the ES is applied until there is no improvement in 25 generations over a percentage $\alpha = 20\%$ of the population of chromosomes (the parameter c of the 1/5-success rule is equal to 0.9); genetic simplification and tuning run over 500 and 1000 generations, respectively. In all cases, the population is formed by 61 chromosomes, the value of the non-uniform mutation parameter b is 5.0, and the crossover and mutation rates are $P_c = 0.6$ and $P_m = 0.1$ (this last one per individual) respectively. The max-min arithmethical crossover parameter a takes the value 0.35.

Finally, we use the *min* t-norm for representing fuzzy implication, and the center of gravity weighted by the matching as defuzzification operator [11].

The results obtained are shown in the tables below. Each table is associated with both of the functions considered. The notation $|R|_x$ stands for the number of fuzzy rules in the RB, while SE_x stands for the mean square error obtained by the current fuzzy model on the corresponding test set at each stage (x is equal to G, S, and T in the genetic construction, simplification, and tuning stages, respectively).

Table 5.1. Results obtained using the three proposed GFIMs for the fuzzy model of the function F_1

GFIM	τ	$|R|_G$	SE_G	$|R|_S$	SE_S	SE_T
1	0.25	108	5.823179	82	2.794654	0.929880
1	0.5	108	5.823179	82	3.423883	1.094663
2	0.25	180	17.448940	119	2.013139	0.994133
2	0.5	180	17.448940	119	2.662488	1.177372
3	0.25	98	2.411402	67	1.779137	0.696869
3	0.5	98	2.411402	73	2.130197	1.118251

Table 5.2. Results obtained using the three proposed GFIMs for the fuzzy model of the function F_2

GFIM	τ	$|R|_G$	SE_G	$|R|_S$	SE_S	SE_T
1	0.25	250	0.398029	181	0.324123	0.290474
1	0.5	250	0.398029	196	0.344089	0.307614
2	0.25	345	0.393328	264	0.283268	0.265746
2	0.5	345	0.393328	275	0.359460	0.327285
3	0.25	346	0.268026	232	0.213960	0.195233
3	0.5	346	0.268026	253	0.232196	0.210177

To illustrate the performance of the proposed GFIM, some of the fuzzy models obtained are shown in the following figures. The two fuzzy models depicted are the ones that best approximate each one of the functions considered, that is, the ones with the lowest mean square error. Figures 5.2 and 5.4 show the fuzzy models for the functions F_1 and F_2; these fuzzy models are obtained by means of the constrained free semantics GFIM using the fitness function Z_3 and $\tau = 0.25$.

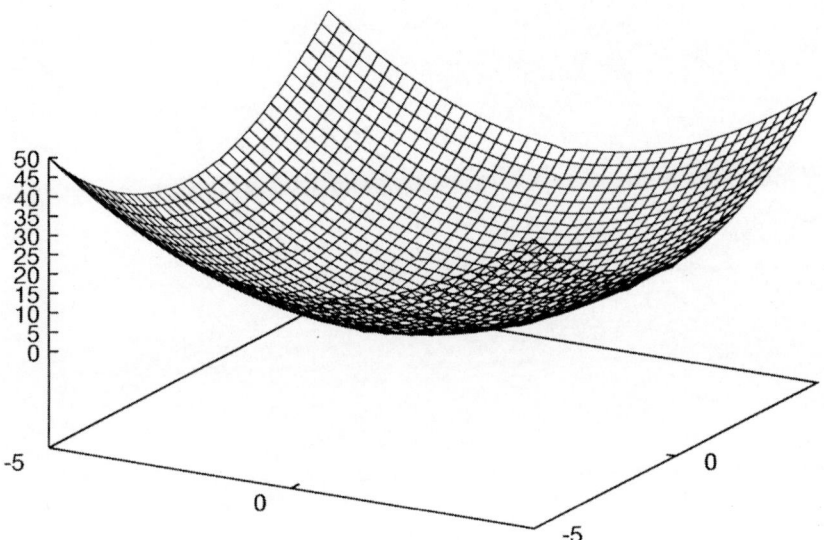

Fig. 5.1. Graphical representation of F_1

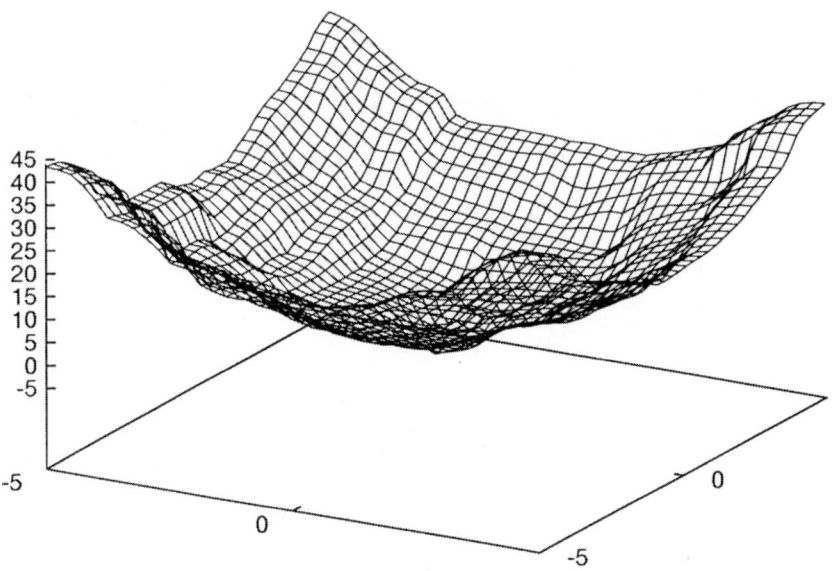

Fig. 5.2. Graphical representation of the fuzzy model obtained for F_1

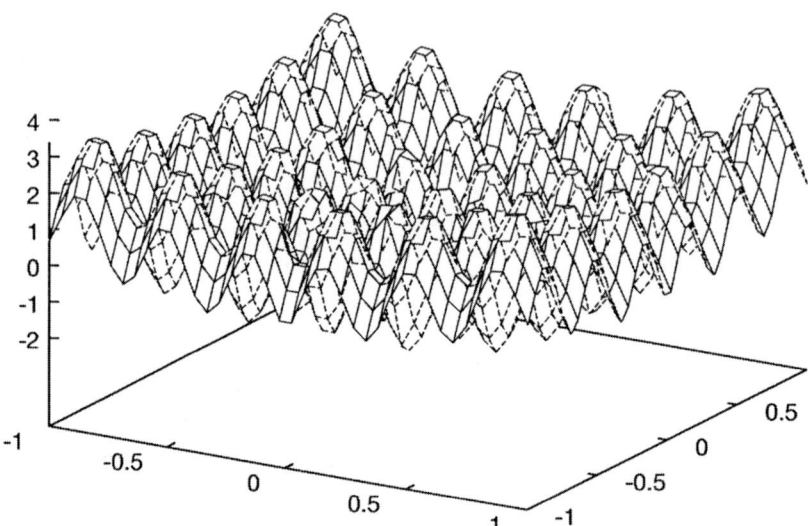

Fig. 5.3. Graphical representation of F_2

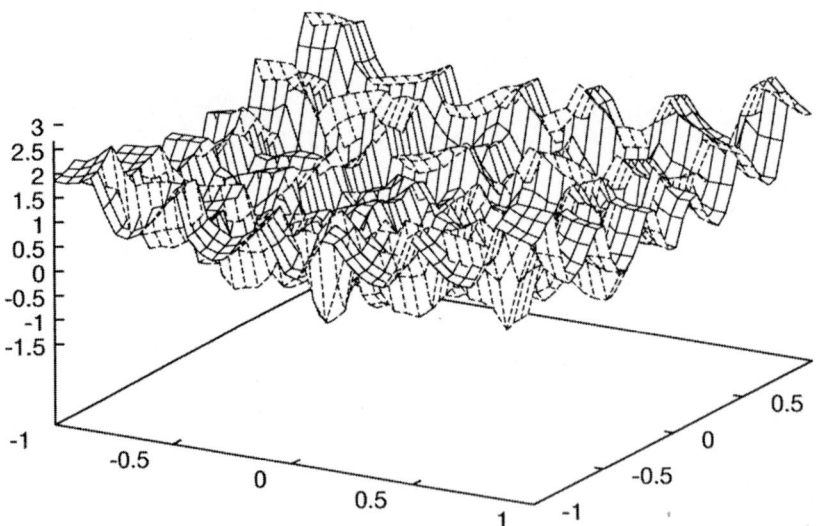

Fig. 5.4. Graphical representation of the fuzzy model obtained for F_2

6. Practical Considerations and Concluding Remarks

This section summarizes the practical aspects of the proposed genetic identification method, pointing out its advantages and drawbacks.

6.1 Use of Prior Knowledge

As observed in Sect. 4, the GFIM can be used in different modes reflecting the knowledge available about the system under identification. This offers the possibility of using both prior expert knowledge and numerical input-output data.

The main advantage of the fuzzy model considered is its ability to deal with complex systems with strong nonlinearities. In [10], the GFIM is used for the purpose of the identification of descriptive and approximative fuzzy models of three-dimensional functions. The approximative approach shows better performance than the descriptive one in the case of more complex functions (such as the function F_2 considered in Sect. 5). The drawback that may be associated with GFIM is the loss of interpretability of the approximative fuzzy model obtained due to the lack of explicit linguistic values. However, this may be justified by the higher modeling accuracy of this type of fuzzy model.

6.2 Model Complexity

Another advantage of the GFIM presented is that the proposed learning process allows the user to obtain the desired tradeoff between fuzzy model accuracy and complexity. The number of fuzzy rules in the RB may be modified by the following factors:

1. The number of linguistic values in the initial fuzzy partitions used in the case of constrained free semantics.
2. The value of the parameter ϵ. The higher the value, the greater the number of fuzzy rules and vice versa.

On the other hand, the identification of the fuzzy model structure is directly guided by the composition of the input-output data set. Therefore, no fuzzy rules are obtained in the regions of the data space that are empty. The necessary number of fuzzy rules is determined by the values of the parameter ϵ. No redundancy may occur due to the low niche interaction rate criterion. Genetic simplification allows us to obtain an adequate interaction rate between the fuzzy rules and avoids overlearning. However, this GFIM working mode may perform badly in case an adequate input-output data set cannot be obtained.

6.3 Robustness of the Identification Method

The robustness of GFIM is highly dependent on the input-output data available. Poor data results in an incomplete fuzzy model structure and a inadequate fuzzy rule interaction level. The presence of noise in the data will have the same effect on the final fuzzy model.

6.4 Real-World Applications

In [21, 23], the proposed GFIM is applied to fuzzy controller design. Recently, the method has been applied to a number of classification problems, obtaining good results on laboratory data sets such as the Iris one [9]. The prediction of economic time series is currently under investigation.

References

1. Babuška, R. (1995): Fuzzy modeling. A control engineering perspective. Proc. of Fourth IEEE International Conference on Fuzzy Systems, Yokohama, Japan, 1897–1902.
2. Bäck, T., Schwefel, H.-P. (1995): Evolution strategies I: Variants and their computational implementation. J. Periaux, G. Winter, M. Galán, P. Cuesta (eds.), Genetic algorithms in engineering and computer science, 111–126. John Wiley and Sons.
3. Baker, J.E. (1987): Reducing bias and inefficiency in the selection algorithm. J.J. Grefenstette (ed.), Proc. Second International Conference on Genetic Algorithms, 14–21. Lawrence Erlbaum, Hillsdale, NJ.
4. Cordón, O., Herrera, F. (1995): A general study on genetic fuzzy systems. J. Periaux, G. Winter, M. Galán, P. Cuesta (eds.), Genetic algorithms in engineering and computer science, 33–57. John Wiley and Sons.
5. Cordón, O., Herrera, F. (1996): A hybrid genetic algorithm-evolution strategy process for learning fuzzy logic control knowledge bases. F. Herrera, J.L. Verdegay (eds.), Fuzzy logic and soft computing, 251–278. Physica-Verlag.
6. Cordón, O., Herrera, F., Lozano, M. (1996): A classified review on the combination fuzzy logic-genetic algorithms bibliography: 1989–1995. E. Sanchez, T. Shibata, L. Zadeh (eds.) Genetic algorithms and fuzzy logic systems. Soft Computing Perspectives. World Scientific.
7. Cordón, O., Herrera, F. (1996): Generating and selecting fuzzy control rules using evolution strategies and genetic algorithms. Proc. Sixth International Conference on Information Processing and Management of Uncertainly in Knowledge-Based Systems (IPMU'96), Granada, Spain, 733–738.
8. Cordón, O., Herrera, F., Lozano, M. (1996): A three-Stage method for designing genetic fuzzy systems by learning from examples. H. M. Voight, W. Ebeling, I. Rechenberg, H. P. Schwefel (eds.), Proc. Fourth International Conference on Parallel Problem Solving from Nature (PPSN IV), Lecture Notes in Computer Science vol. 1141, 720–729, Springer, Berlin.
9. Cordón, O., del Jesus, M.J., Herrera, F., (1996): A fuzzy classification system based on genetic algorithms (in Spanish). Proc. Sixth Spanish Conference on Fuzzy Logic and Technologies (FLAT'96), 95–100, Oviedo, Spain.

10. Cordón, O., Herrera, F. (1996): A three-stage evolutionary process for learning descriptive and approximative fuzzy logic controller knowledge bases from examples. To appear: International Journal of Approximate Reasoning.
11. Cordón, O., Herrera, F., Peregrin, A. (1996): Applicability of the fuzzy operators in the design of fuzzy logic controllers. To appear: Fuzzy Sets and Systems.
12. Deb, K., Goldberg, D.E. (1989): An investigation of niche and species formation in genetic function optimization. Proc. Second International Conference on Genetic Algorithms, 42–50. Lawrence Erlbaum, Hillsdale, NJ.
13. Delgado, M., González, A. (1993): An inductive learning procedure to identify fuzzy systems. Fuzzy Sets and Systems, 55, 121–132.
14. Driankov, D., Hellendoorn, H., Reinfrank, M. (1993): An Introduction to fuzzy control. Springer, Berlin. 2nd ed. 1996.
15. Fogel, D.B. (1995): Evolutionary computation. toward a new philosophy of machine intelligence. IEEE Press.
16. Genetic algorithms archive. URL address: http://www.aic.nrl.navy.mil/galist/.
17. Goldberg, D.E. (1989): Genetic algorithms in search, optimization, and machine learning. Addison-Wesley.
18. Grefenstette, J.J. (ed.) (1994): Genetic algorithms for machine learning. Kluwer Academic.
19. Harris, C.J., Moore, C.G., Brown, M. (1993): Intelligent control: Aspects of fuzzy logic and neural nets. World Scientific.
20. Herrera, F., Verdegay, J.L. (1996): Genetic algorithms and soft computing. Physica-Verlag.
21. Herrera, F., Lozano, M., Verdegay, J.L. (1995): Tuning fuzzy logic controllers by genetic algorithms. International Journal of Approximate Reasoning, 12, 299–315.
22. Herrera, F., Lozano, M., Verdegay, J.L. (1995): Generating fuzzy rules from examples using genetic algorithms. B. Bouchon-Meunier, R.R. Yager, L.A. Zadeh (eds.), Fuzzy logic and soft computing, 11–20. World Scientific.
23. Herrera, F., Lozano, M., Verdegay, J.L. (1995): A learning process for fuzzy control rules using genetic algorithms. Technical Report #95108, Dept. of Computer Science and Artificial Intelligence, University of Granada, Spain.
24. Herrera, F., Lozano, M., Verdegay, J.L. (1996): Tackling real coded genetic algorithms. To appear: Artificial Intelligence Review.
25. Holland, J.H (1975): Adaptation in natural and artificial systems. Ann Arbor. MIT Press 1992.
26. Jang, J.R. (1991): ANFIS: Adaptive-network-based fuzzy inference System. IEEE Transactions on Systems, Man, and Cybernetics, 23 665–685.
27. Lee, C.C. (1990): Fuzzy logic in control systems: Fuzzy logic controller. Parts I and II. IEEE Transactions on Systems, Man and Cybernetics, 20, 404–435.
28. Lee, C.C. (1991): A self-learning rule-based controller employing approximate reasoning and neural net concepts. International Journal of Intelligent Systems, 6, 71–93.
29. Mamdani, E.H., Assilian, S. (1975): An experiment in linguistic synthesis with a fuzzy controler. International Journal of Man-Machine Studies, 7, 1–13.
30. Michalewicz, Z. (1992): Genetic algorithms + data structures = evolution programs. Springer, Berlin. 3rd ed. 1996.
31. Nomura, H., Hayashi, I., Wakami, N. (1992): A learning method of fuzzy inference rules by descent method. Proc. IEEE Conference on Fuzzy Systems, San Diego, 203–210.
32. Pedrycz, W. (1984): Identification in fuzzy systems. IEEE Transactions on Systems, Man, and Cybernetics, 14, 361–368.

33. Schwefel, H.-P. (1995): Evolution and optimum seeking. Sixth-Generation Computer Technology Series. John Wiley and Sons, New York.
34. Spears, W.M., De Jong, K.A., Bäck, T., Fogel, D.B., de Garis, H. (1993): An overview of evolutionary computation. Proc. European Conference on Machine Learning.
35. Takagi, T., Sugeno, M. (1985): Fuzzy identification of systems and its applications to modeling and control. IEEE Trans. on Systems, Man, and Cybernetics, 15, 116–132.
36. Thrift, P. (1991): Fuzzy logic synthesis with genetic algorithms. Proc. Fourth International Conference on Genetic Algorithms, 509–513.
37. Yager, R.R., Filev, D.P. (1994): Essentials of fuzzy modeling and control. John Wiley and Sons, New York.
38. Wang, L.X., Mendel, J.M. (1992): Generating fuzzy rules by learning from examples. IEEE Trans. on Systems, Man, and Cybernetics, 22 1414–1427.
39. Yoshinari, Y., Pedrycz, W., Hirota, K. (1993): Construction of fuzzy models through clustering techniques. Fuzzy Sets and Systems, 54, 157–165.

Optimization of Fuzzy Models by Global Numeric Optimization

V. Vergara and C. Moraga

University of Dortmund, Otto Hahnstr. 16, D-44221 Dortmund, Germany

1. Introduction

In this chapter we deal with the identification of an optimal linguistic fuzzy model via the use of an evolutionary search technique., e.g., genetic algorithms. First some theoretical aspects of linguistic fuzzy models will be presented. Second, the identification problem in the context of fuzzy models will be discussed, and the structure identification and parameter estimation methods that form the subject of this chapter will be presented. Third, we illustrate the use of the identification method proposed by an example and discuss the results obtained. Finally, practical aspects concerning the implementation of the proposed method are examined.

2. Theoretical Aspects of Fuzzy Models

2.1 Fuzzy Models and Identification

A fuzzy model can be considered as a transfer function $f_z(\mathbf{x}(t), y(t))$ between the input variables and the output (Fig. 2.1), where $f_z(\cdot)$ is a nonlinear function and t denotes time. The identification problem is to find this mathematical relationship based on the available input-output data. The approach [16] is to construct the function $\hat{f}_z(t, \rho(t), \theta)$ based on training sets $\rho(t)$ and the unknown parameter vector θ, and to use this function to predict the output of the system. The model output is defined as

$$\hat{y}(t) = \hat{f}_z(t, \rho(t), \theta) \tag{2.1}$$

where $\rho(t)$ is a finite collection of observations (learning, or training set), and θ is the parameter vector that contains the unknown model parameters.

In this chapter we consider a fuzzy model with only one output and several inputs (MISO fuzzy model), because a fuzzy model with more than one output (MIMO fuzzy model) can be constructed by the parallel connection of several MISO fuzzy models as shown below.

A fuzzy model differs from a conventional model by its internal structure and the way its output is computed.

The basic configuration of a fuzzy model is shown in Fig. 2.2 and it comprises three main components [15, 5, 19]:

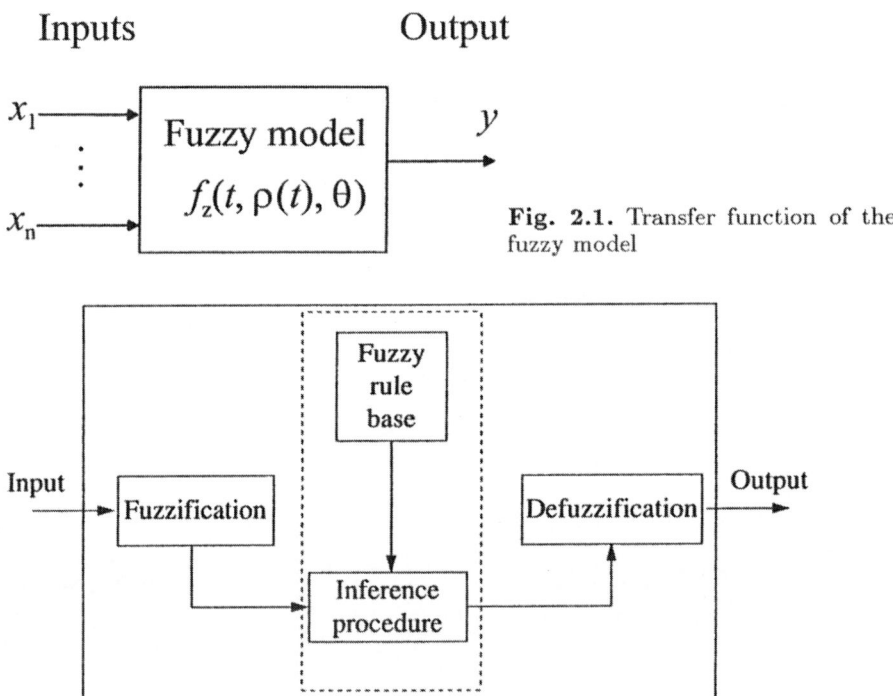

Fig. 2.1. Transfer function of the fuzzy model

Fig. 2.2. Architecture of the fuzzy model

1. A **fuzzification interface:** Performs normalization of the crisp input values and a consequent of fuzzification of these crisp inputs by converting them into fuzzy inputs.
2. The **decision making logic:** Given a normalized, fuzzified input computes the global fuzzy output of the fuzzy model by aggregating the (local) fuzzy output of each fuzzy rule.
3. A **defuzzification interface:** Defuzzifies the global fuzzy output into a global crisp output and then de-normalizes the crisp global output.

A fuzzy model will be represented in this chapter by a collection of if-then fuzzy rules also called the fuzzy rules, or the fuzzy rule base [30]. A fuzzy if-then rule has in general the form

$$R_i : \text{ If } \underbrace{x_1 \text{ is } A_{1,i} \text{ and} \dots \text{ and } x_n \text{ is } A_{n,i}}_{\text{antecedent}} \text{ then } \underbrace{y \text{ is } B_i}_{\text{consequent}} \qquad (2.2)$$

where

- $x_1, x_2, \dots, x_n, y$ denote linguistic variables defined on the corresponding universes of discourse $\mathcal{U}_{x_1}, \dots, \mathcal{U}_{x_n}, \mathcal{V}$.

- $A_{1,i}, A_{2,i}, \ldots, A_{n,i}, B_i$ are the linguistic values taken by the corresponding linguistic variables $x_1, x_2, \ldots, x_n, y$.
- $A_{j,i}(x_j) : \mathcal{U}_{x_j} \to [0, 1]$ are the membership functions defining $A_{j,i}$ on $\mathcal{U}_{x_j}$ where the latter is the universe of discourse of x_j.
- $i \in [1, \ldots, m]$ where m is the number of fuzzy rules constituting the fuzzy rule base.

Let $\mathcal{R}$ denote the fuzzy rule base of a MIMO fuzzy system. This fuzzy rule base can be constructed from a set of r sub-rule bases R_{MISO}^j, $j = [1, \ldots, m]$ where R_{MISO}^j is a fuzzy rule base with m_j fuzzy rules with several inputs and one output as described above. That is,

$$R = \{R_{\mathrm{MISO}}^1, R_{\mathrm{MISO}}^2, \ldots, R_{\mathrm{MISO}}^r\} \tag{2.3}$$

$$= \left\{ \bigcup_{j=1}^{r} \bigcup_{i=1}^{m_j} [(A_{1,i} \times A_{2,i} \times \cdots \times B_i)]_j \right\} \tag{2.4}$$

Thus, a MISO fuzzy system can be considered without any loss of generality. The fuzzy rules are formally interpreted by a fuzzy implication

$$R_i : A_{1,i}(x_1) \times A_{2,i}(x_2) \times \cdots \times A_{n,i}(x_n) \to B_i(y) \tag{2.5}$$

where $A_{1,i}(x_1) \times A_{2,i}(x_2) \times \cdots \times A_{n,i}(x_n)$ are defined on the universes of discourse $\mathcal{U}_1 \times \mathcal{U}_2 \times \cdots \times \mathcal{U}_n$ and $B_i(y)$ is defined on the universe of discourse $\mathcal{V}$. The set of fuzzy implications (2.5) represents the (fuzzy) relation between system inputs and the system output. They constitute the fuzzy model and computation with them reproduces the available input-output data [4, 20, 29].

The fuzzy output of the fuzzy model is obtained as

$$B_i(y) = R_i \circ (A_{1,i}(x_1) \times \cdots \times A_{n,j}(x_n)) \tag{2.6}$$

where $i = 1, 2, \ldots, m$ is the number of fuzzy rules c, $k = 1$ (in the case of one output), and $\circ$ denotes the min-max composition or product composition. We can rewrite (2.6) in the form

$$B_i(y) = \sup_{x_i \in \mathcal{U}_i} [\min(A_{1,i}(x_1), \ldots, A_{n,j}(x_n), R(x_1, x_2, \ldots, x_n, y))] \tag{2.7}$$

in the case of min-max composition, and as

$$B_i(y) = \sup_{x_i \in \mathcal{U}_i} [A_{1,i}(x_1), \ldots, A_{n,j}(x_n), R(x_1, x_2, \ldots, x_n, y)] \tag{2.8}$$

in the case of product-max composition.

The identification problem then consists of the search for a suitable subset of fuzzy rules $\mathcal{R}_i$ ($i \in [1, \ldots, m]$ and $m \leq p$) such that they reproduce the input-output data available. The fuzzy rules are then the fuzzy model structure, and the parameter estimation procedure must search for suitable parameter values for these fuzzy rules, so that the output of the fuzzy model matches the output of the system under identification. Each fuzzy rule consists of an antecedent (if-part) and a consequent (then-part). The antecedent contains a predefined combination of linguistic values for the system inputs and the consequent contains a predefined linguistic value for the system output.

Using the notation and terminology from conventional system identification, the model structure, $\mathcal{M}$, includes the following components:

- The type of model set (in our case a linguistic fuzzy model).
- The size of the model set (the size of a model set $\mathcal{M}^*$ over which the search for a model can be carried out).
- A parameterization of the model set (needed to find the best model from $\mathcal{M}^*$).

After the parameterization of the model set each member of $\mathcal{M}^*$ can be described by a finite-dimensional parameter vector $\theta \in \mathcal{D}_{\mathcal{M}} \subset \mathbb{R}^{d \times 1}$, so that the model corresponding to θ is denoted $\mathcal{M}(\theta)$, and the model structure to which this model belongs is defined by the following relation:

$$\mathcal{M} : \mathcal{D}_{\mathcal{M}} \ni \theta \to \mathcal{M}(\theta) \in \mathcal{M}^*. \tag{2.9}$$

For the fuzzy model, the model structure is given by the set of fuzzy rules, which are parameterized by the membership functions giving the meaning of the linguistic values from these fuzzy rules.

As described in [16], any identification procedure consists of three steps:

1. Determination of the model structure.
2. Calculation of the parameters of the chosen model structure.
3. Validation of the obtained model.

In case of identification of a fuzzy model, the same three steps must be considered, but they are reinterpreted in the following manner:

- **Determination of the model structure:** In our case this is construction of the fuzzy rule base, that is a set of fuzzy rules R_i as well as the membership functions of the linguistic variables from these rules in terms of their support ($suppA = \{x \in \mathcal{U} \mid \mu_A(x) > 0\}$) and shape, for example triangular, trapezoidal, or Gaussian.
- **Determination of the model parameters:** In our case this means the parameters of the membership functions given their support and shape.
- **Validation of the fuzzy model.**

A fuzzy set can be expressed as a mapping $m_A(x)$ [12, 30], with $m_A(x)$ defined as $m_A(x) : x \to [0,1]$. For more than one variable each one will be mapped onto the the interval $[0,1]$, so that a fuzzy model composed of several variables can be seen as a mapping between parts of unit hypercubes [14], that is, a transformation $S : I^n \to I^p$ where $I^j = [0,1]^j$ is the unit hypercube with dimension j, n, and p (the corresponding input and output space dimensions). This type of representation of a fuzzy model will be called a fuzzy associative memory (FAM). A FAM is constituted by a set of associations between fuzzy sets, these associations being the fuzzy rules. Each fuzzy rule R_i will be described then as a couple (A_i, B_i) which associates the fuzzy n-dimensional fuzzy set A_i with the p-dimensional fuzzy set B_i. The FAM encodes and processes in parallel the fuzzy rules R_i (2.5) as a bank of m fuzzy rules (A_i, B_i) [14, 21], $i = 1, \ldots, m$. The fuzzy model is a collection of fuzzy rules (A_i, B_i) that will be processed in parallel.

After the determination of the input and output variables of the fuzzy model, the construction of the fuzzy rules is the next task. At this point the following is done:

- Decompose the input and output domains into fuzzy regions or partitions of the universes of discourse.
- Associate a linguistic term with each of these regions.
- Determine the relations between the different partitions of the universes of discourse.

All the above correspond to the first step of the identification procedure, i.e., the identification of the structure of the fuzzy model.

Once the model structure is determined, the second step is to determine the parameters of the fuzzy model. The parameters depend on the chosen fuzzy model structure, and they affect the output of the fuzzy model. Possible model parameters can be some of the following:

1. The membership functions.
2. The operators realizing the logical connectives (and, or).
3. The inference mechanism (e.g., max-min composition or product-max composition).
4. The type of fuzzy implication.
5. The defuzzification method (e.g., center of area, max criterion).

Normally not all these parameters have a notable influence on the transfer function of the fuzzy model. Tong [27] reports that changes in the elements of the fuzzy rule base and in the membership functions affect strongly the transfer function of a fuzzy model. Here we consider the membership functions as the only parameter of the fuzzy model to be identified. Having made this choice, the parameter identification consists of the following steps:

1. Determine the membership functions (shape and distribution) for the antecedents of the fuzzy rules.
2. Determine the membership functions for the consequents of the fuzzy rules.

Now the structure of the fuzzy model is given by the fuzzy rule base, i.e., the set of fuzzy rules expressed as a FAM-bank; and the parameters for this fuzzy model structure are the membership functions of the different fuzzy set defined on the input and output universes of discourse.

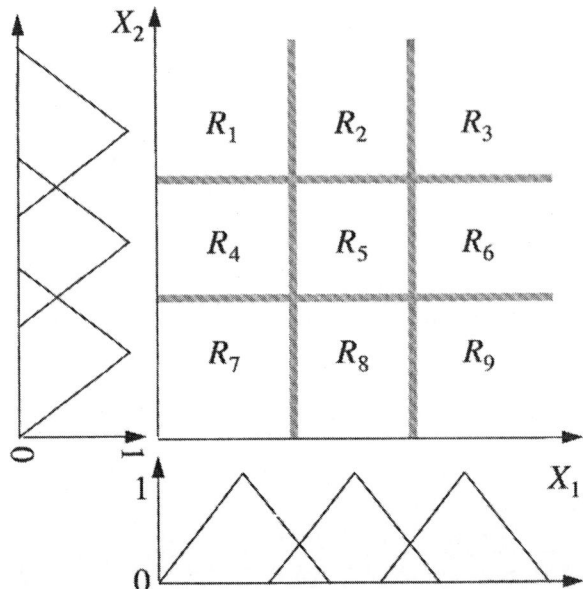

Fig. 2.3. Fuzzy rule base with nine rules

Now the problem is to find a convenient notation for the fuzzy model, so that each fuzzy model can be described by its parameter vector, and selection criteria over the model set can be applied. Both topics will be considered in the next section.

2.2 Notation

The rule base is a suitable combination of all fuzzy partitions. Figure 2.3 shows a normal fuzzy rule base with nine fuzzy rules where the universe of discourse of the input variables is divided in three uniform fuzzy partitions which correspond to the three membership functions.

If some of the parameters of the fuzzy model are fixed prior to the definition of its structure, then the fuzzy model can be represented as a series expansion of a fuzzy basis function (FBF) which is a superposition of fuzzy membership functions. This kind of representation was studied by Wang and Mendel [28] for the approximation of real continuous functions. When the parameters of the fuzzy model are chosen as

1. A disjunctive set of fuzzy rules, i.e., R_i: **If** x is A_i **then** y is B_i;
2. An inference method based on Mamdani implication, i.e.,

$$\textbf{If } x \textbf{ is } A \textbf{ then } y \textbf{ is } B \quad \Rightarrow \quad \mu_R(x,y) = \min(\mu_A(x), \mu_B(y)); \quad (2.10)$$

3. The center of area (COA) as defuzzification method, i.e.,

$$z = \frac{\sum_{j=1}^{m} \bar{z}^j \mu_{A_x} \circ R_j(\bar{z}^j)}{\sum_{j=1}^{m} \mu_{A_x} \circ R_j(\bar{z}^j)}; \tag{2.11}$$

4. Triangular membership functions, i.e., $\mu_{A_i^j}(\mathbf{x})$;

the fuzzy model can be expressed as a FBF in the following form

$$f(\mathbf{x}) = \frac{\sum_{j=1}^{m} \bar{z}^j \left(\prod_{i=1}^{n} \mu_{A_i^j}(x_i) \right)}{\sum_{j=1}^{m} \left(\prod_{i=1}^{n} \mu_{A_i^j}(x_i) \right)} = \sum_{j=1}^{m} p_j(\mathbf{x}) \theta_j, \tag{2.12}$$

where $\bar{z}_j$ is the point in $\mathcal{R}$ at which $u_B^j(z)$ achieves its maximum value. For normal supports the height of the fuzzy set is 1, hence $\max(\mu_B^j(z)) = 1$, and

$$f : \mathcal{U} \subset \mathbb{R}^n \to \mathbb{R}, \ \mathbf{x} \in \mathcal{U} \tag{2.13}$$

where $\mathcal{U}$ is the input universe of discourse and x the input variable.

If the parameters of (2.12) are free design parameters, then the FBF is a nonlinear expansion in the parameters, and a nonlinear optimization technique must be used. On the other hand, if the parameters of $p_j(\mathbf{x})$ are fixed at the beginning of the FBF expansion design procedure, so that the only free design parameters are θ_j, then $f(\mathbf{x})$ is linear in the parameter, and the search will be reduced to

$$\frac{\partial(f(\mathbf{x}) - \hat{f}(\mathbf{x}))}{\partial \theta} = 0. \tag{2.14}$$

The result is a set of parameters for which the error between the system and its model is minimal.

This kind of notation for the fuzzy model is convenient because the only parameters of the fuzzy model are contained in the vector θ, and can be calculated by a conventional parameter estimation method. The fuzzy model consists then of a collection of disjunctive fuzzy rules, which are processed based on Mamdani's implication, and where the output of such a model will be obtained by means the defuzzification COA method, and triangular membership functions give the meaning of the linguistic values from the fuzzy rules.

At this point, only the vector θ must be specified. This vector contains the parameters that characterize each fuzzy model inside the model set. The parameter vector contains also the parameters of each of the membership functions. The parameter vector for a fuzzy model with triangular membership functions is composed by the characteristic points of these membership functions. A fuzzy set A_i defined by a triangular membership function [25] has the form

$$\mu_{A_i}(x) = \begin{cases} \dfrac{x - a_{i-1}}{a_i - a_{i-1}} & \text{if } a_{i-1} < x \leq x_i \\ \dfrac{-x + a_{i+1}}{a_{i+1} - a_i} & \text{if } a_i \leq x \leq a_{i+1} \\ 0 & \text{otherwise.} \end{cases} \tag{2.15}$$

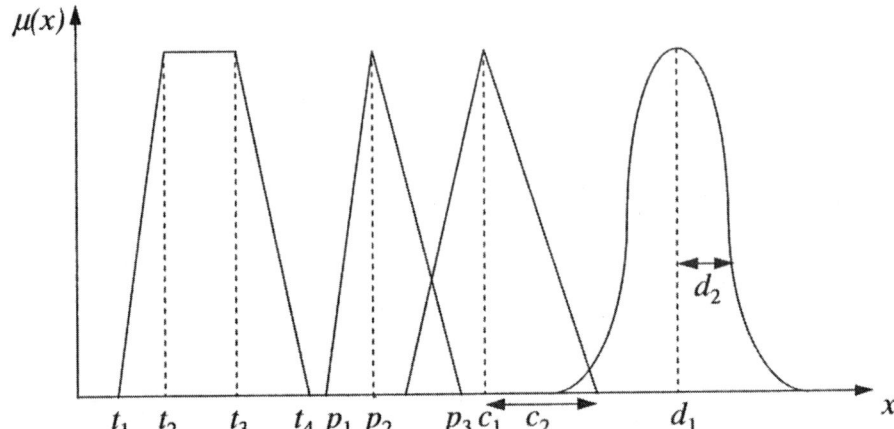

Fig. 2.4. Possible parametrization of membership functions

Usually triangular or trapezoidal membership functions will be used, but for simplicity of calculations a triangular form will be implemented. Several kinds of membership functions are presented in Fig. 2.4. Each of the membership functions will be characterized by the coordinates of the corners, so that a triangular membership function can be represented by the triple $\text{MBF}_i = (a_i, b_i, c_i)$, by the quadruple $\text{MBF}_j = (a_j, b_j, c_j, d_j)$ in the case of a trapezoid, or by $\text{MBF}_k = (m, \sigma)$ in the case of a Gaussian function. A theoretical motivation for the use of a triangular membership function can be found in Pedrycz [22].

The vector θ is then defined by the following equation and will be determined by using the available input-output data

$$\hat{\theta} = \arg \min_{\theta \in D_{\mathcal{M}}} V_N(\theta, Z_N), \tag{2.16}$$

where $\arg \min$ returns the argument that minimizes the loss function, Z_N is the training set, and $V_N(\theta, Z_N)$ is the selection criterion for the search method.

The last step is to determine the output of the fuzzy model, which will be compared with the output of the system under identification (2.14). After the computation of the membership functions, the FAM-bank (A_i, B_i) is completely defined, so that for an input A, the output of the fuzzy model will be computed as follows:

- The input A activates the different stored rules with different intensity. The more A resembles A_i, the more B_i' resembles B_i.
- The fuzzy output B summarizes the partially activated output fuzzy subsets B_j' ($j = 1, \ldots, m$) by taking their combination as a weighted sum (2.17).

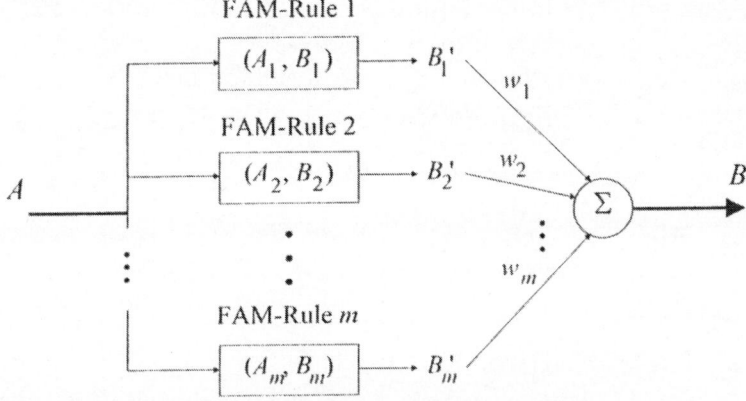

Fig. 2.5. FAM system architecture [11]

- The output fuzzy set B will be defuzzified by means of a chosen defuzzification method to a single numerical value y.

$$B = w_1 B_1' + \cdots + w_m B_m', \tag{2.17}$$

where w_i is the weight of B_i.

This operation is shown in Fig. 2.5, where the input A activates each fuzzy rule in the FAM bank with different intensity, and produces an output fuzzy set B that must be defuzzified.

The last stage of the identification is the validation of the constructed fuzzy models. As for neural networks [9] these characteristic will be measured by the capacity to produce good results with patterns that do not belong to the learning set. Due to the nonlinear nature of the fuzzy models, a special kind of measure is necessary, besides the information criteria (FPE, AIC) [16, 1] that consider also the quantity of searched model parameters (parsimony principle). The validation of the identified fuzzy model will be presented in Sect. 4.

2.3 Dictionary

As a summary, a dictionary is presented to compare the fuzzy terminology with the one usually used by the system identification community [17].

model structure	set of fuzzy rules
model order	number of fuzzy subspaces
parameters	parameters of the membership functions
estimation	training, learning
estimation data	training set
validation	generalization
validation data	generalization set
recursive algorithm	search method
overfit	overtraining

3. The Fuzzy Identification Method

Here genetic search methods will be introduced to find an appropriate model structure and then to estimate the parameters of the membership functions using only the available input-output data. First, we briefly introduce genetic algorithms and their principal characteristics, and then we describe their use for the purpose of identifying a fuzzy model.

3.1 Genetic Algorithms

The underlying principles of genetic algorithms and their mathematical framework were developed in the 1960s by Holland [11]. A genetic algorithm (GA) [7] differs from other search techniques by the use of concepts taken from natural genetics and evolution theory. Over many generations, populations evolve according to the principles of natural selection. The individuals of a population will be reproduced by crossing them with other individuals in the population. This produces new individuals, that share some features from each parent. The least fit members of the population are less likely to get selected for reproduction, so that the whole new population of possible solutions contains a higher proportion of individuals with better characteristics. By directing the mating of the fitter individuals, the most promising regions of the error surface (for example the difference between the system output and the fuzzy model output) will be explored. In this sense the populations will converge to an optimal solution of the problem.

The principal characteristics of the GA are:

- The algorithm works with a population of individuals, each of them representing a possible solution to a given problem, and can search in parallel several peaks of the error surface.
- GAs work with coded parameters, not with the parameters themselves.
- To each individual a fitness measure is assigned that guides the search of the GA.
- The search rules are probabilistic rather than deterministic. The randomized search is guided by the fitness measure of each individual relative to that of the others.

The power of GAs comes from the fact that the technique is robust regarding convergence (if the selection is elitist), and can successfully deal with a wide kind of problems. The GA operates with coded parameters, so that the representation of the problem plays a key role in the convergence of the algorithm.

The use of genetic search methods in the context of fuzzy models will be applied to find the structure and the parameters of the fuzzy model. Normally a number of parameters of the fuzzy model are candidates for optimization:

− The membership functions.
− The fuzzy rules.
− The operators used for the logical connectives, the inference procedure, and defuzzification.

Nevertheless several studies have shown that a simultaneous optimization of all parameters is not convenient [24, 10]. The numeric optimization of a fuzzy system normally consists of two stages:

1. The optimization of the rule base (coarse tuning). In this case the parameters of the membership functions are not affected and are left unchanged.
2. The tuning of the membership function given a predefined rule base (fine tuning).

The above optimization stages correspond to the identification of the structure of the fuzzy model, i.e., the set of fuzzy rules, and the search for suitable parameters for the membership functions.

3.2 Identification of the Model Structure

When applying GAs in order to identify a fuzzy model, the structure and the initial parameters of the initial population of fuzzy models will be determined randomly when no prior knowledge about the system is available. An initial population for the GAs means a set of individuals, also called chromosomes, each one of them coding a particular fuzzy rule base. Our working principle here is that the fuzzy rule base is constructed with standard membership functions, where this means the common use of triangular (and trapezoidal) membership functions with an overlap of 50% (i.e., the height of the intersection of each two successive membership functions equals 0.5).

Figure 3.1 outlines the fuzzy identification procedure. The GA-level determines the policy of parameter estimation by using genetic algorithms.

The GAs operate on a basic randomized initial population of chromosomes, searching the function surface for possible peaks in parallel. This is the surface that represents, in a predefined manner, the fit of all possible models inside the model class. A chromosome represents a binary coded parameter over the search space. The chromosomes will be evaluated by means

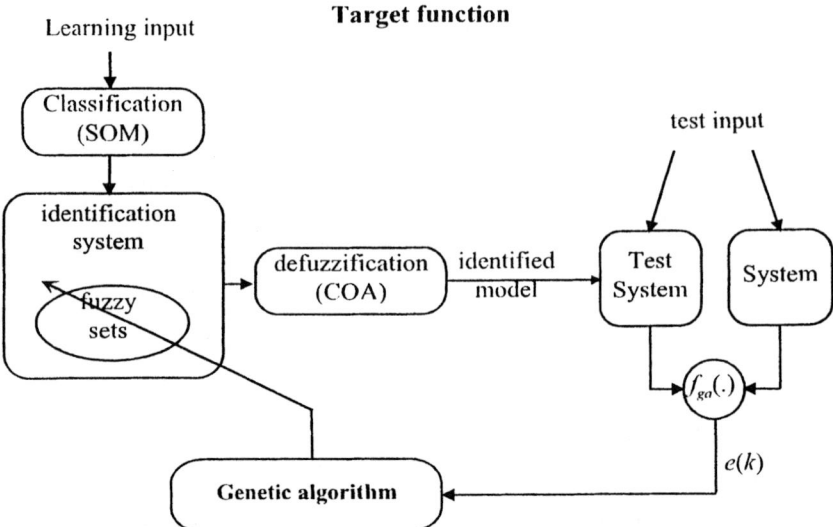

Fig. 3.1. Optimization diagram for an identification process

of a fitness function over a period of time and will be combined with a randomized information exchange by using genetic operators. The simplest of these operators are reproduction, crossover, and mutation [7].

At this stage the construction of the fuzzy model consists mainly of the following tasks that must be carried out:

1. All the input-output pairs of input-output data must be classified into overlapping fuzzy sets. This task can be achieved by applying adaptive clustering algorithms which classify the data in an ordered manner or another classification method [6].
2. The search for a suitable coding of the different fuzzy sets so that the GA can operates with one possible fuzzy rule base.
3. The definition of a target function, also called fitness function, which represents the function that will be minimized.
4. Evaluation of the generated fuzzy rule bases and selection of the one with better fitness.

The first task is the problem of the classification and the corresponding formation (training) of the fuzzy partitions. For our purposes a self-organizing map (SOM) from Kohonen [13] is suitable for the classification of the pattern, because it allows one to realize an independent classification for each variable, and the output is a given number of defined classes that represent the most important operating areas of the system. Each of these classes is associated with a membership function of the corresponding fuzzy sets, i.e., with one of the linguistic labels of the fuzzy relation (2.5).

The cluster algorithm classifies the input data into different fuzzy relations $R_j = (A_j, B_j) \subset I^n \times I^p$, where n and p are the corresponding dimensions of the input-output universes of discourse. Each training data $(\mathbf{x}_i, y_i) \in \mathbb{R}^n \times \mathbb{R}$ belongs to the class R_j, when $\mathbf{x}_i$ and y_i are near the corresponding attraction region of A_j and B_j. The larger the amount of training data that is associated with an attraction region, the stronger will be the weight w_j of the corresponding fuzzy rule R_j. The output of the classification procedure is a set of defined classes that define the fuzzy sets.

The second step is to define a suitable coding of the fuzzy rule base. This coding must be able to represent all possible fuzzy rule bases, i.e., the whole Cartesian product of the universes of discourse. The coding function transforms the fuzzy rule base into a binary string representation and will be defined as

$$f_{ga} : (\text{rule base}) \rightarrow x, x \in \{0, 1\}^{\ell}, \tag{3.1}$$

where $\ell \in \mathbb{N}$ is the string's length (the chromosome).

The relation between the fuzzy subsets will be given by the output of the SOM procedure. The SOM represents the result of a vector quantization algorithm that places a number of reference vectors into a high-dimensional input data space to approximate its data sets in an ordered fashion [13]. The SOM defines a mapping from the input data space into the output data space. Some parameters of the SOM must be defined before clustering. These parameters influence the output of the SOM and are normally set heuristically. By the introduction of the GA at this level of the identification procedure these parameters will be set automatically, avoiding the repetition of experiments to determine them. These parameters are:

- the block size $b \in 1, 2, \ldots, 20000$,
- the learning rate $\eta \in [0.01; 1]$,
- the radius $r \in [0; 9]$,
- the learning factor $c_\eta \in [0.01; 1]$, and
- the radius factor $c_r \in [0, 1]$.

There are also other fixed parameters of the SOM that must be defined before the clustering: the type of lattice used (rectangular or hexagonal), the type of neighborhood function used (step or Gaussian function), and the number of units in each direction (the map dimension).

Each of the parameters (unfixed) will be coded into a binary string that defines a particular SOM, i.e., the structure of the fuzzy model.

After the definition of the coding function the next step is the definition of the fitness function. The natural solution for the fitness function is a function that reflects the difference between the output of the system and the model. This task corresponds to the goal of the identification.

$$F = f(\theta, \mathbf{e}) \tag{3.2}$$

A measure for the fitness function must be then defined like the selection criterion function $V_N(\theta, \mathbf{Z}_N)$ in system identification. This function is a scalar measure of the fit between measured and predicted values:

$$V_N(\theta, \mathbf{Z}_N) = \frac{1}{N} \sum_{k=1}^{N} \ell(y(k) - \hat{y}(k, \theta)), \qquad (3.3)$$

where $\ell(\cdot)$ is a positive scalar-valued function, $N \in \mathbb{N}$ is the number of samples, $y(k)$ is the output of the system under identification, and $\hat{y}(k, \theta)$ is the output of the identified model.

The fitness function for the model structure will be measured in function of the Euclidian distance between the input data and the defined nodes of the SOM

$$\| x - m_c \| = \min_i \| x - m_i \| \qquad (3.4)$$

where $m_i \in \mathbb{R}^n$ is a parameter reference vector and

$$\| x - m_c \| = \sqrt{\sum_i (x_i - m_i)^2}. \qquad (3.5)$$

The better the mapping, the smaller the Euclidian distance for the test pattern will be.

Finally the evaluation of the fuzzy model corresponds to the calculation of the fuzzy output B (2.17) and its defuzzification to a numeric value $y_i \in \mathbb{R}$. Since the defuzzification method will be the COA this requires the use of all the information in the inferred output fuzzy set B to construct the deterministic output of the fuzzy model.

3.3 Estimation of the Model Parameters

After the fuzzy rule base has been determined, the membership functions for the linguistic values of the input and output variables can be optimized, i.e., the search for suitable parameters fuzzy model parameters. Here we will use triangular membership functions with the following initial restrictions:

1. The overlap between two adjacent fuzzy sets must be in the interval [20%–80%].
2. The peak of each normal membership function must be within the support for this membership function.

Both restrictions avoid the building of redundant fuzzy sets, i.e., two or more fuzzy sets are defined over the same interval and their membership functions are defined inside the universe of discourse (see Fig. 3.2).

The membership functions will be defined by means of their ends (corners) and overlap points with their neighbors (2.4).

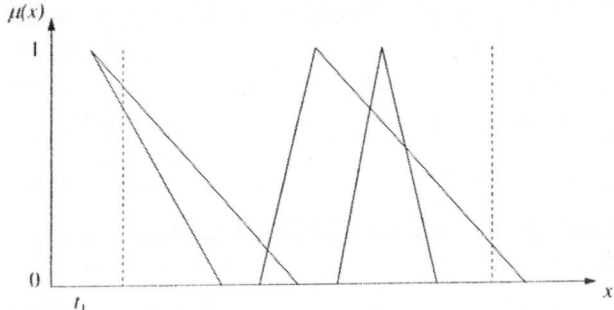

Fig. 3.2. Undesired situation for membership functions

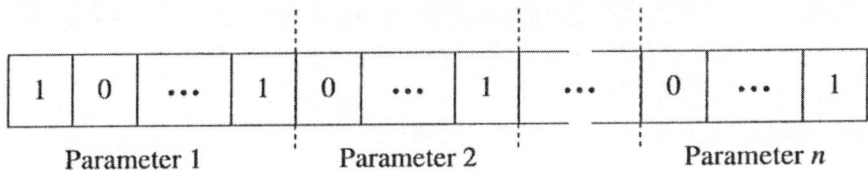

Fig. 3.3. Genetic representation of the rule base

As showed below, the first step of the optimization process with genetic algorithms is the search for a suitable coding function for the parameters of the linguistic values. In this phase, the genetic algorithm is used to optimize the shape and distribution of the fuzzy sets covering the ranges of the input-output variables of (2.5). To genetically code the fuzzy model parameters a chromosome must be defined as a group of parameters that represent the membership functions (see Fig. 3.3). These chromosomes are then linked together to form the entire fuzzy model representation. The GA manipulates this genetic representation according to its fitness. If no priori knowledge about the system exists, the initial position of the membership functions for some individuals will be set to the standard setting, i.e., equally distributed isosceles triangles. Other kinds of membership functions could also be used, but triangular membership functions will be principally implemented with regard to simplicity of calculations, and number of parameters per membership functions (see next paragraph). The introduction of this initial setting biases the initial population of possible solutions, but does not restrict the GA in its search. The advantage of the biased initial population is that if the solution is situated in the neighborhood of the initial population, the GA will converge more quickly.

Another important aspect of the implementation of identification procedure via a GA is the number of parameters used to represent each membership function: the size of the parameter set influences the speed of convergence of the search method. While trapezoidal membership functions can be represented by four parameters, only three parameters are enough for triangular membership functions. If only isosceles triangles are used, they can be represented only by two parameters each, which is also the number of parameters used to represent Gaussian membership functions. Another possibility is to use triangular membership functions with constant overlap. In this case, each membership function can be represented only by its modal value (Fig. 2.4).

Furthermore, the following restrictions are imposed on the membership functions in order to avoid contradictory and redundant information:

1. A triangular membership function with one base-angle greater than 90° is not accepted.
2. Membership functions must be correctly sorted, i.e., membership functions marked with the linguistic value NB must be to the left of those marked NS.
3. Two membership functions shall not cover the same support, because the overlap would be bigger than 80%. This would imply redundant and/or contradictory information.

Some of these restrictions are shown in Fig. 3.2.

Then the codification imposed on the membership functions will take the form

$$f_{ga} : (\text{shape}, \text{position}) \rightarrow x, x \in \{0, 1\}^{\ell+r+c}, \qquad (3.6)$$

where $\ell, r, c \in \mathbb{N}$, and $\ell + r + c$ is the length of the chromosome.

In case of triangular membership functions, each of them will be defined by three points. Each of these points will be coded into a binary string using Gray code (see Fig. 3.3). This presents the advantages that (i) adjacent numbers differ only by a single bit in the string representation of the coded membership functions and (ii) a mutation only causes small perturbation effects.

Now the last step is to determine a fitness function. In case of identification embedded into a GA, the fitness function is defined based on the difference between the fuzzy model and system, that is,

$$F(\varepsilon, \theta) = [q(\varepsilon) + \lambda]^{-1} = \left[\frac{1}{\eta} \sum_{\eta} (y(\eta) - \hat{y}(\eta, \theta))^2 + \lambda \right]^{-1} \qquad (3.7)$$

where λ is a small positive constant to avoid dividing by zero in case of perfect matching, η is the number of patterns, and $q(\cdot)$ is a quadratic function. Thus the minimum of function $F(\varepsilon, \theta)$ will be searched for by the genetic algorithm on the surface defined by $F(\varepsilon, \theta)$. The identification is then treated as a problem of optimization (minimization) of the sum of the distances between the output of the fuzzy model and the system, and is proportional to the inverse of the squared error (3.7).

3.4 Summary of the Identification Steps

This section summarizes the steps necessary to identify a fuzzy model by using GAs and SOM.

Step 1: *Choice of model structure.* At this step the relevant system variables must be determined. After the selection of the relevant system variables, the corresponding universes of discourse are determined, and the number of linguistic values is decided.

Step 2: *Classification of the data.* Now the SOM will be used to form the set of fuzzy relations. Some parameters $(b, \eta, r, c_\eta, c_r,$ type of lattice and type of neighborhood function) must be defined before clustering. The last two are easy to determine, but the other ones must be determined heuristically or with the introduction of GAs as reported in Sect. 3.2.

Step 3: *Design of the GA parameters.* By the introduction of GAs the coded individuals and the fitness function must be defined. Each parameter will be coded into a predefined string length to form the complete fuzzy model representation (Sects. 3.3 and 3.4).

Step 4: *Generate the models.* The fuzzy model will be generated by the introduction of GAs in the identification process as a two-level process. First, the model structure (the fuzzy rule base) will be determined with help of the SOM, and the parameters identified (definition of the distribution and shape of the membership functions). The first level assumes the repetition of Step 3 to find the correct values for the parameters of the SOM. At the second level, the fine tuning will be realized by the GA; some restrictions must be considered at this level to avoid the generation of incorrect membership functions that could be created by the operations of the GA (crossover and mutation). The output is an "optimal" fuzzy model.

Step 5: *Validate the model.* For the validation of the fuzzy model some caution is required because the fuzzy model will be handled as a nonlinear model.

4. Simulation Results

A series-parallel model is implemented here and its structure is shown in Fig. 4.1. This structure has the advantage that the output of the system will be fed back into the model. We use the same examples as in Narendra [18].

For the implementation of the identification method proposed in this chapter, the SNNS (Stuttgarter Neuronale Netz Simulator) tool [31] was used, in which some modifications and additions were made. The module hierarchy of the implemented system is shown in Fig. 4.2. These modules are summarized as follows:

- **NN Module** (neural network): Searches for the parameters of a neural model of the system. Necessary for comparison with Narendra's models.
- **SOM Module** (self-organizing memory): Performs the classification of the learning patterns based on self-organizing maps. The output of this

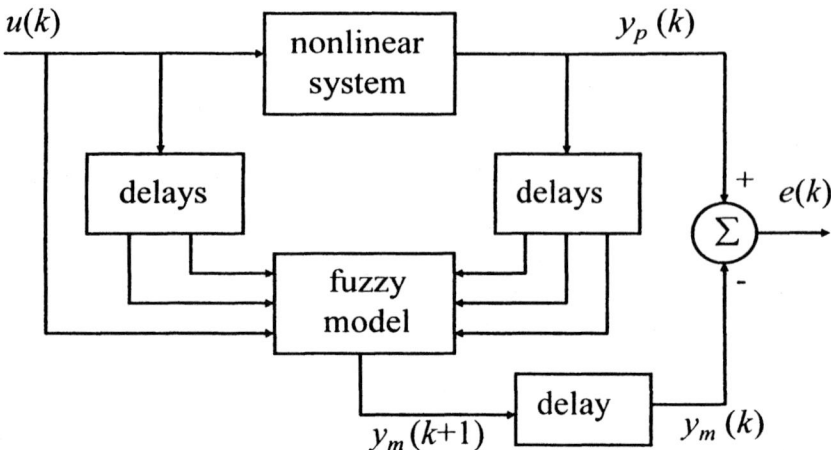

Fig. 4.1. Identification structure

module are the different partitions of the universe of discourse of each system variable. Its implementation is necessary for the fuzzy model.

- **Fuzzy Module**: Searches for the fuzzy model, realizes the fuzzification of the input patterns, implements the inference procedure, and performs defuzzification.
- **KVL Module** (conventional): Search for the parameters of conventional models (in this case only Volterra series).

The main modules manage the control flow of the different programs and use the functions of the I/O module. These functions read the design specifications from the input file with the necessary information for the execution of the experiments. Furthermore, all main modules utilize the system manager in which all the I/O, storage, data processing, and data preparation routines are summarized. In addition to the main modules a procedure to implement a GA (GA simulator) was developed. This realizes the task of introducing new individuals into the population by means of genetical operations (selection, crossover, mutation). These individuals would pass to the corresponding main modules for evaluation, which will be compared with the output of the system to estimate the fitness of each evaluated individual.

A convenient structure of the procedure is shown in Fig. 4.3, although there are several possible ways to embed the identification process into the GA. Here the link between GA and identification procedure is given by the evaluation of the individuals, i.e., decoding the chromosomes and calculating their fitness.

The first implemented example corresponds to the system S_1 which is described by

$$S_1 : y(k + 1) = f_1[y(k), y(k - 1), y(k - \ell), u(k), u(k - 1)], \tag{4.1}$$

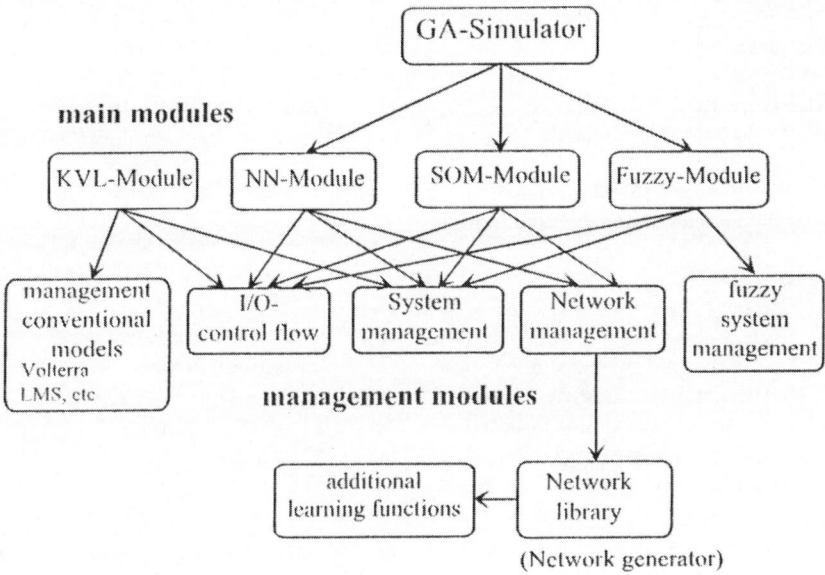

Fig. 4.2. Module hierarchy

where

$$f_1[x_1, x_2, x_3, x_4, x_5] = \frac{x_1 x_2 x_3 x_5 (x_3 - 1) + x_4}{1 + x_3^2 + x_2^2} \qquad (4.2)$$

and

$$x_1 = y(k)$$
$$x_2 = y(k - 1)$$
$$x_3 = y(k - 2)$$
$$x_4 = u(k)$$
$$x_5 = u(k - 1)$$

and the input:

$$u(k) = \begin{cases} \sin(2\pi k/250) & \forall k \leq 500 \\ 0.8 \sin(2\pi k/250) + 0.2 \sin(2\pi k/25) & 500 < k \leq 800. \end{cases} \qquad (4.3)$$

The data consists of 800 pairs of input-output test samples, distributed within the interval $(u(k), y_p(k)) \in [-1, 1] \times [-1, 1]$, where we assume BIBO (bounded input, bounded output) signals. For the classification of the system data, a Kohonen map with rectangular topological structure was used. The final parameters applied to the map were

block size	$b = 11576$
radius	$r = 1.6722$
learning rate	$\eta = 0.2883$
learning factor	$c_\eta = 0.6525$
radius factor	$c_r = 0.4636.$

```
begin
        collect data;
        t := 0;
        initiate generation;                              {ramdomly generated}
        for i := 1 to population size do
                begin
                        decode individual(i);
                        calculate fitness individual(i);
                end;
        repeat
                t := t+1;
                select generation(t) from generation(t-1);   {roulette wheel selection}
                recombination generation(t);                 {crossover, mutation}
                evaluate generation(t);
        until (t > max number of generations)
end
```

Fig. 4.3. Procedure structure

All of these were optimized during the global process by the GA. The size of the population was 20 individuals per generation, the learning set comprised 10^5 patterns and the test set 10^4, all of them uniformly distributed into the input-output universes of discourse.

A SOM with size 10×10 and a uniform distributed learning signal on the interval $[-1, 1]$ of 10^5 pattern was used. This map was the basic structure for the classification of the learning patterns. The SOM classifies the learning set into a predefined number of classes, which represent the working states of the system. Each class will be associated with a membership function, and each membership function will be labeled with a corresponding linguistic value.

The SOM represents these classes and their relationship through weighted vectors that correspond to the distribution of the frequency of the learning patterns. Figure 4.4 shows the learning patterns (a) and the attraction regions for the relations between some of the delayed I/O variables (b, c, d). For this example the dimension of the learning vector is six. The internal topology of the map is not interesting for identification purposes, because only the association capacity of the net is important for the construction of the rule base. The fuzzy model represents the relationship between the input fuzzy cube I^5 and the fuzzy output I ($S : I^5 \times I$ with $I \subset [0, 1]$).

For the estimation of the parameters of the fuzzy model a target function will be minimized. The target function is a measure of the error between the output of the fuzzy model and the output of the system, so that the objective of the search method is to keep this error as low as possible. For the target function the MSE was used and as search method a GA. For the coding of the parameters an indirect coding was used, so that each parameter field in the binary string represents a number from the interval $[0, 1]$. This value will be adapted again to the universe of discourse of the corresponding variable.

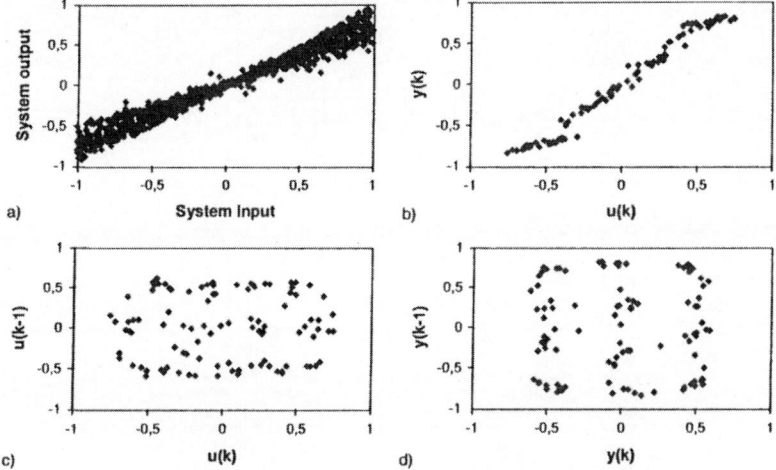

Fig. 4.4. Distribution of the weighted vector of the SOM (10 × 10)

Nevertheless, some restrictions on the search for the optimal parameters of the fuzzy model and the initial position of the membership functions must be considered. These are related to the overlap between the parameterized membership functions and to the covering of the working interval. As reported before, the overlap between two neighboring membership functions is 20–80% (of the area) to avoid on the one side the effect of two membership functions covering the same interval and on the other side uncovered intervals within the universes of discourse. The second restriction refers to the second and fourth overlap, where redundant linguistic terms (which are totally covered by the others) are undesirable. Finally a restriction is introduced for the membership functions that are at the extremes of the interval of discourse. Both membership functions must be defined beyond the end of the interval in order to avoid uncovered regions and to allow learning at the periphery, i.e., near to the extremes.

Figure 4.5 shows the performance of two fuzzy models. The difference between these fuzzy models is the number of linguistic values for each fuzzy variable: the first one has seven linguistic values (NB, NM, NS, Z, PS, PM, PB) and the second eleven. In practice, seven or five linguistic values will be normally used for the characterization of fuzzy variables, but here eleven linguistic values have been used to show the difference between the fuzzy models. Each linguistic value was associated with a triangular membership function. The distribution of these membership functions for the I/O variables of the fuzzy models after the optimization process is shown in Fig. 4.6. The different supports are centered in the middle of the working interval although they are not symmetric. This phenomenon is due to the distribution of the learning pattern.

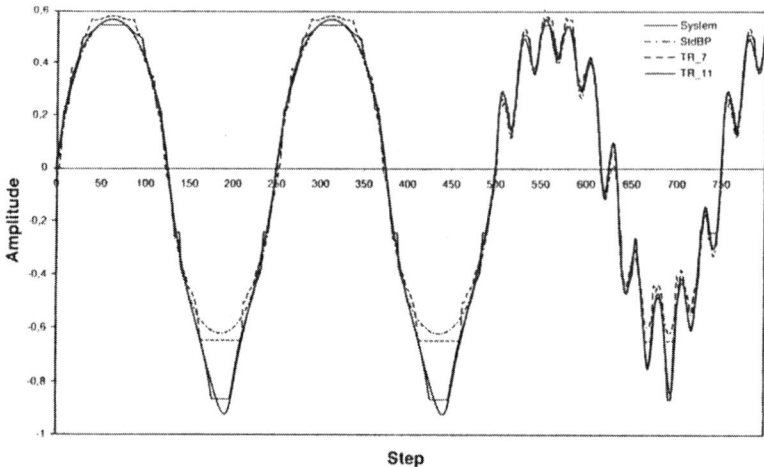

Fig. 4.5. Response of the plant and the identification models

If the system under analysis is linear there exist a number of test available for validating the estimated model, but in the case of nonlinear systems another kind of tests must be applied. Many works [2, 8] show that the introduction of correlation tests and information criteria are sufficient and optimal for validating nonlinear models. It is important, too, to apply performance indices that besides considering the residual terms ($\{\varepsilon(t, \hat{\theta})\}$) also penalize the number of parameters.

Coming back to the fuzzy model, to show how the model performs in the presence of noise and data not belonging to the training set (cross validation), the model was validated with a random noise signal (mean $= 0$, var $= 0.6$). The response of the system and model are presented in Fig. 4.7. If the model is considered adequate, the residue $\varepsilon(t, \hat{\theta})$ will be unpredictable from all combinations of past inputs and outputs, and model validity tests were developed by Billing et al. [2] based on this principle. The correlation values obtaining by applying this validity test to the achieved fuzzy model are shown in Fig. 4.8. The results for the fuzzy model lie below the 5% limit, i.e., the correlation functions are within 95% of the confidence intervals, which shows that the fuzzy model is regarded as adequate. This additional validity test is introduced as a measure of the quality of the fit of the model based on correlation functions.

For the white noise response or generalization capability (Fig. 4.7) of the fuzzy model a residual criterion (MSE) of 0.025 is obtained, which shows that the model has a good response for patterns that are not learned too (the cross validation).

As a complement to the models obtained, the kinds of support for the memberships functions were changed. Trapezoidal and Gauss based membership functions were implemented and tested with seven and eleven linguistic

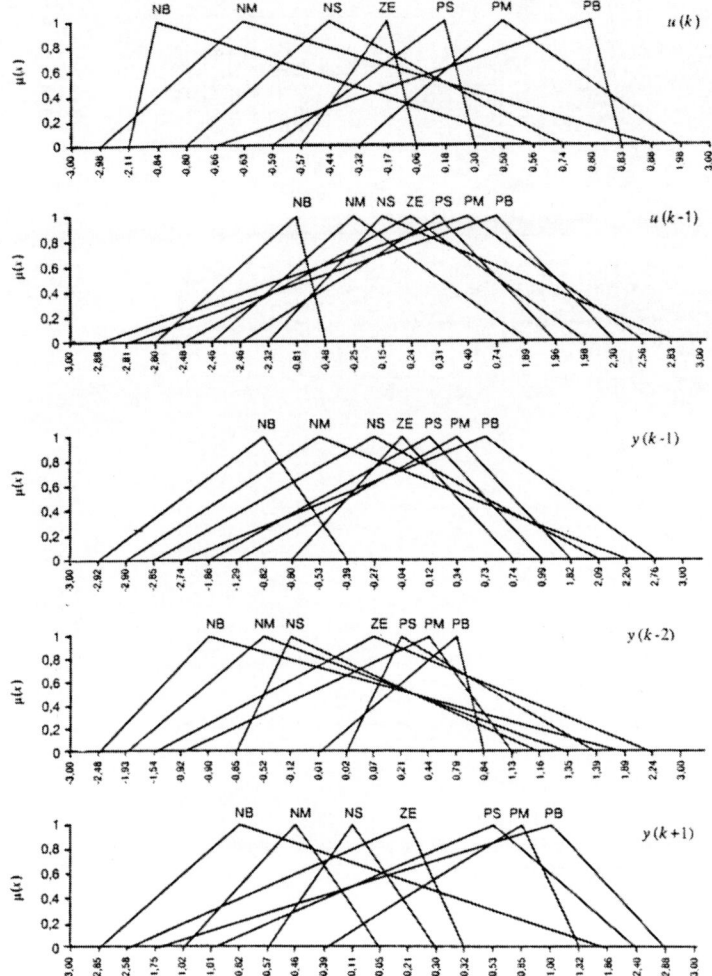

Fig. 4.6. Distribution of the MBF

terms. The results are summarized in Table 4.1, where 'Vol3' corresponds to a Volterra model of cubic order and 'StdBP' to the neuronal network implemented by Narendra in his paper. The different kinds of membership functions are labeled TR for triangular, GS for Gauss, TP for trapezoidal, and 7 or 11 for the number of linguistic terms for each fuzzy variable.

In addition, another system $(S_2 : \mathbb{R}^2 \rightarrow \mathbb{R})$ was identified, its description given by

$$y(k+1) = f_2[y(k), y(k-1)] + u_2(k) \tag{4.4}$$

with

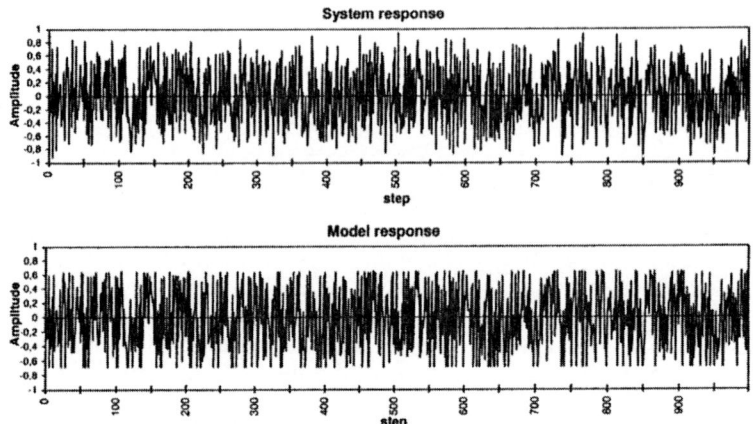

Fig. 4.7. Response for the generalization capability

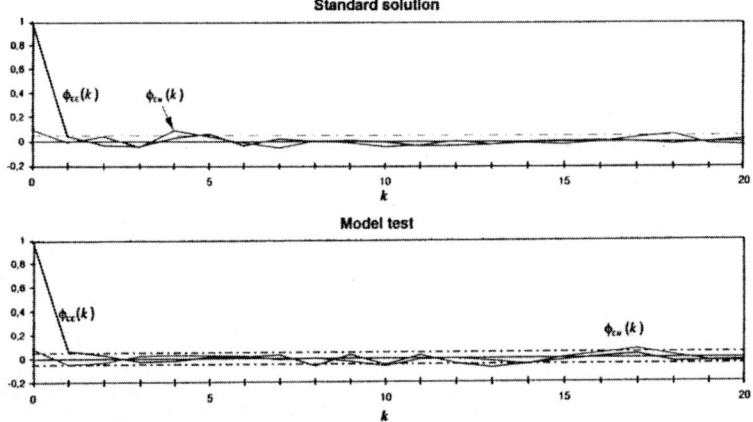

Fig. 4.8. Correlations values

$$f_2(x_1, x_2) = \frac{x_1 x_2 (x_1 + 2.5)}{1 + x_1^2 + x_2^2} \qquad (4.5)$$

and

$$u_2(k) = \sin(2\pi k/25), \quad \forall k \qquad (4.6)$$

From the different experiments it can be concluded that the higher the number of linguistic values, the higher the fitness of the fuzzy model (the smaller the error). A compromise between the number of linguistic values, the desired accuracy of the fuzzy model, and the complexity of the calculations must be found, where the usually used number of seven linguistic values yield a good fitness for the fuzzy model. The kind of support for the membership function does not have an important influence on the efficiency of the fuzzy model, so that the use of a triangular membership function is

Table 4.1. Performance of the different models

Model	Learning capability (MSE)	Generalization capability (FPE)
StdBP	0.00680015	0.00691371
TR11	0.00561018	0.00897263
TR7	0.00465870	0.00773863
GS11	0.00505817	0.00513211
GS7	0.00853049	0.00655820
TP11	0.00475293	0.00845801
TP7	0.00680295	0.00777984
Vol3	0.00167245	0.00652714

totally justified, because the implementation is easier and also the search procedure has a shorter duration. Moreover, the triangular supports implement the fuzzy membership concepts relatively well and are piecewise linear, which allows a compact representation by means of the vertices.

5. Practical Aspects

In this section some practical aspects of the implementation of the identification method presented in this chapter will be discussed.

5.1 Incorporation of Prior Knowledge

Prior knowledge about the system under analysis makes the optimization procedure faster with a corresponding reduction of computation time. The introduction of GAs into the identification process is computationally very expensive. Prior knowledge could be incorporated to reduce the computation time. Usually the initial population of coded models will be randomly determined, by setting some of the initial individuals near to the presumed solution and setting the rest of them randomly to assure the necessary heterogeneity of the population – this will accelerate the search. The initial population will then be biased, which influences the evolution of the GA, but does not influence its final result if the population is large enough and the individuals inside the population are sufficiently heterogeneous.

In case of the model structure the set of fuzzy rules will be determined by means of a clustering procedure like the SOM, whose results will then be optimized by a GA. The available data plays an important role for the purpose of optimization. If the information contained in this data is not sufficient to reflect all possible states of the system under analysis, the SOM will not build the rules that correspond to these states. This could be avoided by the design of good test data sets, or, if this is not possible, by the use of prior knowledge.

Here the prior knowledge can be incorporated easily as linguistic information in the form of fuzzy relations (2.5) to the SOM.

5.2 Model Complexity

The identification method will not generate rules for regions that do not contain sufficient data. This is caused by insufficiently representative learning data or possibly by the nature of the SOM itself, e.g., the size of the lattice in the self-organizing map, and the effect of the attraction regions. This is so because learning from data admits the possibility that the learning patterns do not cover all the regions on the universes of discourse, resulting in an incomplete rule base. The quantity of points (learning samples) at the periphery is small compared with that of points at the center, and these points are attracted to stronger regions (regions containing more samples). Each node in the lattice represents a metric ordered relation of the input samples. If peripheral points outside the best matching node are absorbed by the node next to them, then there is a gap in the function for those peripheral points, which leads to saturation (the wrong rules are fired). This problem can be solved by the use of a SOM with a small grid for the lattice, i.e., a small radius for the attraction regions. This causes an increase of the number of rules in the fuzzy rule base. Unfortunately, this solution does not guarantee the removal of the saturation problem. This is because there are not enough learning patterns at the periphery, since they are attracted to stronger regions. The GA cannot interpret this phenomenon and tries to optimize the architecture of the SOM only considering the available data, and evaluating each model by its fitness function. The GA is a pure numerical procedure and does not consider the semantical aspects.

The saturation of the output of the model due to the effect of strong regions may be observed in Fig. 4.5. In spite of the good fitness achieved by this model the data at the extremes would be weakly weighted so that stronger rules map these samples into wrong values.

Another effect present in this kind of model is redundancy due to membership functions that cover the same interval of the corresponding variable. This is a result of the optimization by the GA and of the cluster projections onto the corresponding intervals of the variables. This phenomenon does not influence the output of the fuzzy model but increases the amount of unnecessary calculations. To avoid higher redundancy, restrictions on the generation of the membership function by the GA were introduced. These restrictions allow the overlap between membership functions only within an acceptable interval.

5.3 Robustness of the Identification Method

Any identification procedure needs suitable data for the implementation of the parameter learning process. This method requires also a suitable data set to build the fuzzy model. The data set was divided into two subsets: the learning set and the test set, because of the computational demand of the GA on one side and the learning process of the SOM on the other side. The two sets are disjoint, and the amount of pattern required to learn the SOM is large and demands expensive calculations steps. Like other types of models, if the samples are not representative, the fuzzy model will perform badly.

The amount of computation required increases with large data sets, the number of variables, and the clustering. The GA operates with many individuals at the same time, so that bigger populations present a better heterogeneity but take more computational time. The quantity of 20 individuals per generation results in an appropriate mass. Experiments on the size of the data sets show that learning data sets of 10^5 patterns did not improve the performance of the clustering method.

References

1. H. Akaike: A new look at the statistical model identification, *IEEE Trans. Automatic Control*, **19**(6), 716–723, 1974.
2. S. A. Billings, W. Voon: Correlation based model validity for non-linear models, *Int. Journal of Control*, **44**(1), 235–244, 1986.
3. S. Chen, S. A. Billings, P.M. Grant: Non-linear system identification using neural networks, *Int. Journal of Control*, **51**(6), 1191–1214, 1990.
4. E. Czogala, W. Pedrycz: *Identification and control problems in fuzzy systems*. TIMS/Studies in the Management Sciences 20, Elsevier, pp. 447–466, 1984.
5. D. Driankov, H. Hellendoorn, M. Reinfrank: *An introduction to fuzzy control*, Springer-Verlag, 1993. 2nd ed. 1996.
6. R. Duda: *Pattern classification and scene analysis*, John Wiley & Sons, 1973.
7. D. Goldberg: *Genetic algorithms in search, optimization, and machine learning*, Addison-Wesley, 1989.
8. R. Haber, H. Unbehauen: Structure identification of nonlinear dynamic systems – A survey on input/output approaches, *Automatica*, **26**(4), 651–677, 1990.
9. J. Hertz, A. Krogh, R. Palmer: *Introduction to the theory of neural computation*, Addison-Wesley, 1991.
10. F. Herrera, M. Lozano, J. Verdegay: Applying genetic algorithms in fuzzy optimization problems, *Fuzzy Systems and Artificial Intelligence*, **3**, 39–52, 1994.
11. J. H. Holland: *Adaptation in natural and artificial systems*, Ann Arbor, MI: Univ. Mich. Press, 1975.
12. G. J. Klir, T. A. Folger: *Fuzzy sets, uncertainty, and information*, Prentice Hall, 1988.
13. T. Kohonen: *Self-organization and associative memory*, Springer-Verlag, 3rd ed., 1989.
14. B. Kosko: *Neural networks and fuzzy systems: a dynamical system approach to machine intelligence*, Prentice Hall, 1992.
15. C. C. Lee: Fuzzy Logic in Control Systems: Part I & II, *IEEE Trans. System, Man, & Cybernetics*, **20**(2), 404–432, 1990.
16. L. Ljung: *System identification, theory for the user*, Prentice Hall, 1987.
17. L. Ljung, J. Sjöberg: *A system identification perspective on neural nets*, Tech. Report LiTH-ISY-I-1373, Linköping University, 1992.

18. K. S. Narendra, K. Parthasarathy: Identification and control of dynamic systems using neural networks, *IEEE Trans. Neural Networks*, **1**(1), 4–27, 1990.
19. W. Pedrycz: *Fuzzy control and fuzzy systems*, John Wiley & Sons, 1989.
20. W. Pedrycz: Processing in relational structures: Fuzzy relational equations, *Fuzzy Sets and Systems*, **40**, 77–106, 1991.
21. W. Pedrycz, K. Hirota, T. Takagi: Fuzzy associative memories: Concepts, architectures, and algorithms. In: *Fuzzy engineering toward human friendly systems*, T. Terano, M. Sugeno, M. Mukaidono, K. Shigemasu (eds.), pp. 163–174, 1992.
22. W. Pedrycz: Why triangular membership functions?, *Fuzzy Sets and Systems*, **64**, pp. 21–30, 1994.
23. E. Sanchez: Resolution of composite fuzzy relation equations, *Information and Control*, **30**, 38–48, 1976.
24. H.-P. Schwefel: *Evolution and optimum seeking*, Springer-Verlag, 1995.
25. T. Sudkamp, R. Hammell: Interpolation, completion, and learning fuzzy rules, *IEEE Trans. Systems, Man, and Cybernetics*, **24**(2), 332–342, 1994.
26. T. Takagi, M. Sugeno: Fuzzy identification of systems and its applications to modeling and control, *IEEE Trans. Systems, Man, and Cybernetics*, **15**(1), 116–132, 1985.
27. R.M. Tong: Analysis of fuzzy control algorithms using the relation matrix, *Int. Journal Man-Machine Studies*, **8**, 679–686, 1976.
28. Li-Xin Wang, M. Mendel: Fuzzy basis functions, universal approximation, and orthogonal least-squares learning, *IEEE Trans. Neural Networks*, **3**(5), 807–814, 1992.
29. R. Yager, D. Filev: Unified structure and parameter identification of fuzzy models, *IEEE Trans. Systems, Man and Cybernetics*, **23**(4), 1198–1204, 1993.
30. L. A. Zadeh: Fuzzy sets, *Information and Control*, **8**, 338–353, 1965.
31. A. Zell: *Simulation neuronaler Netze*, Addison-Wesley, 1994.

Artificial Intelligence

Identification of Linguistic Fuzzy Models Based on Learning

Y. Nakoula, S. Galichet, and L. Foulloy

Université de Savoie, 74016 Annecy Cedex, France

1. Introduction

The learning method presented in this chapter is aimed at the identification of linguistic fuzzy models, also called Mamdani-type fuzzy models ([1, 19, 26, 30, 34]). The learning of this type of fuzzy models is, in general, done by dividing the learning process into two subtasks which are solved independently from each other: the construction of the reference fuzzy sets (fuzzy partition of the input and output spaces) on the one hand, and on the other hand, the generation of the rule base. The partition of the input and output spaces is normally obtained from human experts [27], by a clustering technique [21, 35], or simply by using a classical fuzzy partition such as triangular membership functions [4]. The rule base is then determined, either by performing a matching procedure on the set of training points (training input-output data) [13, 26, 30], by solving a fuzzy relational equation [22, 24], or by training a fuzzy neural network that represents the underlying logic of the input-output relationships in terms of fuzzy if-then rules [12, 25]. However, because of the close interdependence between the fuzzy partion and the rule base, the division of the global learning strategy into two independent subtasks may not be adequate.

In this context, we present an alternative solution to the learning of a fuzzy model. The idea is to improve the learning strategy by performing the tasks of fuzzy partition and the rule base generation simultaneously. The guiding principle of the proposed learning algorithm then consists in the successive improvement of the fuzzy model by adding new linguistic values and new rules at the same time.

In Sect. 2, we present some basic concepts and notations from fuzzy modeling. The fuzzy model identification problem is presented in Sect. 3. Section 4 is devoted to the presentation of the identification method and the global learning algorithm. Experimental results in the context of two examples are presented in Sect. 5. The first example concerns one-dimensional function approximation. The second one treats the problem of the Box and Jenkins time series prediction. The results are compared to those obtained by other methods. Section 6 treats the practical aspects of the proposed identification method and presents some concluding remarks.

2. Basic Concepts and Notation

Here we present some basic concepts and notation used in fuzzy modeling.

2.1 Linguistic Knowledge Representation

A linguistic representation of knowledge about the behavior of a system leads to the notion of a linguistic variable. Such a variable whose values are words or sentences [38] is defined by the quadruplet [6]

$$\langle x_{\text{name}}, \mathcal{L}(X), X, M \rangle, \tag{2.1}$$

where

- x_{name} is the name of the linguistic variable, such as *temperature, age, error,* etc.
- $\mathcal{L}(X) = LX_1, \ldots, LX_n$ is the set of linguistic values that x_{name} can take. For example, $\mathcal{L}(X) = \{$cold, cool, warm, hot$\}$ may be defined for the linguistic variable *temperature*. $\mathcal{L}(X)$ is also called the term-set or reference-set of x_{name}.
- X is the physical domain associated with the linguistic variable x_{name}, also called the universe of discourse. For example, X may be defined as $[0, 150]$ for the linguistic variable *age*.
- M is a semantic function that associates a fuzzy meaning to each linguistic value $LX \in \mathcal{L}(X)$.

Fuzzy modeling involves both numeric and linguistic information. As the inference mechanism uses a linguistic rule base, its link with the system's inputs and outputs requires a clear definition of the "numeric-to-linguistic" and "linguistic-to-numeric" interfaces. If the fuzzy meaning M introduced in the definition of a linguistic variable is implemented as a "linguistic-to-numeric" interface, then the dual notion of "numeric-to-linguistic" becomes irrelevant. However, it was introduced in the early work of Zadeh [36] by means of a fuzzy description of a crisp numeric value. Since this notion will be used in this chapter, we find it useful to reintroduce it here.

Let E be any set; $\mathcal{F}(E)$ will denote the set of fuzzy subsets of E. Classical conventions will be used for the notation of fuzzy subsets [37]. If E is discrete, any element F of $\mathcal{F}(E)$ will be expressed as

$$F = \sum_{e \in E} \mu_F(e)/e. \tag{2.2}$$

In the case when E is continuous, the $\sum$-symbol is replaced by an $\int$-symbol

$$F = \int_{e \in E} \mu_F(e)/e. \tag{2.3}$$

In the context of the above introduced notation, $\mathcal{F}(X)$ represents the set of fuzzy subsets defined on the universe of discourse X. Since X corresponds to a physical domain, any element of $\mathcal{F}(X)$ is a numeric fuzzy subset, that is a fuzzy subset defined on a numeric set with the help of a membership function. In the same way, $\mathcal{F}(\mathcal{L}(X))$ is the set of fuzzy subsets defined on the term-set $\mathcal{L}(X)$. Any element of $\mathcal{F}(\mathcal{L}(X))$ is thus a linguistic fuzzy subset, that is, a fuzzy subset defined on a finite set of linguistic values, and is thus discrete by nature.

2.2 Fuzzy Meaning of a Linguistic Value

The fuzzy meaning of a linguistic value is an injective mapping M from the linguistic term-set $\mathcal{L}(X)$ to the set of numeric fuzzy subsets $\mathcal{F}(X)$, that is,

$$M : \quad \mathcal{L}(X) \to \mathcal{F}(X)$$
$$LX \to M(LX). \tag{2.4}$$

Thus $M(LX)$ is a numeric fuzzy subset whose membership function defines the meaning attributed to the linguistic value LX in the domain X. The fuzzy meaning of a linguistic term (or linguistic value) was defined by Zadeh in [36]. Consider any linguistic term x and let $M(x)$ be its meaning. Then Zadeh states:

> Although x and $M(x)$ are entirely different entities, it is expedient to abbreviate $M(x)$ to x, relying on the context for the determination of whether x stands for a term or for its meaning, $M(x)$. This is what we usually do in everyday discourse, because in such discourse it is rarely necessary to differentiate between x and $M(x)$. On the other hand, it is important to differentiate, or at least to understand, the difference between x and $M(x)$ in the case of programming languages, machine translation of languages, and other areas in which ambiguity of interpretation can lead to serious errors.

In this chapter, in order to prevent any ambiguity, we will use different notations for a linguistic value LX and its fuzzy meaning $M(LX)$. Various notations have already been introduced in the literature, especially in [6] where $M(LX)$ is denoted with $\widetilde{LX}$ or μ_{LX}, i.e., a membership function without an argument. In that case, $\sim$ or μ is a reference to a fuzzy subset, i.e., the fuzzy meaning of the linguistic value LX. Although this type of notation can considerably simplify the writing, it will not be used in this chapter. The general expression $M(LX)$ will indeed be kept without any abbreviation for the fuzzy meaning of the linguistic value LX. This choice can appear as being hard with respect to writing, but it will be justified in the next paragraph.

2.3 Fuzzy Description of a Crisp Value

The fuzzy description of a crisp value as defined in [36] is a mapping from the domain X to the set of linguistic fuzzy subsets $\mathcal{F}(\mathcal{L}(X))$, that is,

$$D : \quad X \to \mathcal{F}(\mathcal{L}(X))$$

$$x \to D(x) = \sum_{LX \in \mathcal{L}(X)} \mu_{D(x)}(LX)/LX. \tag{2.5}$$

Thus $D(x)$ is a linguistic fuzzy subset, discrete by nature, that provides for any term LX the degree to which it describes the crisp numeric value x. The grade of membership of each linguistic value LX in $D(x)$ is equal to the grade of membership of x in $M(LX)$. This leads to the following relationship that links the fuzzy meaning and the fuzzy description:

$$\forall LX \in \mathcal{L}(X), \forall x \in X : \mu_{M(LX)}(x) = \mu_{D(x)}(LX). \tag{2.6}$$

In the same manner that we distinguish between a linguistic value LX and its fuzzy numeric meaning $M(LX)$, we find it important to differentiate between a crisp numeric value x and its fuzzy linguistic description $D(x)$. The problem of notation has again to be considered. How can we abbreviate or at least simplify $D(x)$? Choosing $\widetilde{LX}$ or μ_{LX} for representing $M(LX)$, as envisaged in the previous paragraph, would lead to the use of $\tilde{x}$ or μ_x for denoting $D(x)$. However, in the literature this notation generally refers to a numeric fuzzy subset that translates the linguistic concept of "around x". This is the reason why the fuzzy meaning of the linguistic value LX and the fuzzy description of the numeric value x will be expressed as $M(LX)$ and $D(x)$ through the rest of this chapter.

2.4 Example

Let us consider the variable error and let its physical domain be denoted as E. Let the set of the linguistic values of error be defined as $\mathcal{L}(E) = \{N, Z, P\}$, where N denotes Negative, Z denotes Zero, and P denotes Positive. The fuzzy meanings of these linguistic values are given by the mapping M as shown in Fig. 2.1a. The fuzzy description of any crisp value of error may then be derived by applying (2.6). For example, Fig. 2.1b illustrates the fuzzy description of the crisp value 0.3 of error obtained according to the following

$$\mu_{D(0.3)}(N) = \mu_{M(N)}(0.3) = 0,$$
$$\mu_{D(0.3)}(Z) = \mu_{M(Z)}(0.3) = 0.25,$$
$$\mu_{D(0.3)}(P) = \mu_{M(P)}(0.3) = 0.75. \tag{2.7}$$

3. The Identification Problem

3.1 The Linguistic Rule Base

The form of the linguistic rule base will be first illustrated by a small example. Let us consider a system with two inputs $\langle x_1, \mathcal{L}(X_1), X_1, M \rangle$,

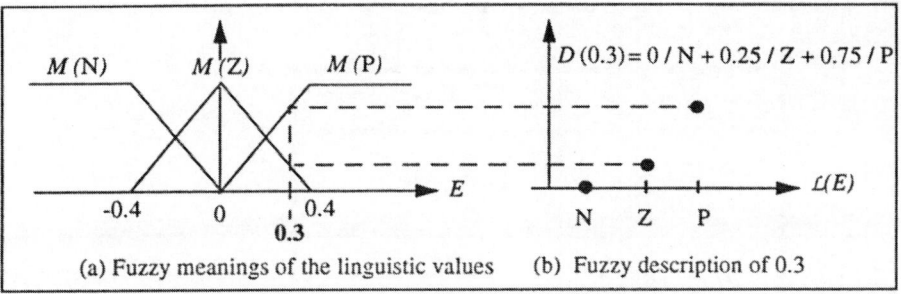

Fig. 2.1. Link between fuzzy meanings and fuzzy description

$\langle x_2, \mathcal{L}(X_2), X_2, M \rangle$, and one output $\langle y, \mathcal{L}(Y), Y, M \rangle$, where $\mathcal{L}(X_1) = \{$Small, Large$\}$, $\mathcal{L}(X_2) = \{$Neg, Pos$\}$, and $\mathcal{L}(Y) = \{$Decrease, Nothing, Increase$\}$. The linguistic model of such a system will be based on a collection of weighted rules of the form [19]:

If x_1 is Small **and** x_2 is Pos **then** y is Nothing	with weight 1,
If x_1 is Small **and** x_2 is Neg **then** y is Increase	with weight 0.7,
If x_1 is Small **and** x_2 is Neg **then** y is Nothing	with weight 0.3,
If x_1 is Large **and** x_2 is Pos **then** y is Decrease	with weight 0.8,
If x_1 is Large **and** x_2 is Pos **then** y is Nothing	with weight 0.2,
If x_1 is Large **and** x_2 is Neg **then** y is Decrease	with weight 1.

The weight associated with each rule is a real number in the interval $[0, 1]$. It represents the degree of confidence or validity of a rule. Each rule can also be expressed in a simplified form. For example, the first rule of the previous rule base can be written as

    If x_1 is Small **and** x_2 is Pos **then** y is 1/Nothing.

In this case, the symbol / is related to the notation of discrete fuzzy subsets as specified in (2.2). Furthermore, when several rules have the same antecedent, we can regroup them in a single compact rule. The formulation with a sign + will also be borrowed from the conventional notation introduced in (2.2). By applying these simplifications, the original rule base becomes:

    If x_1 is Small **and** x_2 is Pos **then** y is 1/Nothing,
    If x_1 is Small **and** x_2 is Neg **then** y is 0.7/Increase + 0.3/Nothing,
    If x_1 is Large **and** x_2 is Pos **then** y is 0.8/Decrease + 0.2/Nothing,
    If x_1 is Large **and** x_2 is Neg **then** y is 1/Decrease.

The above rule base can also be represented in a table form: see Table 3.1.

Let us now consider the more general case where $\mathcal{L}(X_1) = \{A_1, A_2, \ldots, A_I\}$, $\mathcal{L}(X_2) = \{B_1, B_2, \ldots, B_J\}$, and $\mathcal{L}(Y) = \{C_1, C_2, \ldots, C_K\}$. The linguistic rule base becomes a collection of weighted rules of the form:

Table 3.1. The table form of a fuzzy rule base

$$x_2$$

		Neg	Pos
x_1	Small	0.7/Increase + 0.3/Nothing	1/Nothing
	Large	1/Decrease	0.8/Decrease + 0.2/Nothing

$$\text{If } x_1 \text{ is } A_i \text{ and } x_2 \text{ is } B_j \text{ then } y \text{ is } C_k \text{ with weight } \omega_{ijk}, \qquad (3.1)$$

where $i \in \mathcal{I}$, $j \in \mathcal{J}$ and $k \in \mathcal{K}$.[1]

In the compact form, the linguistic rule base can also be expressed as a reduced set of rules:

$$\text{If } x_1 \text{ is } A_i \text{ and } x_2 \text{ is } B_j \text{ then } y \text{ is } \sum_{C_k \in \mathcal{L}(Y)} \omega_{ijk}/C_k. \qquad (3.2)$$

The weight ω_{ijk} involved in the "summation" is related to the rule "**If** x_1 **is** A_i **and** x_2 **is** B_j **then** y **is** C_k." When this rule has no relevance for the system under identification, its weight is taken as zero, i.e., $\omega_{ijk} = 0$. This notational convention authorizes the "summation" over $\mathcal{L}(Y)$ in the consequent of each rule.

Although this formalism may be easily extended to deal with any number of inputs, for the sake of simplicity and without loss of generality, the rest of this presentation will only consider systems with two inputs.

3.2 Computational Architecture

Linguistic models based on a set of rules of the form (3.1) or (3.2) are considered here. The entire computational architecture (or the model's internal implementation) consists of a linguistic fuzzification function, a linguistic inference mechanism, and a linguistic defuzzification function. These three components are considered in this paragraph (for more details see [7]).

3.2.1 Linguistic Fuzzification. It transforms a crisp numeric input into a linguistic fuzzy subset over the set of associated linguistic values. In the proposed method the two crisp inputs x_1 and x_2 are fuzzified independently using the fuzzy description function D.

Let us come back to the previous example with $\mathcal{L}(X_1) = \{\text{Small, Large}\}$, $\mathcal{L}(X_2) = \{\text{Neg, Pos}\}$. By defining the fuzzy meaning of these linguistic values

---

[1] For sake of convenience we define $\mathcal{I} = \{1, 2, \ldots, I\}$, $\mathcal{J} = \{1, 2, \ldots, J\}$, $\mathcal{K} = \{1, 2, \ldots, K\}$, $\mathcal{M} = \{1, 2, \ldots, M\}$, and $\mathcal{N} = \{1, 2, \ldots, N\}$.

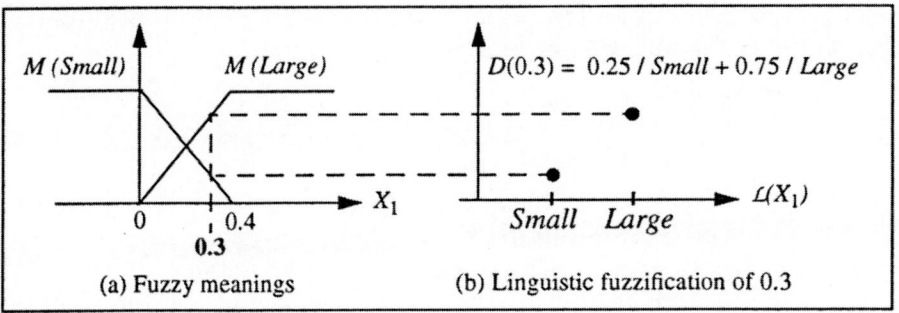

Fig. 3.1. Linguistic fuzzification on X_1

as shown in Fig. 3.1a and Fig. 3.2a, the linguistic fuzzification of any crisp value x_1 or x_2 can be achieved by means of the fuzzy description D. Figure 3.1b illustrates the result of the linguistic fuzzification of $x_1 = 0.3$, that is 0.25/Small + 0.75/Large. The same is carried out to fuzzify the second variable x_2 as shown in Fig. 3.2b for $x_2 = 2$. The obtained result is 0.4/Neg + 0.6/Pos.

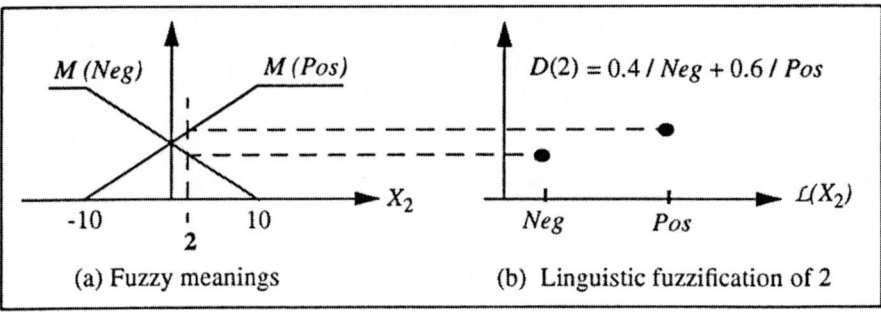

Fig. 3.2. Linguistic fuzzification on X_2

3.2.2 Linguistic Inference. Let R be the fuzzy relation on $\mathcal{L}(X_1) \times \mathcal{L}(X_2) \times \mathcal{L}(Y)$ induced by the rule base (3.2). This relation may be defined as:

$$\forall i \in \mathcal{I}, \forall j \in \mathcal{J}, \forall k \in \mathcal{K} : \mu_R(A_i, B_j, C_k) = \omega_{ijk}. \tag{3.3}$$

The image of an input fuzzy subset E defined on $\mathcal{L}(X_1) \times \mathcal{L}(X_2)$, i.e., $E \in \mathcal{F}(\mathcal{L}(X_1) \times \mathcal{L}(X_2))$, by the relation R, is the fuzzy subset F of $\mathcal{L}(Y)$, i.e., $F \in \mathcal{F}(\mathcal{L}(Y))$. This result is obtained by Zadeh's compositional rule of inference [37], directly applied on the linguistic version of the rule base rather than on its numeric version (see [7] for details). The inferred fuzzy subset F is then given by the following expression:

$$\forall C_k \in \mathcal{L}(Y):$$
$$\mu_F(C_k) = \underset{(A_i, B_j) \in \mathcal{L}(X_1) \times \mathcal{L}(X_2)}{\perp} T(\mu_E(A_i, B_j), \mu_R(A_i, B_j, C_k)) \tag{3.4}$$

where $\perp$ and T represent a t-conorm (projection operator) and a t-norm (combination operator) respectively.

The input fuzzy subset E can be constructed as being the conjunction (a t-norm) of the two independent fuzzy subsets obtained at the fuzzification step. That is,

$$\forall A_i \in \mathcal{L}(X_1), \forall B_j \in \mathcal{L}(X_2):$$
$$\mu_E(A_i, B_j) = T(\mu_{D(x_1)}(A_i), \mu_{D(x_2)}(B_j)). \tag{3.5}$$

Using (3.3) to (3.5), the final inferred F for a given linguistic value C_k is

$$\forall k \in \mathcal{K}: \mu_F(C_k) = \underset{(i,j) \in \mathcal{I} \times \mathcal{J}}{\perp} \mu_{D(x_1)}(A_i) \, T \, \mu_{D(x_2)}(B_j) \, T \, \omega_{ijk}. \tag{3.6}$$

Using (2.6), this result may also be expressed as

$$\mu_F(C_k) = \underset{(i,j) \in \mathcal{I} \times \mathcal{J}}{\perp} \mu_M(A_i)(x_1) \, T \, \mu_M(B_j)(x_2) \, T \, \omega_{ijk}. \tag{3.7}$$

Let us now illustrate the inference mechanism on the example previously introduced. From the rule base presented in Table 3.1, the fuzzy relation R can be determined according to (3.3). The fuzzy relation obtained is illustrated in Table 3.2.

Table 3.2. The fuzzy relation

		x_2	
		Neg	*Pos*
x_1	*Small*	μ_R(*Small, Neg, Decrease*) = 0 μ_R(*Small, Neg, Nothing*) = 0.3 μ_R(*Small, Neg, Increase*) = 0.7	μ_R(*Small, Pos, Decrease*) = 0 μ_R(*Small, Pos, Nothing*) = 1 μ_R(*Small, Pos, Increase*) = 0
	Large	μ_R(*Large, Neg, Decrease*) = 1 μ_R(*Large, Neg, Nothing*) = 0 μ_R(*Large, Neg, Increase*) = 0	μ_R(*Large, Pos, Decrease*) = 0.8 μ_R(*Large, Pos, Nothing*) = 0.2 μ_R(*Large, Pos, Increase*) = 0

In this table, each square regroups the expression of the three linguistic relational equalities that can be derived from the corresponding compact rule of Table 3.1. A zero value of μ_R means that the corresponding rule was not present in the original rule base.

After the fuzzy relation is available, we can now proceed with the inference according to (3.6). Consider the two inputs $x_1 = 0.3$ and $x_2 = 2$. By applying the linguistic fuzzification interface described in Fig. 3.1b and Fig. 3.2b, the following linguistic fuzzy inputs have to be considered:

$$D(0.3) = 0.25/\text{Small} + 0.75/\text{Large} \text{ and } D(2) = 0.4/\text{Neg} + 0.6/\text{Pos}. \quad (3.8)$$

For each output linguistic value, we obtain:

$$\mu_F(\text{Decrease}) = \bot(\mu_{D(0.3)}(\text{Small}) \text{ T } \mu_{D(2)}(\text{Neg}) \text{ T } \mu_R(\text{Small}, \text{Neg}, \text{Decrease}),$$
$$\mu_{D(0.3)}(\text{Small}) \text{ T } \mu_{D(2)}(\text{Pos}) \text{ T } \mu_R(\text{Small}, \text{Pos}, \text{Decrease}),$$
$$\mu_{D(0.3)}(\text{Large}) \text{ T } \mu_{D(2)}(\text{Neg}) \text{ T } \mu_R(\text{Large}, \text{Neg}, \text{Decrease}),$$
$$\mu_{D(0.3)}(\text{Large}) \text{ T } \mu_{D(2)}(\text{Pos}) \text{ T } \mu_R(\text{Large}, \text{Pos}, \text{Decrease})).$$
$$\mu_F(\text{Decrease}) = \bot(0, 0, 0.75 \text{ T } 0.4 \text{ T } 1, 0.75 \text{ T } 0.6 \text{ T } 0.8).$$
$$\mu_F(\text{Decrease}) = \bot(0.75 \text{ T } 0.4, 0.75 \text{ T } 0.6 \text{ T } 0.8).$$
$$\mu_F(\text{Nothing}) = \bot(\mu_{D(0.3)}(\text{Small}) \text{ T } \mu_{D(2)}(\text{Neg}) \text{ T } \mu_R(\text{Small}, \text{Neg}, \text{Nothing}),$$
$$\mu_{D(0.3)}(\text{Small}) \text{ T } \mu_{D(2)}(\text{Pos}) \text{ T } \mu_R(\text{Small}, \text{Pos}, \text{Nothing}),$$
$$\mu_{D(0.3)}(\text{Large}) \text{ T } \mu_{D(2)}(\text{Neg}) \text{ T } \mu_R(\text{Large}, \text{Neg}, \text{Nothing}),$$
$$\mu_{D(0.3)}(\text{Large}) \text{ T } \mu_{D(2)}(\text{Pos}) \text{ T } \mu_R(\text{Large}, \text{Pos}, \text{Nothing})).$$
$$\mu_F(\text{Nothing}) = \bot(0.25 \text{ T } 0.4 \text{ T } 0.3, 0.25 \text{ T } 0.6 \text{ T } 1, 0, 0.75 \text{ T } 0.6 \text{ T } 0.2).$$
$$\mu_F(\text{Nothing}) = \bot(0.25 \text{ T } 0.4 \text{ T } 0.3, 0.25 \text{ T } 0.6, 0.75 \text{ T } 0.6 \text{ T } 0.2).$$
$$\mu_F(\text{Increase}) = \bot(\mu_{D(0.3)}(\text{Small}) \text{ T } \mu_{D(2)}(\text{Neg}) \text{ T } \mu_R(\text{Small}, \text{Neg}, \text{Increase}),$$
$$\mu_{D(0.3)}(\text{Small}) \text{ T } \mu_{D(2)}(\text{Pos}) \text{ T } \mu_R(\text{Small}, \text{Pos}, \text{Increase}),$$
$$\mu_{D(0.3)}(\text{Large}) \text{ T } \mu_{D(2)}(\text{Neg}) \text{ T } \mu_R(\text{Large}, \text{Neg}, \text{Increase}),$$
$$\mu_{D(0.3)}(\text{Large}) \text{ T } \mu_{D(2)}(\text{Pos}) \text{ T } \mu_R(\text{Large}, \text{Pos}, \text{Increase})).$$
$$\mu_F(\text{Increase}) = \bot(0.25 \text{ T } 0.4 \text{ T } 0.7, 0, 0, 0).$$
$$\mu_F(\text{Increase}) = \quad 0.25 \text{ T } 0.4 \text{ T } 0.7.$$

If we use the product t-norm (i.e., $T(u, v) = u \cdot v$) and the bounded sum t-conorm (i.e., $\bot(u, v) = \min(1, u + v)$) as fuzzy conjunction and disjunction operators respectively, then the output fuzzy subset is finally described by:

$$\mu_F(\text{Decrease}) = 0.3 + 0.36 = 0.66,$$
$$\mu_F(\text{Nothing}) = 0.03 + 0.15 + 0.09 = 0.27,$$
$$\mu_F(\text{Increase}) = 0.07,$$

that is, $F = 0.66/\text{Decrease} + 0.27/\text{Nothing} + 0.07/\text{Increase}$.

3.2.3 Linguistic Defuzzification. Linguistic defuzzification aims at transforming the inferred fuzzy subset $F \in \mathcal{F}(\mathcal{L}(Y))$ into a crisp numeric value $y \in Y$. In order to implement this "linguistic-to-numeric" interface, we have first to define the fuzzy meaning attributed to each output linguistic value

C_k, $k \in \mathcal{K}$, that is, $M(C_k)$. Any numeric value y of the universe of discourse Y which verifies $\mu_{M(C_k)}(y) = 1$ will be called a modal value[2] of $M(C_k)$.

The height defuzzification method [6] can be easily adapted for the purpose of linguistic defuzzification (see [7] for details). If we consider that the fuzzy meaning of each output linguistic value C_k has a single modal value denoted as χ_k,[3] the following expression is then obtained:

$$y = \frac{\sum_{k=1}^{K} \mu_F(C_k) \cdot \chi_k}{\sum_{k=1}^{K} \mu_F(C_k)}, \qquad (3.9)$$

where $\mu_F(C_k)$ is given by (3.6) or (3.7).

Let us now illustrate the proposed linguistic defuzzification method on the example previously introduced. After the inference step, the output fuzzy subset obtained was $F = 0.66/\text{Decrease} + 0.27/\text{Nothing} + 0.07/\text{Increase}$ (see Fig. 3.3a). By defining the fuzzy meaning of the output linguistic values as shown in Fig. 3.3b, the following output can be derived from (3.9):

$$y = \frac{-5\mu_F(\text{Decrease}) + 0\mu_F(\text{Nothing}) + 5\mu_F(\text{Increase})}{\mu_F(\text{Decrease}) + \mu_F(\text{Nothing}) + \mu_F(\text{Increase})}, \qquad (3.10)$$

where -5, 0, and 5 represent the modal values associated with the linguistic values Decrease, Nothing, Increase respectively. The crisp output value is finally computed as

$$y = \frac{-5 \cdot 0.66 + 0 \cdot 0.27 + 5 \cdot 0.07}{0.66 + 0.27 + 0.07} = -2.95. \qquad (3.11)$$

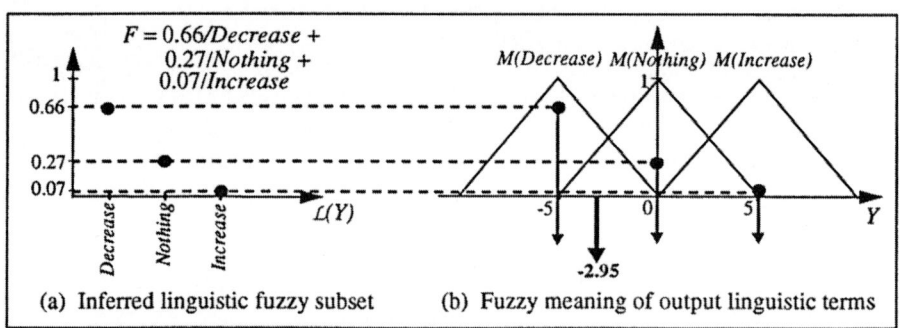

(a) Inferred linguistic fuzzy subset   **(b)** Fuzzy meaning of output linguistic terms

Fig. 3.3. Height defuzzification method

---

[2] The designation "modal value" has been introduced in [5] and is currently used in the literature (e.g. in [23]). However, the same concept is also called "peak value" in [6] or "core value" in [1].

[3] By convention the modal value associated with a linguistic term is noted with the corresponding Greek letter. For example, the modal value associated with the linguistic term A_i (resp. B_j, C_k) is denoted by α_i (resp. β_j, χ_k).

It should be noted here that the fuzzy model considered here may also be put in the relational fuzzy model category. Indeed, from equation (3.3), it is easily established that the identification of the rule weights ω_{ijk} is equivalent to the identification of the fuzzy relation R that links the input and output linguistic values. However, when considering the global model, that is, the rule base together with the fuzzification and defuzzification interfaces, it may be viewed as a linguistic fuzzy model (or a Mamdani-type fuzzy model).

3.3 Representation of a Dynamic System

The class of fuzzy systems that will be used for the modeling task is now completely determined. Its external form is the collection of weighted rules introduced in Sect. 3.1. Its internal (computational) implementation is carried out according to the linguistic fuzzification, inference, and defuzzification steps described in Sect. 3.2. The problem is now to choose the linguistic variables that will be part of the antecedent and consequent of each rule. In other words, how can the fuzzy model represent the system's dynamics? The first solution is based on the following line of reasoning.

A time discrete dynamic SISO system of n-th order with single input $u(k)$ and single output $y(k)$ can be expressed as

$$y(k+1) = f[u(k), u(k-1), \ldots, u(k-n), y(k), y(k-1), \ldots, y(k-n)] \quad (3.12)$$

where f is a nonlinear function. As a universal approximator [14, 20, 26, 31], a fuzzy system with $2(n+1)$ inputs[4] can approximate any nonlinear $2(n+1)$-dimensional mapping, and in particular the function f from (3.12). A dynamic SISO system may thus be modeled by a fuzzy MISO system whose rule base is derived from (3.12). Consequently, each rule has to be expressed as:

If $u(k)$ is A_0 and $u(k-1)$ is A_1 and ... and $u(k-n)$ is A_n and
$y(k)$ is B_0 and $y(k-1)$ is B_1 and ... and $y(k-n)$ is B_n
then $y(k+1)$ is C with weight ω,   (3.13)

where $A_\ell, \ell = 0, \ldots, n$, are linguistic values used to characterize the input variable u at the time instant $k - \ell$; $B_\ell, \ell = 0, \ldots, n$, are also linguistic values, but this time related to the output variable y at the time instant

---

[4] The term *input* for a fuzzy system is different from the term *input* for the plant. In point of fact, inputs for the plant are real physical inputs. In the SISO case, the only input is then the control u. In the opposite, for the fuzzy model we call input every past instance of the variables u and y that appear in the f-function expression (3.12). Consequently, $u(k)$, $u(k-1)$, $\ldots$, $u(k-n)$, $y(k)$, $y(k-1)$, $\ldots$, $y(k-n)$ are the fuzzy model inputs that will be considered in the rule antecedents. A fuzzy model with multiple inputs (variables involved in the rule antecedent) and single output (variable involved in the rule consequent) will be called a fuzzy MISO system.

$k - \ell$; and C is the corresponding linguistic value of the output variable y at the time instant $k + 1$.

A fuzzy model based on a set of rules of the form (3.13) implements the nonlinear regression equation (3.12). It can thus be viewed as a NARX (Nonlinear AutoRegressive with eXogenous input) model. In the case of a dynamic MISO systems, the same time discretization may be applied to each input variable, for example u_1, u_2, ..., u_m. The representation of a system with m inputs and of the n-th order leads then to the use of a fuzzy MISO model with $(m + 1)(n + 1)$ inputs. Although there is no real obstacle to implementing such a fuzzy model for large values of m and n, it is clear that the rule base readability is lost and the model complexity becomes very high (see the discussion on practical aspects in Sect. 4.2). The case of a dynamic MIMO system can also be handled with the proposed fuzzy formalism, but this is not considered in this chapter. However, in order to deal with multiple output variables, for example y_1, y_2, ..., y_r, the fuzzy model has to be considered as a collection of r MISO fuzzy models, where each fuzzy model provides an approximation for a given output. This division may induce undesirable effects, such as different fuzzy partitions for the same input variable with respect to the particular output considered [8].

The choice of a NARX structure to represent a dynamic system is not the only possible one. A state-space model may also be considered. However, in that case, the fuzzy model should implement a system of difference equations using an auxiliary state vector. Once again, a collection of fuzzy MISO systems has to be considered, where each fuzzy system provides an approximation of a particular state vector component. As previously mentioned, the problem of keeping a unique partitioning of the fuzzy inputs arises. This type of model is not dealt with in this chapter.

In the paragraphs that follow, a fuzzy MISO model represents a fuzzy system with multiple inputs and a single output. According to the meaning attributed to each input involved in the rule antecedent, this model can be implemented as a NARX model of a dynamic SISO or MISO system.

3.4 The Type of Identification

From a conceptual point of view, system identification follows a general methodology characterized by two phases: model structure selection and parameter estimation. In [16], the model structure selection (identification) is decomposed into the following three steps:

Step 1: *Choice of the model type.* This involves the selection of linear or nonlinear models, input-output, or state space models.

Step 2: *Choice of the model size.* This involves the issue of selecting which variables to include in the model description. This task is related to the determination of the process order and the time delay.

Step 3: *Choice of a model parameterization.* This consists of finding a suit-able model structure according to the chosen model type and size.

After having determined a model structure, a parameter estimation can be carried out by maximizing a chosen performance criterion. Due to the nature of fuzzy systems, a fuzzy model parameterization depends on the number of rules. When assuming an "and" connective in the rule antecedent and a complete rule base, the number of rules directly depends on the number of linguistic values defined for each input variable. In the same way, the parameterization of the rule consequent is linked to the number of linguistic values defined for the output variable. Finally, in the assumed fuzzy modeling framework, the structure identification task consists of making the following four choices:

1. Model type.
2. Model size.
3. Number of linguistic values defined for each input.
4. Number of linguistic values defined for the output.

The associated parameter identification task consists of the following two steps:

5. Determination of the membership functions associated with each input and output linguistic value.
6. Determination of the weights of all possible rules.

In the identification method presented here, no particular method is pro-posed to solve tasks 1 and 2 from the structure identification part. In fact, a NARX input-output model is assumed. Furthermore, a linguistic fuzzy model with weighted rules is used to implement it. In the same way, it is assumed that the variables involved in the function f from (3.12) have been selected according to a preliminary analysis. This assumption directly translates in a fixed structure for the rules (3.13) and also implies the composition of the training data. For example, in the case of a SISO process, the choice of $u(k-2)$ and $y(k)$ as past input-output data leads to considering the training set as a collection of $(u(k-2), y(k), y(k+1))$ vectors. In this case, the general notation (x_1, x_2, y) used in the rules (3.1) and (3.2) has to be understood as

$$x_1 \equiv u(k-2),$$
$$x_2 \equiv y(k),$$
$$y \equiv y(k+1). \tag{3.14}$$

Assuming that the rule form has been chosen, the proposed identifica-tion method deals with model parameterization and parameter estimation (tasks 3 to 6). However, in contrast to conventional identification approaches (Fig. 3.4a), the proposed method does not handle the structure selection and parameter identification in a single pass. The global fuzzy model is con-structed in an incremental way according to Fig. 3.4b. An initialization phase produces a minimal fuzzy model and this model is then improved during a refining phase.

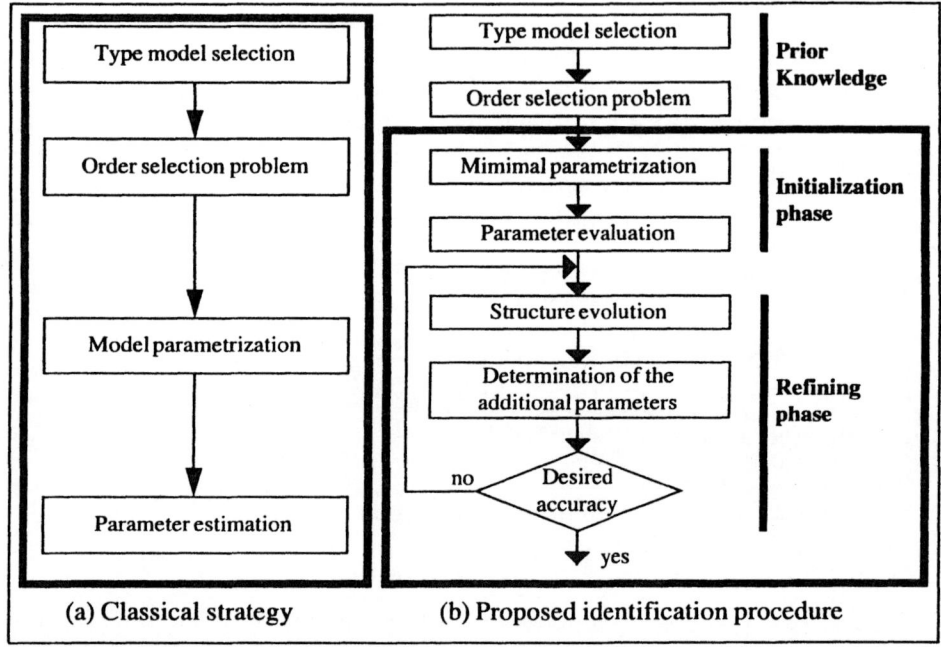

Fig. 3.4. Structure and parameter identification

The minimal parameterization involved in the initialization phase is based on the definition of only two linguistic values for each fuzzy variable. The parameter estimation is then achieved with respect to the distribution of the training input-output data.

At each iteration of the refining phase, the number of linguistic terms is increased (structure evolution). One more linguistic value is inserted in the term-set associated with each input and output variable. The membership function associated with each new linguistic value and the weight of the corresponding additional rules are then simultaneously determined in such a way that the discrepancy between the model and the modeled process decreases (determination of the additional parameters). It should be noticed that a strict[5] triangular partition of the different universes of discourse is permanently maintained. This constraint reduces the number of parameters needed for the characterization of a membership function to a single one, i.e., its

---

[5] Given any crisp numeric value $x \in X$, its fuzzy description on the set of linguistic values $\mathcal{L}(X)$ verifies $\sum_{LX \in \mathcal{L}(X)} \mu_{D(x)}(LX) = 1$.

modal value. The interest of such an iterative construction rests in the fact that the addition of new linguistic values does not modify the parameters already identified. This property results from the intrinsic nature of a fuzzy model. Indeed, a fuzzy model may be viewed as a collection of local models which remain valid when a new local model (new rule) is introduced. The developed identification strategy is finally a combination of the *brute-force* and *smart* approaches proposed in [1]. The former suggests increasing the number of linguistic values involved in the fuzzy model without taking into consideration their distribution. The latter suggests placing the membership functions of a given number of linguistic values in such a way that the modeled system does not significantly diverge from the model obtained by linear interpolation between the rules. The method described here employs an increasing number of linguistic values together with an appropriate distribution of these linguistic values. This can be viewed as an automatic zoom capacity.

4. The Fuzzy Identification Method

4.1 Introduction

In this paragraph, we present the identification algorithm. The identified fuzzy model is implemented with the product t-norm (i.e., $T(u, v) = u \cdot v$) and the bounded sum t-conorm (i.e., $\perp(u, v) = \min(1, u + v)$) according to equations (3.7) and (3.9). These fuzzy operators have been chosen because they induce multilinear interpolation between modal values [9].

As already stated, the identification process is composed of two phases. The first one is an initialization phase during which an initial model based on selected training points is constructed. The second is an iterative phase that refines the initial partitions and rule base, by adding new linguistic values and new rules. At each iteration, the constructed fuzzy model is tested over the training set in order to determine the training point that generates the maximal approximation error. For each training point the model output is computed by applying equations (3.7) and (3.9) and the approximation error is determined. The training point that provides the maximal approximation error is then chosen to generate a new linguistic value in each partition and the corresponding new rules. The iterative phase continues until the maximal error over the training set reaches a desired value.

The two proposed phases, i.e., initialization and refining, are concerned with a common task which is the generation of new linguistic values and rules from a given training point. The approach proposed here is to create the new linguistic values and rules such that the modified model perfectly reproduces the considered training point, that is, the new fuzzy model should produce a zero approximation error for the learned point. This construction constraint guarantees a global convergence of the algorithm on the training set. Indeed, even if new rules do not always decrease the maximal approximation error, it becomes null when all the training points are learned. Given a set of N training points, by learning a different training point at each iteration, a maximal of N iterations is then sufficient to ensure a zero approximation error on the training set.

In order to clarify this principle before going into details, we will first illustrate it with a fuzzy SISO example in Sect. 4.2. The learning process will then be presented in details for a fuzzy MISO system with two inputs and one output, without any loss of generality. Although the learning algorithm involves the sequence of first initialization and then refining, it was chosen here, for the sake of presentation, to introduce first the refining phase in Sect. 4.3 and then the initialization phase in Sect. 4.4. The global learning algorithm is developed in Sect. 4.5. Finally, Sect. 4.6 summarizes the identification procedure from data preparation to model validation.

4.2 Illustrative Example

Consider the training set consisting of the data (x, y) given in Fig. 4.1. The aim of the identification problem is then to determine a fuzzy model that accurately represents the relationship $y = f(x)$.

In the initialization phase two linguistic values will be created on the input space (A_1 and A_2) and on the output space (B_1 and B_2). This choice corresponds to a minimal parameterization of the fuzzy model. Assuming a strict triangular partitioning, the parameters to identify are: (i) α_1 and α_2, i.e., the modal values attributed to A_1 and A_2, (ii) β_1 and β_2 i.e., the modal values attributed to B_1 and B_2), and (iii) the weights of all possible rules "**If** x is A_i **then** y is B_j with weight ω_{ij}", $i = 1, 2$, $j = 1, 2$.

Regarding the membership functions, the parameter assessment is achieved in the following way:

$$\alpha_1 = x_{\min} \qquad \alpha_2 = x_{\max}$$
$$\beta_1 = y_{\min} \qquad \beta_2 = y_{\max}$$

where $x_{\min}$, $x_{\max}$ (resp. $y_{\min}$ and $y_{\max}$) represent the minimum and maximal x-coordinates (resp. y-coordinates) extracted from the training set. The induced membership functions are illustrated in Fig. 4.1a. The weight parameters are then determined so that the fuzzy model precisely reflects the training points with minimum and maximal x-coordinates: $P^1(x_{\min}, y^1)$ and $P^2(x_{\max}, y^2)$ (see Fig. 4.1a). In fact, it is required to learn these points. After identification, we obtain $\omega_{11} = 0.76$, $\omega_{12} = 0.24$, $\omega_{21} = 0$, and $\omega_{22} = 1$. The initial corresponding rule base consists of the two rules

> **If** x is A_1 **then** y is $0.76/B_1 + 0.24/B_2$,
> **If** x is A_2 **then** y is $1/B_2$.

It should be noted that "$0.76/B_1 + 0.24/B_2$" is the fuzzy description of the y-coordinate of the training point P^1, that is, $D(y^1)$. In the same way,

"$1/B_2$" is the fuzzy description of the y-coordinate of the training point P^2, that is, $D(y^2)$. The rule weights can thus be derived from the fuzzy description of the y-coordinate of the points to learn (P^1 and P^2) and the rule base may be rewritten as

If x is A_1 then y is $D(y^1)$,
If x is A_2 then y is $D(y^2)$.

Using the identified rule base and applying linguistic fuzzification, linguistic inference, and linguistic defuzzification operations, the result is a linear interpolation between the two learned points P^1 and P^2.

At this point, the refining iterative phase may be started. The training point that produces the maximal approximation error, that is, P^3, is now used to improve the initial fuzzy model. Two new linguistic values A_3 and B_3 are created respectively on the X and Y spaces. This structure evolution is carried out in order to learn the point $P^3(x^3, y^3)$. The identification of the new parameters is thus achieved so that the increased fuzzy model satisfies this constraint. The modal value of A^3 (resp. B^3) is defined as being x^3 (resp. y^3) and the expanded rule base becomes

If x is A_1 then y is $0.2/B_2 + 0.8/B_3$,
If x is A_2 then y is $1/B_2$,
If x is A_3 then y is $1/B_3$.

The first rule seems to have changed. However, from a formal point of view, the consequents of the already existing rules have not been modified and are still defined as $D(y^1)$ and $D(y^2)$. Indeed, a modification of the Y partition has implied a modification in the description of y's crisp values. In particular, the description of the y-coordinate of the point P^1 has become $D(y^1) = 0.2/B_2 + 0.8/B_3$. In the same way, we have $D(y^2)/B_2$ and $D(y^3)/B_3$. The new partitions and the corresponding model are illustrated in Fig. 4.1b. It should be noted that the points P^1, P^2 and P^3 are now precisely represented by the model.

For the second iteration, the point P^4 which generates the maximal error is used to create the linguistic values A_4 and B_4 and the new rule "If x is A_4 then y is $1/B_4$." The fuzzy description of the y-coordinate of point P^1 is once more changed, and the first rule becomes "If x is A_1 then y is $0.46/B_3 + 0.54/B_4$." The new partitions and the resulting model are shown in Fig. 4.2a. In the next iterations, the structure evolution is continued. The generation of new linguistic values and rules is thus continued in order to obtain a model that fits the system with the desired accuracy. Figure 4.2b illustrates the fuzzy model after the eighth iteration.

4.3 Refining Phase

Let us now come back to the general case of a fuzzy MISO system with two inputs (x_1, x_2) and one output (y). The learning process is performed on

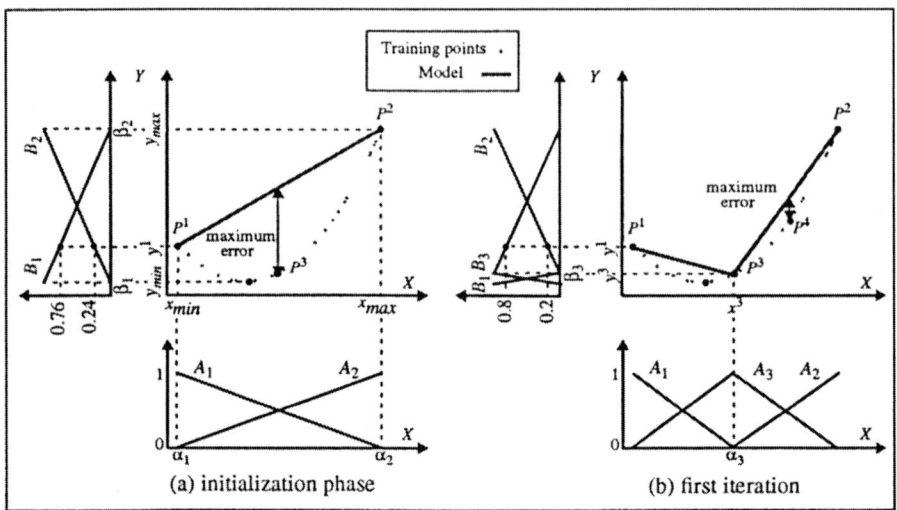

Fig. 4.1. Initialization phase and first iteration

a set of training points $T = \{P^n(x_1^n, x_2^n, y^n), \; n \in \mathcal{N}\}$. Let us also assume that, at a given time, M linguistic values have been defined for each input and output variable, that is, $\mathcal{L}(X_1) = \{A_1, \ldots, A_M\}$, $\mathcal{L}(X_2) = \{B_1, \ldots, B_M\}$ and $\mathcal{L}(Y) = \{C_1, \ldots, C_M\}$. For sake of convenience, the enumeration of the different linguistic values corresponds to increasing order of their associated modal values. Let us also suppose that the fuzzy meanings of these linguistic values are triangular and that the modal values of the linguistic values $m-1$ and $m+1$ are respectively equal to the left and right-hand limits of the support of the linguistic value m (see Fig. 4.3 where $\alpha_1, \ldots, \alpha_M$ represent the modal values of the symbols $A_1, \ldots, A_M$). In this case, the fuzzy description of any crisp value involves only one or two linguistic values, and the sum of the corresponding membership degrees is equal to 1 (strict partition of the universes of discourse).

Furthermore, a rule base of the form (3.1) or (3.2) is assumed. For example, Table 4.1 shows a table representation of a rule base of type (3.1) or (3.2) for $M = 3$, where $F1$–$F9$ are linguistic fuzzy subsets defined on $\mathcal{L}(Y)$, that is elements of $\mathcal{F}(\mathcal{L}(Y))$. For example, $F1$ may be defined as $F1 = 0.2/C_1 + 0.8/C_2$. In that case, the corresponding compact rule of form (3.2) is

If x_1 is A_1 and x_2 is B_1 then y is $0.2/C_1 + 0.8/C_2$.

The fuzzy model is assessed over the training data set and the point $P^*(x_1^*, x_2^*, y^*)$ that produces the maximal approximation error will now be learned. The learning strategy consists of adding new input and output linguistic values (structure evolution) and consequently new rules, so that the

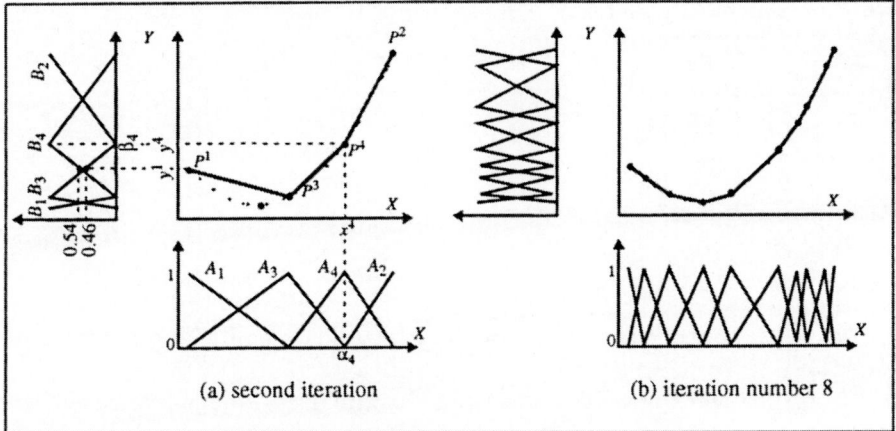

(a) second iteration   (b) iteration number 8

Fig. 4.2. Further iterations

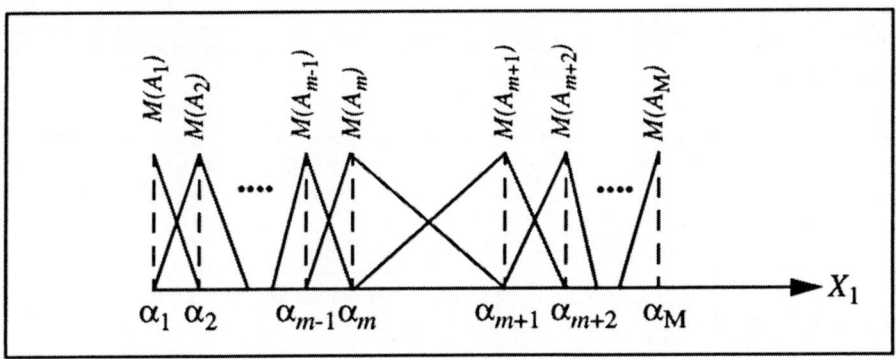

Fig. 4.3. Fuzzy meanings of the M linguistic values over the X_1 space

modified fuzzy system precisely models the new point $P^*(x_1^*, x_2^*, y^*)$. This objective can be attained by an adequate evaluation of the additional parameters of the expanded fuzzy model. It can be carried out according to the four steps presented in Sects. 4.3.1–4.3.4.

4.3.1 Identification of the New Input Linguistic Values. This first step consists in defining one new linguistic value for each input variable x_1 and x_2 and in modifying the associated fuzzy partitions. Denoting by A^* the newly generated linguistic value of x_1, the fuzzy partition of the universe of discourse X_1 is updated according to the principle illustrated in Fig. 2.1. The original fuzzy partition of the X_1 space is shown in Fig. 2.1a, and the modified fuzzy partition is presented in Fig. 4.4b. Let us assume that $\alpha_m < x_1^* < \alpha_{m+1}$, where α_m and α_{m+1} are respectively the modal values of A_m and A_{m+1}. Then we have that

$- x_1^*$ becomes the modal value of A^*

Table 4.1. Rule base table for $M = 3$

B_3	F2	F8	F4
B_2	F5	F7	F9
B_1	F1	F6	F3
	A_1	A_2	A_3

– $[a_m, a_{m+1}]$ becomes the support of the triangular membership function of $M(A^*)$,
– $[a_{m-1}, x_1^*]$ becomes the support of the triangular membership function of $M(Am)$,
– $[x_1^*, a_{m+2}]$ becomes the support of the triangular membership function of $M(Am+1)$.

The definition of the new linguistic value B^* for x_2 variable is made in the same manner.

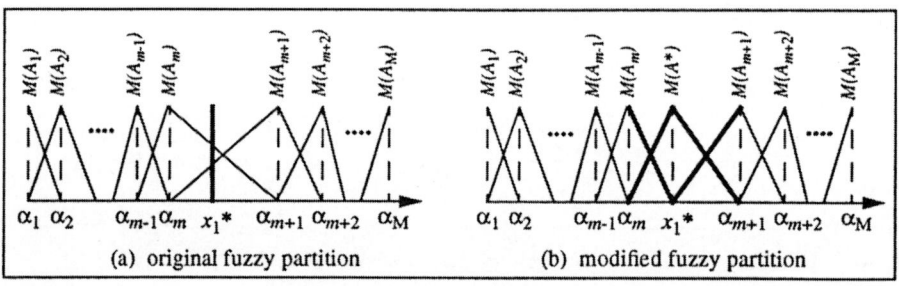

Fig. 4.4. Modification of the fuzzy partition by adding a new linguistic value

4.3.2 Identification of an Output Linguistic Value and a New Rule.
The second step of the learning strategy of the considered training point $P^*(x_1^*, x_2^*, y^*)$ consists in defining a new rule whose antecedent will be "**If** x is A^* **and** y is B^*." A new output linguistic value C^* is generated in the same way as for the input variables. Particularly, y^* is chosen as the modal value of C^*. In that case, the definition of the new rule is

$$\text{If } x_1 \text{ is } A^* \text{ and } x_2 \text{ is } B^* \text{ then } y \text{ is } 1/C^*. \tag{4.1}$$

The above rule is represented by the shaded cell in Table 4.2. From equation (3.7), the linguistic output of the modified fuzzy model for the crisp inputs x_1^* and x_2^* fulfills

$$\forall m \in \mathcal{M} : \mu_F(C_m) = 0 \text{ and } \mu_F(C^*) = 1. \tag{4.2}$$

By applying (3.9) for the purpose of linguistic defuzzification, the final result obtained is the modal value of C^*, that is y^*. As required, the modified fuzzy model, with new linguistic values A^*, B^*, and C^* and the new rule (4.1), has learned exactly the desired training point (x_1^*, x_2^*, y^*).

Table 4.2. Addition of a new rule

B_3	F2	F8		F4
B_2	F5	F7		F9
B^*			1/C*	
B_1	F1	F6		F3
	A_1	A_2	A^*	A_3

If one of the new linguistic values A^*, B^*, or C^* has already been created in previous iterations, there is no need to create it again. In this case, the model may have a different number of linguistic values for the different input and output spaces. Although this statement is incompatible with the previous assumption of having the same number of linguistic values for each space, its violation does not cause any difficulty. In fact, this restrictive assumption was only made for sake of convenience.

4.3.3 Identification of the Consequent of the Additional New Rules.

The creation of the linguistic values A^* and B^* has produced additional new rules whose consequents have not been determined, that is

$$\forall i \in \mathcal{M} : \textbf{If } x_1 \textbf{ is } A_i \textbf{ and } x_2 \textbf{ is } B^* \textbf{ then } y \textbf{ is ???},$$

$$\forall j \in \mathcal{M} : \textbf{If } x_1 \textbf{ is } A^* \textbf{ and } x_2 \textbf{ is } B_j \textbf{ then } y \textbf{ is ???}. \tag{4.3}$$

The above rules are represented by empty cells in the rule base table from Table 4.2. They correspond to 2D fuzzy regions of the forms $M(A^*) \times M(B_j)$ and $M(A_i) \times M(B^*)$. For example, the empty cell (A^*, B_j) is characterized by the fuzzy region $M(A^*) \times M(B_j)$ which has the modal point $P_{(*,j)}$. The coordinates of $P_{(*,j)}$ are respectively the modal values of A^* and B_j, that is, $P_{(*,j)}(x_1^*, \beta_j)$ (see Fig. 4.5).

Let T' be the projection of the training set T on $X_1 \times X_2$, namely,

$$T' = \text{Proj}_{X_1 \times X_2} T = \{P'^n(x_1^n, x_2^n), \quad n \in \mathcal{N}\}. \tag{4.4}$$

Consider now the empty cell (A^*, B_j) in the rule base. It should be noted that, in general, the modal point $P_{(*,j)}$ characterizing the 2D fuzzy region $M(A^*) \times M(B_j)$, does not belong to T'. In order to fill the empty cell, the training example that best characterizes the associated 2D fuzzy region is then used.

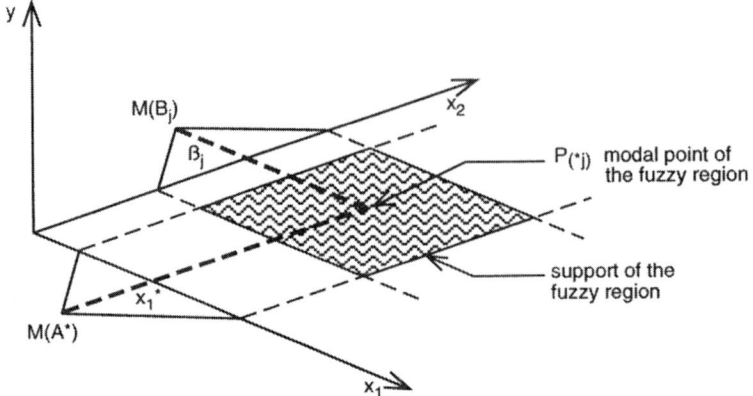

Fig. 4.5. 2D fuzzy region associated with a fuzzy cell

The proposed solution consists in using the training example $P^\ell(x_1^\ell, x_2^\ell, y^\ell) \in T$, whose projection on $X_1 \times X_2$, $P'^\ell(x_1^\ell, x_2^\ell)$, has the maximal membership degree to the involved 2D fuzzy region, that is,

$$\mu_{M(A^*) \times M(B_j)}(x_1^\ell, x_2^\ell) = \max_{n=1,\ldots,N} \mu_{M(A^*) \times M(B_j)}(x_1^n, x_2^n)$$

$$= \max_{n=1,\ldots,N} \mathrm{T}(\mu_{M(A^*)}(x_1^n), \mu_{M(B_j)}(x_2^n)), \quad (4.5)$$

where $P'^n(x_1^n, x_2^n) \in T'$.

The y-coordinate of P^ℓ, y^ℓ, can be viewed as the best candidate to represent the output of the model for the considered 2D fuzzy region. The idea is thus to fill the empty cell so that the resulting model produces the output y^ℓ for the input pair (x_1^*, β_j). One way to satisfy this requirement could be to generate a new output linguistic value with y^ℓ as a modal value and a corresponding new rule as done in the former step. However, since we want to keep the rule base readable, the number of output linguistic values has to be limited. The conclusion of the new rule has therefore to be expressed by means of output linguistic values already defined. This leads to the synthesis of the following compact rule:

If x_1 is A^* and x_2 is B_j then y is $D(y^\ell)$, $\qquad (4.6)$

where $D(y^\ell)$ represents the fuzzy description of y^ℓ over the actual set of linguistic values $\mathcal{L}(Y)$ as defined in (2.5). It can be shown that the result effectively models the point (x_1^*, β_j, y^ℓ). By applying the described strategy for all empty cells, the rule base is filled as illustrated in Fig. 4.6, where $y^{10}, \ldots, y^{15}$ represent the y-coordinates of the training points verifying the equation (4.5) for the involved fuzzy regions. When $\mathcal{L}(Y)$ changes in next iterations, that is when new linguistic values C^* are generated, the fuzzy description $D(y^\ell)$ may be modified if y^ℓ belongs to an output fuzzy region whose support is modified by the creation of a new output linguistic value (see illustrative example of Sect. 4.2).

B_3	F2	F8	$D(y^{14})$	F4
B_2	F5	F7	$D(y^{13})$	F9
B^*	$D(y^{10})$	$D(y^{11})$	$1/C^*$	$D(y^{15})$
B_1	F1	F6	$D(y^{12})$	F3
	A_1	A_2	A^*	A_3

Fig. 4.6. Initial rule base after completion

4.3.4 Completion of the Rule Base. Unfortunately, the proposed solution is not able to fill an empty 2D fuzzy region of the form $M(A^*) \times M(B_j)$ or $M(A_i) \times M(B^*)$ when there exists no training example associated with it, that is,

$$\forall P'^n(x_1^n, x_2^n) \in T' : \mu_{M(A^*) \times M(B_j)}(x_1^n, x_2^n) = 0, \tag{4.7}$$

or

$$\forall P'^n(x_1^n, x_2^n) \in T' : \mu_{M(A_i) \times M(B^*)}(x_1^n, x_2^n) = 0. \tag{4.8}$$

Thus the rule base remains incomplete and now the problem is how to complete it. In [26], Sudkamp et al. proposed two methods for rule base completion based on the notion of similarity between fuzzy cells. The more similar a filled cell is to an empty cell, the more it affects the value assigned to the empty cell. This kind of completion method is independent of the training set. Only the current rule base, that is, the points already learned, are involved in the completion process. The same principle of interpolating by similarity is applied here. However, in order to take into account the wholeness of the training set, and not only the learned points, the similarity of all the training points with the empty cell is evaluated. In fact, for each empty cell, the method consists in finding its "nearest" training point to represent it.

Consider an empty cell (A^*, B_j) in the rule base. The Euclidean distance d between its modal point $P(*, j)$ and all the points of T' is evaluated. Let $P'^\ell(x_1^\ell, x_2^\ell)$ be the point of T' with minimum distance, that is

$$d(P'^\ell, P_{(*,j)}) = \min_{n \in \mathcal{N}} d(P'^n, P_{(*,j)}). \tag{4.9}$$

The y-coordinate of the corresponding training point $P^\ell(x_1^\ell, x_2^\ell, y^\ell) \in T$ is then used to fill the empty cell (A^*, B_j) by generating the compact rule (4.6).

This iterative phase described in the Sects. 4.3.1–4.3.4 is repeated until the maximal error reaches a given threshold.

4.4 Initialization Phase

In this phase, a set of two linguistic values is assumed for each of the input and output variables. In the considered case the initial term-sets are then $\mathcal{L}(X_1) = \{A_1, A_2\}$, $\mathcal{L}(X_2) = \{B_1, B_2\}$ and $\mathcal{L}(Y) = \{C_1, C_2\}$. This assumption leads to a minimal parameterization of the initial fuzzy model. The parameter identification is then carried out according to the next two steps.

4.4.1 Identification of Input and Output Linguistic Values.

The purpose of this step is to determine the membership functions associated with each linguistic term. To sum up, it consists in evaluating the modal value of the fuzzy meaning of A_1, A_2, B_1, B_2 and C_1, C_2. Indeed, by assuming strict triangular partition, the membership functions can be unequivocally reconstructed from the knowledge of the modal values alone. The proposed strategy starts by extracting the maximal and minimum crisp values of all input and output variables over the training data set. In the present fuzzy MISO system, the chosen values are x_{1min}, x_{1max}, x_{2min}, x_{2max}, y_{min}, and y_{max}. The modal value of the fuzzy meaning of the first linguistic value (resp. second) is then initialized with the minimum (resp. maximal) value of the considered variable. The initial partitions of the three spaces X_1, X_2, and Y are shown in Fig. 4.7. It may be noted that the strict partitioning requirement causes a distortion of the extreme membership functions and defines a plateau outside of the minimum and maximal limits.

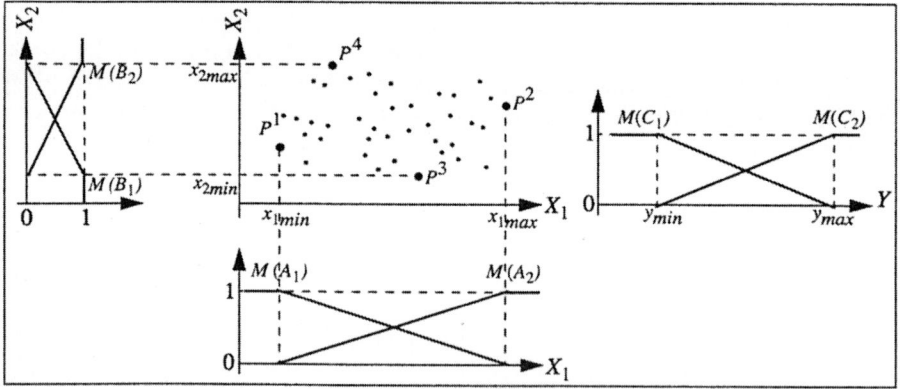

Fig. 4.7. Initial partitions

4.4.2 Identification of the Initial Rule Base.

The creation of the initial linguistic values produces an initial rule base with no determined consequents as illustrated in Fig. 4.8a. Our purpose is now to identify the undetermined rule consequents. This is equivalent to the assessment of the weights of all the possible rules. This task is achieved by using the two steps of the learning strategy presented in Sects. 4.3.3 and 4.3.4. The rule base obtained is shown in Fig. 4.8b, where $F1$, $F2$, $F3$, and $F4$ are the fuzzy descriptions of the y-coordinate of the training points that best represent the considered input fuzzy regions.

If x_1 is A_1 and x_2 is B_1 then y is ???

If x_1 is A_1 and x_2 is B_2 then y is ???

If x_1 is A_2 and x_2 is B_1 then y is ???

If x_1 is A_2 and x_2 is B_2 then y is ???

B_2	$F2$	$F4$
B_1	$F1$	$F3$
	A_1	A_2

(a)   (b)

Fig. 4.8. Initial rule base

4.5 Global Learning Algorithm

The already described identification method can be represented as the following global learning algorithm:

Data   /* T is the training data set */
$$T = (x_1^1, x_2^1, y_1), (x_1^2, x_2^2, y_2), \ldots (x_1^N, x_2^N, y^N)$$

Begin

/***** Initialization phase *****/

- Determine the maximal and minimum crisp values of each x_1, x_2, and y, i.e., $x_{1\min}$, $x_{1\max}$, $x_{2\min}$, $x_{2\max}$, $y_{\min}$, and $y_{\max}$.
- Generate two linguistic values for each input and output with the previously determined limits as modal values, i.e.,

$$\mathcal{L}(X_1) = \{A_1, A_2\}, \qquad \mathcal{L}(Y) = \{C_1, C_2\},$$
$$\mathcal{L}(X_2) = \{B_1, B_2\}, \qquad M = 2.$$

- Construct the initial rule base using steps 3 and 4 of the learning strategy from Sect. 4.3.

/***** Refining phase *****/

Repeat

    For each point $(x_1^i, x_2^i, y^i) \in T$ **do**
        /* Apply the model to the input (x_1^i, x_2^i) and get the corresponding output $\hat{y}^i$ */

$\hat{y}^i = \text{Model}(x_1^i, x_2^i)$

/* Compute the error err^i between the model output and the training output y^i */

$\text{err}^i = |y^i - \hat{y}^i|$

Endfor

Find j such that $\text{err}^j = \max_{i \in [1,N]}(\text{err}^j)$

If $\text{err}^j > \text{err}^{\text{th}}$ **then**   /* Desired accuracy not obtained */

/* Apply the learning strategy from Sect. 4.3 on the training point (x_1^j, x_2^j, y^j) */

    1. Generate a linguistic value A^* with modal value x_1^j.
       $\mathcal{L}(X_1) = \mathcal{L}(X_1) \cup A^*$.
    2. Generate a linguistic value B^* with modal value x_2^j.
       $\mathcal{L}(X_2) = \mathcal{L}(X_2) \cup B^*$.
    3. Generate a linguistic value C^* with modal value y^j.
       $\mathcal{L}(Y) = \mathcal{L}(Y) \cup C^*$.
    4. Generate the new rule **If** x_1 is A^* **and** x_2 is B^* **then** y is $1/C^*$.
    5. Generate other rules covering 2D fuzzy regions of the form $M(A^*) \times M(B_j)$ and $M(A_j) \times M(B^*)$, $j = 1, \ldots, M$.
    6. Complete the empty cells remaining after step 3.
    Renumber the linguistic values, $M = M + 1$.

Endif

Until $(\text{err}^j \le \text{err}^{\text{th}})$

/* repeat the refining phase until the maximal error reaches err^{th} */

End

4.6 Summary of the Identification Procedure

This section summarizes the presented identification method.

Step 1: *Prepare the training data.* After having chosen the relevant variables for the fuzzy model, this step consists in generating a training data set based on the available input-output data. For example, when $u(k-2)$ and $y(k)$ are chosen as antecedent variables for the fuzzy model, and $y(k+1)$ is chosen as output variable, the training set is composed of points of the form $[u(k-2), y(k), y(k+1)]$.

Step 2: *Construct an initial model.* In this step, extreme points are extracted from the training data set. From these points, two linguistic values are configured on each input and output space and an initial rule base is identified.

Step 3: *Evaluate the model.* In this step, the current model is tested over the whole training set. The training point that induces the maximal approximation error is determined. It is the next candidate for the learning process. If the maximal approximation error is less than a predefined threshold then the model synthesis is finished (go to step 7).

Step 4: *Modify the model structure.* One new linguistic value is created on each input and output space. This structure evolution is performed so that the fuzzy model can easily learn the training point which produced the worst performance during the model evaluation.

Step 5: *Determine the additional parameters.* In this step, the modal values associated with the new linguistic values are identified. The rule base is increased in consequence. After this procedure is completed, the training point candidate for the learning process has been learned.

Step 6: *Go to step 3.*

Step 7: *Validate the model.* In this step, the linguistic model obtained is validated over a test set. Usually it consists in evaluating a performance index that indicates the model relevance for the system under consideration. The test set can be generated from random inputs when the system allows it. In other cases, the available data are divided into two subsets, a training set and a test set. When the available data set is small, statistic learning methods such as cross validation [29] can be used.

5. Numeric Examples

The identification method described in this chapter is tested on two numeric examples. The first one deals with function approximation. The second one uses the Box and Jenkins numeric data set [3] as an example for dynamic system modeling. All the results have been obtained by using the product as t-norm and the bounded sum as t-conorm in (3.7) and (4.5).

5.1 Function Approximation

The aim of this example is to generate a fuzzy model that constitutes an approximation $\hat{f}$ of a function f given by a set of training points composed of N input-output data.

In [11], an analytical one-dimensional function was proposed in order to evaluate a gradient descent method for optimizing fuzzy rules of the Takagi–Sugeno type. This function was defined as:

$$
\begin{aligned}
x \in [0, \tfrac{1}{3}]: &\quad y = -9x^2 + 3x + \tfrac{1}{3} \\
x \in [\tfrac{1}{3}, \tfrac{1}{2}]: &\quad y = 2x - \tfrac{1}{3} \\
x \in [\tfrac{1}{2}, 1]: &\quad y = \frac{0.03}{(0.03 + (4.5x - 3.85))^2}.
\end{aligned}
\tag{5.1}
$$

The above function presents three difficulties: a discontinuity of the derivative at $x = \tfrac{1}{3}$, a discontinuity of the function at $x = \tfrac{1}{2}$, and a peak with high curvature zone around $x = 0.85$. Because of these particularities, we choose the same function to illustrate the capacity of the learning method to isolate linear behaviors and to decompose nonlinear behaviors in terms of linear ones. A training set of $N = 100$ input-output data points of the form $(x, y = f(x))$, where x is randomly distributed over the interval $[0, 1]$ has been generated. At the *initialization phase*, the identification method performs the following tasks:

- Extraction of the extreme crisp values of the input and output variables. We find $x_{\min} = 0.021$, $x_{\max} = 0.968$, $y_{\min} = 0.012$, and $y_{\max} = 0.970$.
- Generation of two linguistic values A_1 and A_2 whose modal values are $x_{\min}$ and $x_{\max}$ on the X-space and similarly two linguistic values B_1 and B_2 whose modal values are $y_{\min}$ and $y_{\max}$ on the Y-space. The fuzzy meanings of these linguistic values constitute an initial partition of the input and output spaces. (See Fig. 5.1 for illustration.)
- Application of step 2 of the initialization phase. This results in the learning of the two points $P_1(x_{\min}, y_{x\,\min})$ and $P2(x_{\max}, y_{x\,\max})$ and in the generation of the two rules:
   If x **is** A_1 **then** y **is** $D(y_{x\,\min}) = 0.6/B_1 + 0.4/B_2$,
   If x **is** A_2 **then** y **is** $D(y_{x\,\max}) = 0.9/B_1 + 0.1/B_2$.
   where $y_{x\,\min} = 0.393$ and $y_{x\,\max} = 0.105$ are the y-coordinates of the learned points $(x_{\min}, y_{x\,\min})$ and $(x_{\max}, y_{x\,\max})$, $D(y_{x\,\min})$ and $D(y_{x\,\max})$ their fuzzy descriptions over the output space. The membership degrees of the two linguistic values B_1 and B_2 to these fuzzy descriptions are illustrated as bold points on the output partition of Fig. 5.1. The obtained model is a line segment between these two learned points (see Fig. 5.1).

At each *iteration of the refining phase* the following tasks are performed:

- Determination of the point (x^*, y^*) that provides the maximal approximation error.
- Generation of a linguistic value A^* with modal value x^* and a linguistic value B^* with modal value y^*, and insertion of their fuzzy meanings into the input and output partitions.
- Generation of the rule: **If** x **is** A^* **then** y **is** $1/B^*$.

At this level there is a rule for each input linguistic value, so the generated rule base is complete and there is no need for steps 3 and 4 of the learning strategy from Sect. 4.3. In fact, these two steps are only implemented for fuzzy MISO systems.

This iterative phase is repeated until the maximal error reaches a predetermined threshold err^{th}.

Figure 5.2 represents the modeled function, the model obtained, and the training data points for $N = 100$ and a rule base of ten rules. We notice that the segment between $x = \frac{1}{3}$ and $x = \frac{1}{2}$ was isolated with only two linguistic terms in the input fuzzy partition. In fact, the model is exactly equivalent to the original function when the latter has linear behavior. Nonlinear parts of the function are approximated by linear behavior. The method was also able

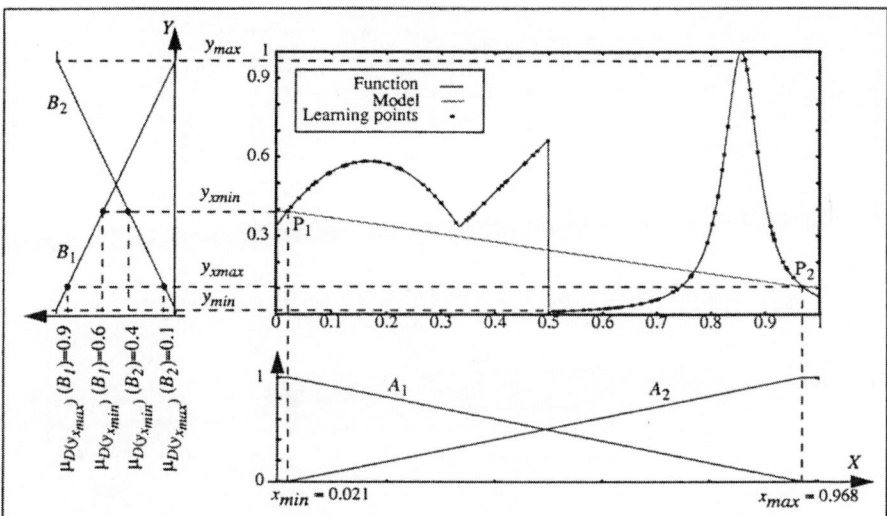

Fig. 5.1. Initial partitions and initial model

to recognize the function discontinuity at $x = \frac{1}{2}$ and to model it with two close input linguistic values.

In order to evaluate the performance of the model, a test set of 100 points was also generated. Figure 5.3a shows the variations of the square error E with the number of rules. We notice that the insertion of new linguistic values decreases the error. This statement is not always true, but it is valid for a sufficiently large number of linguistic values. In fact new linguistic values create finer fuzzy regions that better represent the local behavior of the function. With 20 rules (20 input linguistic values) the square error is 0.0027 for the training set and 0.0062 for the test set. Comparing these results with those obtained by the TS-C method of [11] which also performs a linear interpolation, we can see that with a small number of rules (10–18) our method is much more efficient. This may be explained by the good placement of the modal values of the linguistic values on the input space X. For a larger number of rules our method still gives good results, but the improvement is less significant.

Figure 5.3b illustrates the influence of the number of training samples on the model accuracy in terms of the mean square error $\mathrm{MSE} = 2 \cdot E / psamples$, where *psamples* is the number of samples in the training data set. Here the number of rules is set to 20. The MSE for the training data set is nearly equal to zero for any $N \in [25, 500]$, so the MSE curve is merged with the horizontal axis. For the test set, the MSE is satisfactory for $N > 100$.

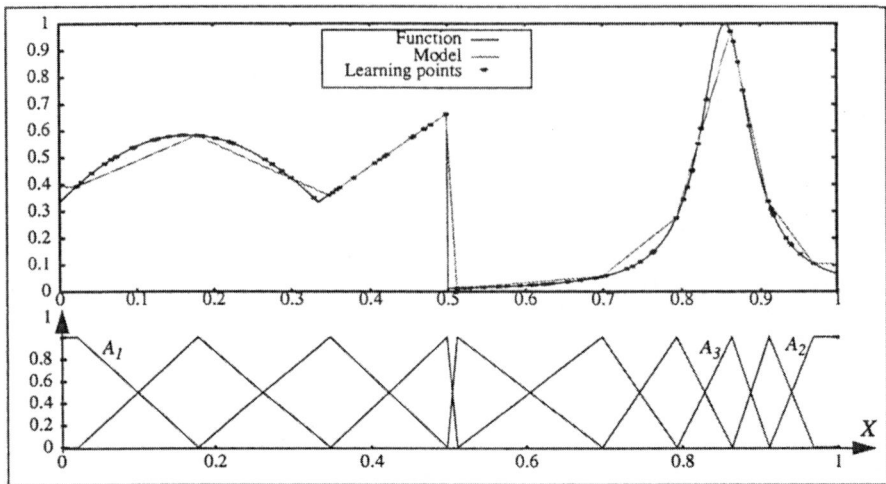

Fig. 5.2. Simulation results and membership functions for 10 rules

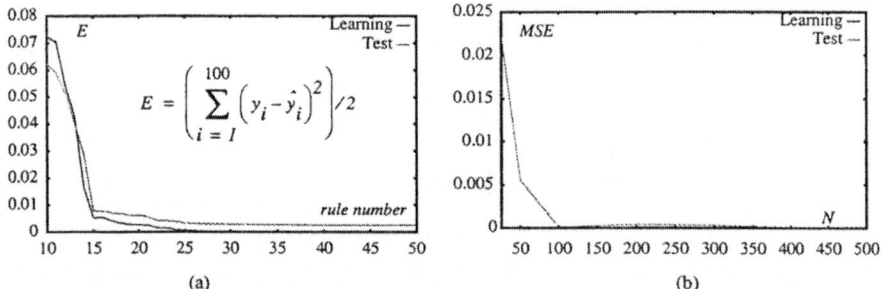

$$E = \left(\sum_{i=1}^{100} \left(y_i - \hat{y}_i \right)^2 \right) / 2$$

(a)   (b)

Fig. 5.3. Error evaluation

The case of noisy data is not studied here. We discuss this issue in Sect. 6 concerning the practical aspects of the proposed identification method.

5.2 The Furnace Model of Box and Jenkins

This example deals with the widely used benchmark furnace model of Box and Jenkins [3]. The modeled system consists of a gas furnace in which air and methane are combined to form a mixture of gases containing CO_2 (carbon dioxide). Air fed to the furnace is kept constant, while the methane feed rate can be varied in any desired manner. Following that, the resulting CO_2 concentration is measured in the exhaust gases at the outlet of the furnace.

The time series used for identification purposes consists of 296 successive pairs of observations of the form $(u(t), y(t))$, where $u(t)$ represents the methane gas feed rate at the instant t and $y(t)$ represents the concentration of CO_2 in the outlet gases.

Using a time-discrete formulation, the dynamics of the system is represented by a relationship that links the predicted system state $y(k + 1)$ to the previous input $u(k - \tau)$ and the previous system state $y(k - \tau_1)$, that is, $y(k + 1) = f(u(k - \tau), y(k - \tau_1))$ where τ and τ_1 are restricted to integers. This benchmark is used in [4, 15, 21, 25, 28, 35]. In [15, 35] the authors try to determine the best values of τ and τ_1. They conclude that $\tau = 2$ and $\tau_1 = 0$ give the best results in terms of model performance. From the 296 data pairs we constructed a training data set of $N = 293$ points of the form (x_1, x_2, y) where, $x_1 = u(k - 2)$, $x_2 = y(k)$, and $y = y(k + 1)$.

The model is then generated using the presented identification method. The model performance J is expressed by the mean square error (MSE) between the data and the model outputs, that is,

$$J = \frac{1}{N} \sum_{i=1}^{N} (y^i - \hat{y}^i). \tag{5.2}$$

We present here the different steps of the model construction. At the initialization phase, the identification method performs the following tasks:

- Extraction of the extreme values of the input and output variables. The extracted values are $x_{1min} = -2.716$, $x_{1max} = 2.835$, $x_{2min} = 45.6$, $x_{2max} = 60.5$, $y_{min} = 45.6$, and $y_{max} = 60.5$.
- Generation of two linguistic values A_1 and A_2 whose modal values are x_{1min} and x_{1max} on the X_1 space, two linguistic values B_1 and B_2 whose modal values are x_{2min} and x_{2max} on the X_2 space and two linguistic values C_1 and C_2 whose modal values are y_{min} and y_{max} on the Y space. The fuzzy meanings of these linguistic values form an initial partition of the input and output spaces (see Fig. 5.4).
- Generation of the four initial rules
    If x_1 is A_1 and x_2 is B_1 then y is $D(y^1)$,
    If x_1 is A_1 and x_2 is B_2 then y is $D(y^2)$,
    If x_1 is A_2 and x_2 is B_1 then y is $D(y^3)$,
    If x_1 is A_2 and x_2 is B_2 then y is $D(y^4)$,
    where $y^1 = 52.8$, $y^2 = 60.4$, $y^3 = 45.6$, $y^4 = 57.8$ are the y-coordinates of the training points that have maximal membership degrees to the 2D fuzzy regions $M(A_1) \times M(B_1)$, $M(A_1) \times M(B_2)$, $M(A_2) \times M(B_1)$, and $M(A_2) \times M(B_2)$. Furthermore, $D(y^i)$, $i = 1, \ldots, 4$, is the fuzzy description of y^i on the linguistic set $\mathcal{L}(Y) = \{C_1, C_2\}$. The initial model obtained here is shown in Fig. 5.5.

At each iteration of the refining phase the following tasks are performed:

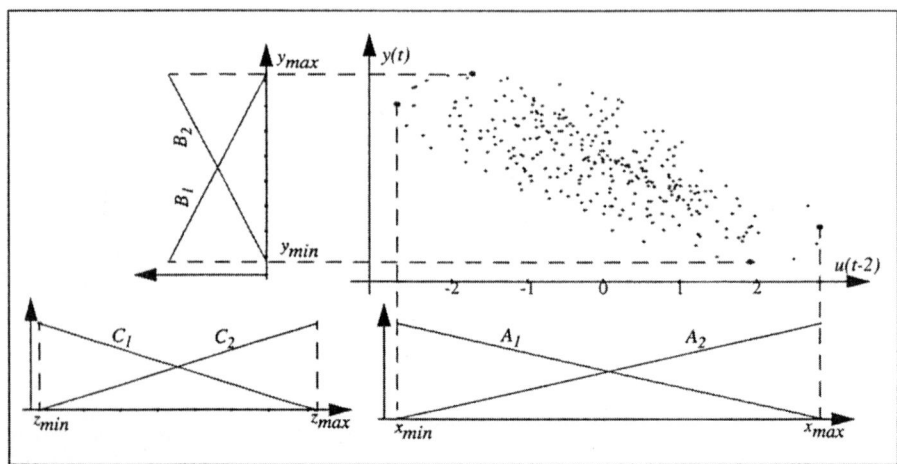

Fig. 5.4. Initialization phase

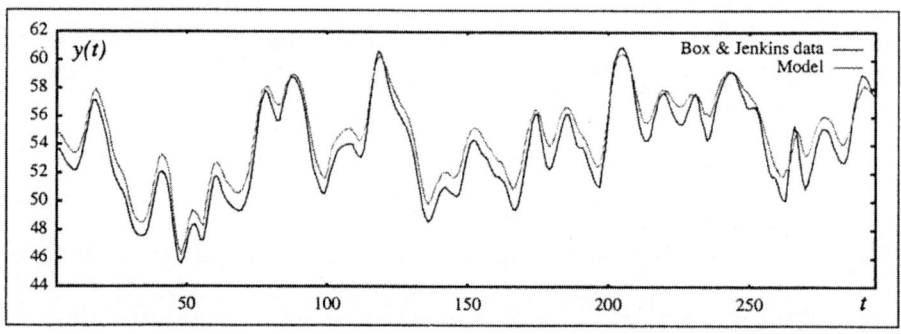

Fig. 5.5. Initial model with 2 linguistic values on each space

- Determination of the point (x_1^*, x_2^*, y^*) which provides the maximal approximation error.
- Generation of a linguistic value A^* with the modal value x_1^*, a linguistic value B^* with the modal value x_2^*, and a linguistic value C^* with the modal value y^*, and insertion of their fuzzy meanings into the input and output partitions.
- Generation of the rule **If** x_1 is A^* **and** x_2 is B^* **then** y is $1/C^*$. The shaded cell in Fig. 5.6 represents such a rule at the first iteration.
- Generation of other rules corresponding to 2D fuzzy regions of the form $M(A^*) \times M(B_j)$ and $M(A_j) \times M(B^*)$. For example, these rules are those of the A^* column and B^* row in the rule base of Fig. 5.6. They are generated by steps 3 and 4 of the learning strategy described in Sect. 4.3.

This iterative phase is repeated until the maximal error reaches a predetermined threshold err^{th}.

B_2	$D(y^2)$	$D(y^7)$	$D(y^4)$
B^*	$D(y^5)$	$\boldsymbol{1/C^*}$	$D(y^8)$
B_1	$D(y^1)$	$D(y^6)$	$D(y^3)$
	A_1	A^*	A_2

Fig. 5.6. Generation of new rules at the first iteration

As the number of linguistic values increases, the performance index J decreases. Figure 5.7 shows the variations of J as a function of the number of iterations c.

For $c = 10$ the X_1, X_2, and Y universes of discourse are respectively decomposed into 10, 9, and 10 linguistic terms. At each iteration a new point is learned and the corresponding new linguistic values are created. If one or more of the new linguistic values has already been created in a previous iteration, there is no need to create it again. This is why the fuzzy model may have a different number of linguistic values for the different input/output spaces. The resulting value of the performance index is $J = 0.175$. So we accept this result as a satisfactory one. The obtained model is visualized in Fig. 5.8.

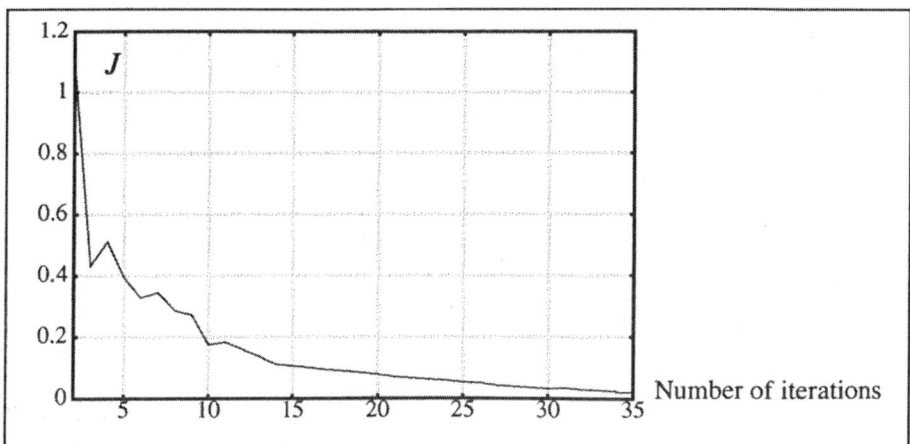

Fig. 5.7. Evolution of the performance index J with c

When we add new linguistic values, the performance index continues to decrease ($J = 0.052$ for $c = 25$), but the model has a great number of rules

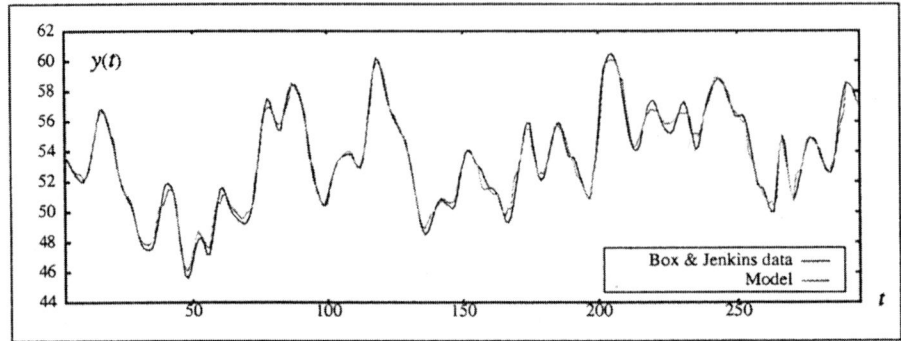

Fig. 5.8. Data versus results of modeling for $c = 10$

(550 for $c = 25$) and becomes unreadable. It should be noted that the J index is an optimistic performance index. Indeed, it is evaluated from the same data as the ones used to build the model, but nothing guarantees that the same performance will be obtained with new data. Especially, overspecialization or overfitting of the model to the data can appear and induce bad generalization performance. However, as classically practised in the literature dealing with the Box and Jenkins benchmark (see Fig. 5.9), we restrict ourselves to the J index.

One must be very careful in interpreting the results given in Fig. 5.9. Indeed, the different methods use neither the same approach to fuzzy modeling nor the same number of inputs or rules in the fuzzy model. Furthermore, the form of the rules may be completely different from one method to another. So the number of rules does not always have the same meaning and cannot be considered as a comparison criterion for the readability of the rule base.

6. Practical Aspects and Concluding Remarks

In this section we will analyze how the proposed identification method relates to some practical aspects of model identification in general. We also present some concluding remarks.

6.1 Use of Prior Knowledge

Whatever the identification procedure, the construction of a dynamic model requires prior knowledge about the plant. The proposed method does not escape this need. Especially, the process order and time-delay has to be estimated to determine a fixed rule form.

Identification method	Index J
Box & Jenkins 70 [3]	0.710
Tong 80 [28]	0.469
Pedrycz 84 [21]	0.320
Xu & Lu 87 [33]	0.328
Costa Branco & Dente 93 [4]	0.312
Yoshinari, Pedrycz & Hirota 93 [35]	0.299
Lee, Hwang & Shih 94 [15]	0.211
Pedrycz 95 [25]	0.395
Wang & Langari [32]	0.066
Proposed method	0.175

Fig. 5.9. Fuzzy modeling results

A major advantage of fuzzy modeling is that it provides a global model in the form of a collection of local models. This property facilitates the integration of new knowledge in the model under construction. This is particularly interesting in the present case because of the linguistic representation of the knowledge. Hence, the proposed fuzzy linguistic model is able to combine the advantages of numeric approximation (global model) and linguistic formulation (local models).

Let us briefly examine how to make use of prior knowledge during model identification in the context of the method proposed in this chapter. Two levels may be considered: on the one hand, the introduction of heuristic knowledge during the refining phase, and on the other, the development of a knowledge-based initialization phase.

As suggested in [2], the construction of a model might be approached as an incremental process, using the current model for guiding the extraction of additional knowledge. The iterative nature of the learning strategy is particularly well adapted to this mechanism. Indeed, at the end of each iteration of the refining phase, some new knowledge may be introduced. Furthermore, the search for additional knowledge may be easily guided by the obtained current model. If the filling methods described in Sect. 4.3. (steps 3 and 4) are not applied, some fuzzy cells remain empty. These cells correspond to local unknown behaviors of the system that are to be learned. Two basic solutions are then possible:

– An expert is able to fill the rule base.
– Some new test signals are generated in order to extract additional training
  points describing the unknown system behavior.

If it is interesting to make use of heuristic knowledge during the refining
step of the model synthesis, it is certainly still more desirable to integrate all
available pieces of knowledge from the initial phase of the learning process.
This may be done during the construction of the initial fuzzy model. When a
linear numeric model has already been identified, it can be directly translated
in terms of a rule base and a fuzzy partitioning of the different universes
of discourse [9]. In the same way, the initial fuzzy model may be directly
expressed by using human knowledge as in the case of the design of a fuzzy
controller [17, 18]. Furthermore, as illustrated in [10], for control purposes the
construction of an initial fuzzy model can successfully involve the knowledge
of a nonlinear numeric law augmented with linguistic knowledge.

6.2 Model Complexity

In fuzzy modeling, the model complexity is often defined as the size of the
rule base. For the proposed method, the model complexity thus depends on
both the number of fuzzy input variables and the number of linguistic terms
per variable. The total number of rules may be roughly evaluated as M^r,
where M represents the number of linguistic values defined over each input
space, and r is the number of fuzzy input variables. Although the proposed
method allows one to reduce M by a good placement of the linguistic values, it does not deal with the major problem of exponential growth in the
number of rules as r increases. The curse of "rule explosion" stems from the
and-decomposition assumed in the antecedents of the rules. However, the
non-parametric form of the proposed model provides a hope for a possible
extension of the method, that could directly deal with the inputs as a whole,
thus avoiding the antecedent decomposition. Each corresponding input vector may then be qualified by a linguistic value whose fuzzy meaning has to
be defined over a multi- dimensional space. However, it should be noted here
that dealing directly with multi-dimensional input vectors can diminish the
readability of the rule base if the grouping of the inputs has no physical
meaning.

An a priori judgement of the model complexity with respect to a given
system remains a critical question. Assuming a given number r of fuzzy input
variables (prerequisite for the identification method), the rule base size evaluation simply depends on the number M of linguistic values generated for
each fuzzy input. In the linear case, an accurate model can be obtained with
only two linguistic values ($M = 2$). When nonlinearities are introduced, M
increases and the nonlinearities are reflected in the distribution of the identified membership functions. The increase in M then depends on the number
and nature of the nonlinearities (see Fig. 5.2).

6.3 Robustness of the Identification Method

The robustness of the proposed identification method with respect to a given problem is related to the training data set. It is clear that the developed algorithm is not robust with regard to noisy data. The sensitivity to noise is caused by the need for precise learning of some particular training points. If these points are noisy, however, because they are precisely learned, the noise is also carried with them. Furthermore, aberrant training points are undoubtedly learned due to the fact that they provide important approximation errors and are then good candidates for the learning process. Therefore, efficient use of the method needs pre-processing of the data in order to eliminate the undesirable noise. The noise sensitivity of the method is counter to its ability to deal efficiently with high nonlinearities as illustrated in Fig. 5.2.

The robustness of the identification method is also linked with the informative content of the training data set. The model produced will perform correctly in generalization only if the training data is representative of the global behavior of the system. In particular, the identified model has no extrapolation capabilities outside the limits given by the training data.

References

1. R. Babuska, H.B Verbruggen, A new identification method for linguistic fuzzy models, *Proc. of FUZZ IEEE'95*, pp. 905–912, Yokohama, Japan, 1995.
2. R. Babuska, Fuzzy modeling, a control engineering perspective, *Proc. of FUZZ IEEE'95*, pp. 1897–1902, Yokohama, Japan, 1995.
3. G.E. Box, G.M. Jenkins, *Time series analysis, forecasting, and control*, (Holden Day, San Francisco, CA, 1970).
4. P.J. Costa Branco, J.A. Dente, A new algorithm for on-line relational identification of nonlinear dynamic systems, *FUZZ'IEEE 93*, pp. 1173–1178.
5. D. Dubois, H. Prade, *Possibility theory – An approach to computerized processing of uncertainty*, Plenum Press, New York, 1987.
6. D. Driankov, H. Hellendoorn, and M. Reinfrank, *An introduction to fuzzy control*, Springer-Verlag, 1993, Chap. 2, pp. 37–102. 2nd ed. 1996.
7. L. Foulloy, S. Galichet, Typology of fuzzy controllers. In: *Theoretical aspects of fuzzy control*, Eds H. Nguyen, M. Sugeno, R. Tong, and R. Yager, Wiley and Sons, New York, 1995.
8. S. Galichet, L. Foulloy, State Feedback Fuzzy Controllers, *EUFIT'94*, Aachen, Germany, 1994, pp. 1161–1167.
9. S. Galichet, L. Foulloy, Fuzzy controllers: synthesis and equivalences, *IEEE Trans. Fuzzy Systems*, **3**(2)(1995)140–148.
10. S. Galichet, L. Foulloy, M. Chebre, J. Beauchene, Fuzzy logic control of a floating level in a refining tank, *Proc. of the 3rd IEEE Conf. on Fuzzy Systems (FUZZ-IEEE'94)*, Orlando, USA, 1994, pp. 1538–1542.
11. F. Guely, P. Siarry, A centred formulation of Takagi–Sugeno rules for improved learning efficiency, *Fuzzy Sets and Systems*, **62**(1994)277–285.
12. J.S. R. Jang, ANFIS: Adaptive-network-based fuzzy inference system, *IEEE Trans. Systems, Man, and Cybernetics*, **23**(3)(1993)665–685.

13. B. Kosko, *Neural networks and fuzzy systems: A dynamical systems approach to machine intelligence*, Prentice Hall, 1992.
14. B. Kosko, Fuzzy systems as universal approximators, *Proc. 1st IEEE Conference on Fuzzy Systems*, San Diego, CA, 1992, pp. 1153–1162.
15. Y.C. Lee, C. Hwang, and Y.P. Shih, A Combined approach to fuzzy model identification, *IEEE Trans. Systems, Man, and Cybernetics*, 24(5)(1994)736–744.
16. L. Ljung, *System Identification – Theory for the User*, Prentice Hall, 1987.
17. Mamdani E.H., Assilian S., An experiment in linguistic synthesis with a fuzzy logic controller, *Int. Journal of Man-Machines Studies*, 7(1975)1–13.
18. Mamdani E.H., Advances in the linguistic synthesis of fuzzy controllers, *Int. Journal of Man-Machines Studies*, 8(1975)669–678.
19. Y. Nakoula, S. Galichet, L. Foulloy, Learning of a fuzzy symbolic rule base, *Proc. EUFIT'95*, 1995, Aachen, Germany, pp. 594–598.
20. H.T. Nguyen, V. Kreinovich, On approximation of controls by fuzzy systems, *Proc. 5th IFSA World Congress*, Seoul, Korea, 1993, Vol. II, pp. 1414–1417.
21. W. Pedrycz, An identification algorithm in fuzzy relational systems, *Fuzzy Sets and Systems*, 13(1984)153–167.
22. W. Pedrycz, On Generalized Fuzzy Relational Equations and Their Applications, *Journal of Mathematical Analysis and Applications*, Vol. 107, pp. 520–536, 1985.
23. W. Pedrycz, *Fuzzy Control and Fuzzy Systems*, John Wiley & Sons Inc., 1989.
24. W. Pedrycz, Fuzzy Relational Modeling. In: *Theoretical Aspects of Fuzzy Control*, H. Nguyen, M. Sugeno, R. Tong, and R. Yager, (eds.) Wiley and Sons, New York, 1995.
25. W. Pedrycz, P.C.F. Lam, A.F. Rocha, Distributed Fuzzy System Modeling, *IEEE Trans. Systems, Man, and Cybernetics*, 25(5)(1995)769–780.
26. T. Sudkamp, R.J. Hammell II, Interpolation, Completion, and Learning Fuzzy Rules, *IEEE Trans. Systems, Man, and Cybernetics*, 24(2), 1994.
27. T. Takagi, M. Sugeno, Derivation of fuzzy control rules from human operator's control actions, *Proc of the IFAC on Fuzzy Information*, pp. 55–60, Marseille, France, 1983.
28. R.M. Tong, The evaluation of fuzzy models derived from experimental data, *Fuzzy Sets and Systems*, 4(1980)1–12.
29. J. S. Urbram Hjorth, *Computer Intensive Statistical Methods, Validation Models Selection and Bootstrap*, Chapman and Hall, 1994.
30. L-X. Wang, J.M. Mendel, Generating Fuzzy Rules by Learning from Examples, *IEEE Trans. Systems, Man, and Cybernetics*, 22(6)(1992)1414–1427.
31. L.-X. Wang, Fuzzy systems are universal approximators, *Proc. 1st IEEE Conference on Fuzzy Systems*, San Diego, CA, 1992, pp. 1163–1169.
32. L. Wang and R. Langari, Building Sugeno-Type Models Using Fuzzy Discretization and Orthogonal Parameter Estimation Techniques, *IEEE Trans. Fuzzy Systems*, 3(4)(1995)454–458.
33. C.W. Xu, Y.Z. Lu, Fuzzy Model Identification and Self-Learning for Dynamic Systems, *IEEE Trans. Systems, Man and Cybernetics*, 17(4)(1987)683–689.
34. R.R. Yager, D.P. Filev, *Essentials of Fuzzy Modeling and Control*, Wiley and Sons, New York, 1994.
35. Y. Yoshinari, W. Pedrycz, K. Hirota, Construction of fuzzy models through clustering techniques, *Fuzzy Sets and Systems*, 54(1993)157–165.
36. Zadeh L.A., Quantitative fuzzy semantics, *Information Sciences*, 3(1971)159–176.

37. Zadeh, L.A., Outline of a new approach to analysis of complex systems and decision process, *IEEE Trans. Systems, Man, and Cybernetics*, **3**(1)(1973)28–44.
38. Zadeh L. A, The concept of a linguistic variable and its application to approximate reasoning, *Information Sciences*, Part I: **8**(1975)199–240, Part II: **8**(1975)301–375, Part III: **9**(1975)43–80.

Springer
and the
environment

At Springer we firmly believe that an international science publisher has a special obligation to the environment, and our corporate policies consistently reflect this conviction.

We also expect our business partners – paper mills, printers, packaging manufacturers, etc. – to commit themselves to using materials and production processes that do not harm the environment. The paper in this book is made from low- or no-chlorine pulp and is acid free, in conformance with international standards for paper permanency.